T0229900

Technology

The focus of this series is engineering, broadly construed. It covers techno-logical innovation from a range of periods and cultures, but centres on the technological achievements of the industrial era in the West, particularly in the nineteenth century, as understood by their contemporaries. Infra-structure is one major focus, covering the building of railways and canals, bridges and tunnels, land drainage, the laying of submarine cables, and the construction of docks and lighthouses. Other key topics include develop-ments in industrial and manufacturing fields such as mining technology, the production of iron and steel, the use of steam power, and chemical processes such as photography and textile dyes.

The Life of Sir William Crookes

In 1870, Sir William Crookes (1832–1919) travelled to Gibraltar to observe the solar eclipse. He kept a diary and produced beautiful accounts of the expedition – alongside altogether more specific observations, including the 656 steps down a local cliff face, and every item in his luggage. It is with the same meticulous approach and cheerful prose that he records, in letters, journal articles and reports, the successes and failures of the vast range of projects in which he was involved. Although initially trained as a chemist, Crookes worked across the spectrum of the sciences, from consulting on preventative measures against cattle plague through to investigating spiritualism. Opening with a foreword by the physicist Sir Oliver Lodge, this biography by Edmund Edward Fournier d'Albe (1868–1933), first published in 1923, explores a remarkable life of enquiry through a host of first-hand sources.

Cambridge University Press has long been a pioneer in the reissuing of out-of-print titles from its own backlist, producing digital reprints of books that are still sought after by scholars and students but could not be reprinted economically using traditional technology. The Cambridge Library Collection extends this activity to a wider range of books which are still of importance to researchers and professionals, either for the source material they contain, or as landmarks in the history of their academic discipline.

Drawing from the world-renowned collections in the Cambridge University Library and other partner libraries, and guided by the advice of experts in each subject area, Cambridge University Press is using state-of-the-art scanning machines in its own Printing House to capture the content of each book selected for inclusion. The files are processed to give a consistently clear, crisp image, and the books finished to the high quality standard for which the Press is recognised around the world. The latest print-on-demand technology ensures that the books will remain available indefinitely, and that orders for single or multiple copies can quickly be supplied.

The Cambridge Library Collection brings back to life books of enduring scholarly value (including out-of-copyright works originally issued by other publishers) across a wide range of disciplines in the humanities and social sciences and in science and technology.

The Life of
Sir William Crookes

EDMUND EDWARD FOURNIER D'ALBE
WITH A FOREWORD BY OLIVER LODGE

CAMBRIDGE UNIVERSITY PRESS

Cambridge, New York, Melbourne, Madrid, Cape Town,
Singapore, São Paolo, Delhi, Mexico City

Published in the United States of America by Cambridge University Press, New York

www.cambridge.org
Information on this title: www.cambridge.org/9781108061599

© in this compilation Cambridge University Press 2013

This edition first published 1923
This digitally printed version 2013

ISBN 978-1-108-06159-9 Paperback

THE LIFE OF
SIR WILLIAM CROOKES

SIR WILLIAM CROOKES, O.M., 1911.

ÆT. 79.

THE LIFE OF SIR WILLIAM CROOKES

O.M., F.R.S. *By* E. E. FOURNIER D'ALBE, D.Sc., F.Inst.P.

WITH A FOREWORD BY
SIR OLIVER LODGE

T. FISHER UNWIN LTD
LONDON: ADELPHI TERRACE

First published in 1923

FOREWORD

M R. FISHER UNWIN and Dr. Fournier d'Albe have both asked me to write an Introduction or Foreword to a " Life of Crookes " : and I am very glad to have an opportunity of saying something in memory of my old friend.

In many respects Crookes was unique in my experience of scientific men. He belonged, I suppose, to the class which has done such first-rate work in England, a class each member of which began as a scientific amateur and forced his way to the front by sheer ability and brilliance of discovery. These constitute a group of whom, I suppose, Joule and Cavendish and Huggins are the most typical instances ; though I intend no sort of comparison between any of them. They may differ from each other in every other particular.

Crookes was not a learned man in the professional sense. He had brilliant ideas—inspirations they might be called—and he worked them out systematically and pertinaciously. It was his orderly system and pertinacity. continued into old age, which must have impressed everyone. And the brilliance of the resulting discoveries forced him into the highest Official position which a scientific man can attain, in spite of the controversy and hostility which some of his investigations had evoked. I must confess that I myself used to share in this ignorant hostility to some slight extent. For when he lived in Mornington Road, close to my lodgings in Delancey Street, I never remember calling at his house, or seeing any of the work that was going on there. I first met him personally at the early meetings of the Physical Society of London, when he was at work on high vacua, and when, with the help of a skilled glass-blowing assistant, he was able to exhaust vessels more perfectly than they had ever been exhausted before. The observations which he made of the electric discharge under those circumstances were of an illuminating character, and led to theoretical and practical achievements which ultimately had prodigious consequences. When my senior friend Professor Carey Foster became President of the Physical Society of London in 1876, and was contemplating beforehand his Presidential Address, I remember his telling me that he was proposing to take as its theme the work and discoveries of Crookes. He did not actually carry out this half-formed intention : but the possibility of it, coming from a man of the universally recognised sound judgement of Carey Foster, impressed me, who at the age of twenty-five was inclined to be hypercritical, and made me think more of Crookes than I had previously.

Before that, however, I remember hearing about his attempt to detect the pressure of light. And I remember, when, or soon after, Professor Fleming and I were both students in Professor Frankland's Advanced Chemical Laboratory at South Kensington, we used to discuss this matter, and rigged up an oblong of pith on a torsion wire inside a bulb in order to repeat Crookes's early experiment. We thought it must be a heat effect ; and so indeed it turned out to be, not a real pressure of light at all, but a reaction from the molecular bombardment on a slightly warmed surface. The pressure of light was too small to be detected by the means then available ; it was masked by this other molecular non-etherial effect.

Crookes, however, continued his researches, and brought them to a climax by his neat invention of the radiometer, which he exhibited at an early meeting of the Physical Society ; incidentally making us wonder by what thaumaturgic skill he and his assistant had managed to get the revolving vanes into the bulb.

However, Crookes did many more wonderful things than that, inside vacuum bulbs ; and later, in the year 1879 at Sheffield, he demonstrated before an Evening Meeting of the British Association the remarkable properties of what for the first time he called cathode rays. It was a little before this time that he regarded his beam of cathode rays as corpuscular light. And it was now that he introduced his prophetic phrase, " Matter in a fourth state " ; that is to say, he claimed that he had obtained, as we now perceive and admit, matter in a state neither solid, liquid, nor gaseous ; no longer consisting of atoms as heretofore known, but split up into its apparently ultimate ingredients, now known to be the units of electric charge.

Before all these brilliant vacuum investigations, I had only known Crookes as the discoverer of Thallium and Editor of the *Chemical News* : though there were rumours that he was also trying to interest leaders of science in certain obscure and apparently superstitious investigations into mysterious and incredible phenomena. He had not then learnt the full force of scientific conservatism. With the enthusiasm of comparative youth, he must have thought that scientific magnates, no matter how orthodox, would be glad to welcome new and startling phenomena, if they could be rigidly and metrically demonstrated. He had the advantage of a powerful medium, Daniel Home ; he witnessed the most astounding things ;

he described some of them with natural enthusiasm, but occasionally with less than scientific restraint ; and he tried to reduce the puzzling phenomena which frequently occurred in Home's presence, to their rudiments, by getting him to tilt a board suspended at one end on a spring-balance, with the other end resting on a table, and to effect this in an obviously non-mechanical manner. But it was just this apparent absence of any possibility of mechanical explanation which prevented high scientific Authorities from coming to see it. They felt, no doubt, in sympathy with the position which Faraday had laid down—that we must make up our minds as to what is possible and what impossible, and not waste time in trying to look for the impossible !

After further fruitless efforts to interest the scientific world in what he considered a whole region of unexplored territory, and after drawing upon himself a great amount of fierce criticism and ridicule—so that he was even accused of being practically insane on one side of his brain, while sane enough on the other—Crookes appears to have realised that the time was not ripe for recognition of facts of this novelty, that there were many other branches of Physical and Chemical research to which he might devote his undivided attention, and that the attempt to storm the citadel of orthodoxy had better be postponed. He felt no doubt that the truth would make its way in due time. He never abandoned his full conviction of the reality of the facts which he had observed. He may or may not have varied his theoretical views concerning them ; for the time for useful theory was not yet. But he refrained from arousing any fresh hostility by pressing home his discoveries on a reluctant world. They had been observed, he knew, by others. They would continue to be studied, although perhaps in a less striking and sensational manner, by those who had less reputation to lose : and he was content to bide his time till the truth should make its way by its own momentum. Before he died he had the pleasure of seeing these psycho-physical phenomena taken up with some vigour on the continent of Europe, where, in spite of opposition, more and more scientific and medical men have become convinced of the reality of these puzzling things ; though in this country they still remain under a cloud of suspicion.

But though Crookes, and others since Crookes, have perceived the wisdom of not pressing truth prematurely on those unwilling to receive it, Societies have grown up with the object of taking a rational interest in these things, and reducing them to law and order, preparatory to their being gradually absorbed by scientific orthodoxy. Among such Societies, Crookes's name is held in high honour all over the world, not in spite, but because of his unpopular researches ; and in the years 1896–9 Crookes was President of the chief of these Societies. In all the talks I had with him, there was never the least sign of his abandoning any of the positions he had taken up. On the contrary, in a congenial atmosphere, he and Lady Crookes would narrate some of their extraordinary experiences ; some of which were beyond anything which, so far as I know, have been seen since ; though the general trend of them, and their more elementary aspects, have been verified up to the hilt again and again. The subject however, has been pursued in England less on the physical than on the mental side ; and on the more purely psychical aspect of the phenomena, Crookes did not feel himself specially competent, though he ratained his vivid interest to the end.

No longer, however, is it considered essential, by those who pursue these studies, to undergo more than a very mitigated martyrdom, in the fruitless task of trying to force recognition on the unwilling. Like Galileo, Crookes, who had certainly suffered to a moderate degree, came to the conclusion that the truth would make its way in due time, and that meanwhile he could afford to wait in patience.

The other discoveries of Crookes are of world-wide fame, and will doubtless be dealt with in their due sequence in the " Life " which is now being written, but which I have not read. It seems possible, perhaps in consonance with the wishes of his family, that only undeniable and uncontroversial discoveries will be fully emphasised. But it would not be fair to the memory of Crookes to omit respectful mention of a branch of inquiry in which hereafter he will be hailed as one of the pioneers. And even now, half, or some smaller proportion, of the civilised world is more interested in these puzzling and obscure phenomena than it is in discoveries of any but the first magnitude in Chemistry and Physics.

Crookes was a great experimental Chemist. His fractionation and separation of the rare earths was a model of systematic procedure. His spectroscopic investigations laid the foundation for much of the brilliant work that is being done in our own day. The vivacity and thoroughness with which he examined any new fact, or any peculiar mineral, is well worthy of imitation. But not to many are given the intuitions which lead straight into the heart of the subject, and yield a clue for further advances. Among these intuitions, I have mentioned his views on the nature of the cathode rays, before the discovery and isolation of electrons. I might also mention his anticipation of the possibilities of Wireless Telegraphy, based on seeing a few developments of the Hertzian demonstration of electric waves in space. The Article will be found in the *Fortnightly Review* for February 1892 (p. 173), and is a rather wonderful example of scientific forecast.

When he was President of the British Association in 1898, he was tempted to resurrect his unpopular researches. But what were considered wiser councils prevailed, and he devoted the greater part of his Address to the Problem of the Wheat Supply of the World—an Address which has often been referred to by those competent to deal, in a practical and political way, with such a topic.

In later investigations by others, connected with the intimate structure of Atoms, Crookes's method of observing for the first time the splash or luminosity excited by each single atom, when projected with sufficient energy, is found of the utmost service. For it has enabled newly formed, or at least newly expelled, atoms of different kinds to be recognised ; discrimination between them being possible by their range, or length of path, through air. In that way the breaking-up of atomic nuclei has been verified, and the conversion of one element into another confirmed. Here again, Crookes's intuition was not at fault. He had an idea, and he mooted it at the British Association in Birmingham in 1896, that the family relationship existing among the series of chemical elements could only be accounted for on the principles of evolution. He made a sort of speculative family-tree, suggesting that the elements came into being in regular order and succession ; and he surmised that the atomic weights were not such simple and definite things as had been thought, but might be a kind of average, about which elements of slightly different weight but similar chemical character, could group them-selves,—a surmise which the discovery of Isotopes in recent years has amply and conspicuously justified.

In every way, then, Crookes was a great scientific man, of surprising perseverance and experimental skill, with flashes of intuition and insight, which though at the time they might seem wildly speculative, were soon justified by the orderly progress of science.

His personality was not specially impressive. In his presence one did not feel the worshipful enthusiasm which some of the great men of science have aroused. It was easy to be critical. And yet, when one visited him at his house in Kensington Park Gardens, and saw his fine library with the well-ordered though small laboratory opening out of it, and admired the neatness of his records and his untiring industry, one felt that here was a workshop from which phenomena of stimulating novelty might at any time emerge. One would not compare it with Faraday's at the Royal Institution—that home of immortal dis-coveries,—still less with such a hive of industry, guided and stimulated by mathematical theory, as the Cavendish Laboratory, Cambridge. But it was a peaceful and secluded home of quiet research. There friendly people used to assemble for weekly converse ; and there, in his later years, after a lifetime spent in exploring the secrets of Nature, lived and died the pioneer, William Crookes.

September 1923   Oliver J Lodge

PREFACE

THE LATE SIR WILLIAM CROOKES had on several occasions planned to write his Reminiscences and to issue a collection of his Scientific Papers. Had he been able to carry out his plans, the world would no doubt have been enriched by a work of great charm and of permanent interest. But the hand of death was laid upon him before he could find time to attempt such a formidable task. The negotiations begun during his lifetime by the present publishers were continued after his death, and, after some vicissitudes, the large collection of letters and other documents came into my hands. The work of sifting these documents, numbering altogether some 40,000, has been somewhat arduous, and it may be that a few of much personal or public interest have failed to find a place in this " Life." But my aim has been to present Crookes as he appeared to himself and to his best friends, and, if possible, to allow him to express his ideas and aims in his own words. That must be my excuse for publishing many of his letters *in extenso*. They express the man and his character better than could be done by any narrative in the third person.

I wish to return thanks to the following persons for the loan of valuable documents : Sir William Barrett, F.R.S., Miss Gladstone, and Mr. A. G. Ionides ; and to the following for supplying much interesting information : Sir Herbert Jackson, F.R.S., Sir Dugald Clerk,

PREFACE

F.R.S., Mrs. Ionides of Witham, Mr. J. W. Gordon, K.C., Mr. A. C. Ionides, Mr. A. E. Franklin, Mr. A. Campbell Swinton, F.R.S., Mr. Percy Spielmann, and Mr. James H. Gardiner. To Mrs. Cowland (Sir William's surviving daughter) my special thanks are due for much practical assistance and advice.

E. E. FOURNIER D'ALBE

CONTENTS

CONTENTS

xvi

CONTENTS

ILLUSTRATIONS

The Life of Sir William Crookes

CHAPTER I

INTRODUCTORY

But in his peaceful cell, the Sage draws significant circles,
Ponders the hidden plan, thinks the Creator's thought ;
Tests elementary powers, the loves and hates of the magnets,
Follows the sound through the air, tracks the ethereal ray,
Traces the friendly Law among Chance's terrible wonders,
Gropes for the constant pole guiding the flight of the world.
                                                    SCHILLER.

THE LATE PROFESSOR POYNTING, of Birmingham, used to say of himself that he was generally taken for a prosperous farmer. His chubby face and square figure did, indeed, effectively counterpoise the massive brain and penetrating eye of the renowned physicist. But the scientific type is difficult to identify. Of the four greatest living English physicists, one might easily be taken for a futurist painter, another for a Labour leader, a third for a thriving member of the Stock Exchange, and yet another for a Hebrew prophet. We have to go back to the time of Franklin, Priestley, or Davy before we arrive at the figure of the typical " natural philosopher."

The gulf which separates the old " natural philosopher " (" very much wrapped up about the throat," as Heaviside remarks) from the modern representative of physical and chemical science is effectively bridged by two vivid personalities—Michael Faraday, the illustrious successor of Humphry Davy at the Royal Institution, and William Crookes, the subject of this biographical essay. And though the history of Faraday, the bookbinder's apprentice who became the foremost discoverer of his time, is full of human as well as scientific interest, there is no doubt that Crookes, more than any other man, represents the popular idea of the scientific investigator of the nineteenth century.

For to the general public the man of science is a man of mystery, a man of inhuman and somewhat unaccountable tastes. Not everyone goes so far as to maintain that he is a freak because he indulges in an activity " with no money in it." But it seems to be generally agreed that the " scientist " is a being living outside ordinary human spheres, not amenable to ordinary human standards, a being who is usually harmless but may quite conceivably become dangerous—a sort of well-meaning revolutionary who plays with fire and other forces without much regard for the safety and stability of existing institutions, and who is therefore best kept at work along recognised lines under the general supervision of the more orthodox leaders of the community.

My own youthful idea of a *savant* was largely based upon Jules Verne's romances. The amiable and enthusiastic scientific leader of impossible expeditions figuring in those wonderful stories appealed irresistibly to my imagination, and I figured myself as a recluse, working mysteriously for years in a wonderful private laboratory of my own at some world-shaking problem and finally bursting forth with a *fait accompli* involving far-reaching consequences for the whole of mankind. I saw myself tackling the most fundamental and formidable problems, exploring the hidden mysteries of the universe with the lamp of science, clearing away encumbrances and obstacles to knowledge, advising and guiding my fellow-beings, and growing old in wisdom and honours.

Translate this dream into actual fact in modern England and you get—William Crookes.

Crookes is the very type and symbol of English science at its best. His career embodies the emergence of the scientific man as a force in English life. He was not of the governing class. His education could not be epitomised as of " Eton and Christchurch," or " Rugby and Trinity." According to early Victorian standards, he had no rightful part to play in English polity at all. His long life of eighty-seven years saw the advance of science from the humble rank it held in 1832 to the all-embracing position to which it attained in 1919, and his own activities and achievements contributed not a little to the astonishing transformation.

Crookes was no linguist. He had no university education, nor did he hold a professorship. His work was partly journalistic and partly that of a consultant. He stood primarily for the widening and dissemination of chemical knowledge and its application to the manifold problems of human life. But his outlook went far beyond that narrow range. He was a keen fighter, but the enemies he preferred to fight were the enemies of the human race. Thus he enlisted in the ranks against cattle plague and cholera when these pests were raging in England, and rapidly gained the higher command by his industry and keen insight. He studied disinfectants and water supply and sewage disposal, leaving a mark on his generation in the shape of a substantial reduction of the death-rate. He threw his whole weight into the development of the photographic art, devised new processes, invented new apparatus, and applied the art to

3

the investigation and recording of scientific phenomena such as meteorological changes and solar eclipses. He even attempted—somewhat prematurely—to photograph a projectile fired from a heavy gun and thus to record its trajectory.

All this was done in the first flush of youthful ardour. But as his experience ripened and his resources increased we find him seeking for hidden treasures at deeper levels. Selenium—an element destined for a career of exceptional interest—had fascinated him from his earliest student days. In examining its spectrum with the newly invented "spectroscope," he found a beautiful green line which nobody had seen before. It belonged to an unknown element, and the story of how Crookes tracked down this substance and triumphantly established its elementary nature in the face of much criticism forms one of the most romantic chapters of his life.

But the strangest interlude in Crookes's career occurred in 1870, when he commenced his four years' investigation of "the phenomena called spiritual." He was then thirty-eight, and had been married fourteen years. It was quite a new departure for him. He threw himself into the investigation with his usual energy and resource, and achieved the same prominence as a psychical researcher as he had done as a chemist and physicist. His results, taken at their face value, are the most amazing things ever obtained by a trained man of science, and, if fully accepted by the scientific world, would bring about a revolution in our views of the universe such as has not been witnessed since the days of Copernicus.

It was but an interlude. Crookes found himself struggling with unfamiliar conditions in a turbulent atmosphere, which made it impossible to convince his colleagues of the reality of phenomena which he had established to his own satisfaction in his laboratory. Such a state of things is not unknown even in purely physical investigations, but it almost always signifies some flaw in the reasoning, some neglected source of error. Crookes did not feel justified in sacrificing more years to such a fruitless investigation—fruitless of that appreciation by his peers which is so powerful an incentive towards scientific endeavour. And so Crookes closed that chapter, regretfully perhaps, but fully determined to devote all his strength to ultimate problems of a nature open to accepted scientific methods.

And then came that wonderful chapter of researches in high vacua, leading to " radiant matter," the Radiometer, and the " Crookes tube," which incidentally solved the problem of electric lighting, and is now universally represented by the electric lamp found even in humble homes.

Here we find Crookes at the very height of his career. He was, indeed, the outstanding discoverer of the day and of his generation. Working mostly alone and apart in his great laboratory, he wrested many a secret from Nature and laid bare his hard-won treasures before an astonished world. For the next thirty years honours fell thick upon him from all sides. The presidency of the Chemical Society, of the Institution of Electrical Engineers, of the British Association for the Advancement of Science, and finally of the Royal Society fell to him in succession, thus giving him some of the most coveted distinctions open to science in England. The knighthood conferred upon him in 1897 was but the Royal assent to a full measure of recognition already earned and received. With a mind untrammelled by dogmatic preconception and a position independent of academic punctilios, he was free to seek the truth wherever it was likely to be found and to promulgate it without fear or favour. As the Victorian era approached its end, and science gradually advanced to a position of greater influence and importance, the position and authority of a man like Crookes assumed an ever-increasing prominence. When finally the European War fell upon the world, the silent revolution completed itself dramatically. The world stood face to face with stark reality. Rhetorical subtleties lost their charm, and men and systems were valued according to the extent to which they harmonised with fundamental truths. It was inevitable that the scientific ideal and outlook should then prevail. For the true man of science worships but one god—truth. He despises the ecclesiastic for teaching half-truths for the sake of moral influence ; the politician for dressing up truth in a partisan guise ; and the business man for subordinating truth to personal gain. Science represents a new aristocracy based upon a new power, but an ascendancy *ouverte aux talents* and attainable by anyone who will go through the necessary mental and physical labour. Like all aristocracies, it is " tempered by revolution," and one is not surprised to hear the " tyranny of experts " denounced as the worst tyranny of all. But any such revolution could only replace one expert by another expert of greater reputation for knowledge and honesty, and therefore endowed to a greater extent with the ideal attributes of the scientific mind. We thus get a close approximation to that " benevolent despotism " which has been advocated as the only cure for the evils of democratic government.

Not that Crookes even remotely resembled a despot. In his later years he gave one the impression of a shrewd and kindly personality, venerable with his white hair and beard. His married life was serene and fruitful, and culminated in a diamond wedding. Up to the last years of his long life he worked away at scientific problems in the hope of a solution which would benefit mankind, and when he attained a solution he gave it freely to the world. While his discoveries created, or assisted in creating, vast industries, he was content to do the work of the pioneer, who, as a rule, is notoriously ill rewarded. And therein he represented but another aspect of the scientific discoverer. While statesmen are at pains to control the forces which together make up the life of the nation, the silent investigator created new forces of incalculable import. While the agitator talks of revolution by violence or by stoppage of work, the chemist or physicist in his " peaceful cell " creates or destroys the work of millions of men, and fundamentally alters the status of both labour and capital. The discoverer is the real arbiter of the destinies of the world. The powers that ruled the Middle Ages knew better than to let him loose. And now it is too late, and we must accept the discoverer's revolutionary activity as we accept the earthquake and the tornado, and statesmen and politicians and Labour leaders must needs dance to the tune of the discoverer's pipe. The alternative is to stop the activities of the discoverer. This, indeed, has often been attempted, by bribery and intimidation and what not else. But it is futile, for the new aristocracy is in itself more democratic than any other human institution. There is no central authority—all scientific authority is provisional. It only exists until displaced by wider knowledge and deeper research. Unlike ecclesiastical authority, it is adaptable and amenable to new truth. It is intensely alive by that very fact, and is indestructible so long as it retains that adaptability. Kings and armies and financial combinations must bow before it. The new despotism is calmly accepted almost everywhere. It therein somewhat resembles the despotism of fashion —which, presumably, is also a " tyranny of experts."

INTRODUCTORY

Sir William Crookes would probably have objected to being classed as a revolutionary. Standing at the head of the staircase at Burlington House and extending a Presidential welcome to the Fellows and friends of the Royal Society, he seemed to embody the dignity of an ancient, well tried and conservative institution, the conservatism of which had been painfully brought home to him during the spiritualistic interlude. It is fortunate for mankind that the youthful impetuosity of its budding " despots " is tempered by the mature wisdom of the elder men. The body of scientific workers has, so to speak, automatically secreted a cortex as the sap does in a tree, so as to add weight and permanence to the general structure. Crookes in his career passed through all the stages from sap to cortex, but when the Great War overtook him at the age of eighty-two he showed a return to his earlier stages, and served his country right through, retiring from the field only when his country had achieved victory and the land that bore him was ready to clasp him to her breast.

That was the man. And how he arose and lived and fought and won his laurels I shall endeavour to show forth in the chapters here following.

CHAPTER II

EARLY DAYS

SIR WILLIAM CROOKES was born at 143, Regent Street, London, on June 17, 1832, and died on April 4, 1919, at 7, Kensington Park Gardens, in the eighty-seventh year of his age.

His father, Joseph Crookes, was born in 1792 in the little town of Masboro', Yorkshire, as the son of a local tailor, who apprenticed his most promising son—one out of a large family—to the trade he knew best. So young Joseph dutifully took to "the board," devoting his scanty leisure to the pleasure of roaming the country to within sight of Sheffield, or joining the young manhood of Masboro' in "holding the bridge" over the Rother against the high-handed and impudent aggression of the lads of Rotherham, the neighbouring town.

But Joseph Crookes had ambitions which could not be compassed within the limits of a Yorkshire townlet. While yet in his teens, he came by coach to London, visiting on the way an uncle, who gave him two guineas as a start in life. Having already paid his fare, he had this sum intact on his arrival in London.

Joseph Crookes promptly entered the service of a West End tailor in Regent Street. Here young Joseph underwent the metamorphosis necessary to convert a Yorkshire country tailor into the master craftsman who dwells within the shadow of the throne.

9

Joseph Crookes took root in London. Combining prudent foresight with great industry, he soon became sufficiently wealthy to lay the foundation of that fortune which his illustrious son put to such excellent use. His savings were mostly invested in ground rents, and subsequent events showed the wisdom of such a choice. He married twice. Of the five children of the first marriage, Henry Crookes, the eldest son, became a bookseller and Sir William's choice companion.

Joseph Crookes then married Mary Scott, who came from Aynhoe, in Northamptonshire, and of her he had no less than sixteen children, the eldest and longest-lived of whom is the subject of this biography. When Crookes was himself sixty-five years old, he wrote to *The Times* to say that his family memory reached back to the time of the Great Plague. He wrote :

My father, Joseph Crookes, died in 1884, aged ninety-two. I have frequently heard him relate how, when a boy, he was interested in hearing from his great-grandmother, Mrs. Lound, then over one hundred, anecdotes and incidents connected with the Great Plague of 1665, which had been told her by her grandfather, a participator in and eye-witness of the events of that year. The narrator, my great-great-great-great-grandfather, was born about the year 1639, and lived at Staveley in Derbyshire, where the Plague was brought in 1665 by refugees from London. He was one of the few who took the Plague and recovered, although it settled in his hip and made him lame. He was employed with a few others in going from house to house to bring out the dead and put them on horses and sledges, when they were taken to Marston for burial. He died in 1729, aged ninety. His granddaughter, born in 1710, married a Mr. Lound, and occupied a farm a few miles from Staveley. She died in 1814, aged 105, in full possession of all her faculties.

The following figures show that these lives overlapped, in one case nineteen years, and in the other case twenty-two years—more than enough to render the above statement probable :

My g-g-g-g-g-father (name unknown), b. 1639, d. 1729, æt. 90.
My g-g-g-mother, Mrs. Lound, b. 1710, d. 1814, æt. 105.
My father, Joseph Crookes, b. 1792, d. 1884, æt. 92.

It will be interesting to know if there are many instances where a similar bridge of two arches is sufficient to carry us back from almost the present time to that of the Great Plague of 1665.

It is interesting to note that Mary Scott's sister Martha married Mr. Geeves, a Regent Street bookseller. This probably accounts for the strong influence exercised by everything connected with printing, illustrating, and publishing upon the mind of young William Crookes.

A diary called *Marshall's Ladies' Daily Remembrancer* for 1832 has been piously preserved. It belonged to Mary Crookes, and under date of Sunday, June 17th, we find an entry in her handwriting to the following effect :

Was taken ill between one and two o'clock in the morning. Baby born half-past five in the evening.

A few days after this auspicious event there came from Mrs. Geeves a quaint and touching little letter, which has also been preserved. It says :

Aunt Martha's kind love to her dear little nephew, and begs his acceptance of the *World with its Animals*, likewise a small purse which she thinks will be large enough to hold his money for the present anyhow.

This letter was probably the first missive received by William Crookes, and the beginning of a vaster correspondence, perhaps, than ever fell to the lot of a scientific man.

The Geeves family lived next door to the house where William Crookes was born. The prosperous bookseller's shop was, in fact, at 141, Regent Street. The following letter, addressed to Mr. Geeves, has been preserved :

> CONNAUGHT PLACE,
> *Augt 8th,* 1834.

SIR,

    In consequence, I believe, of the failure of Mrs. Stockdale, or her Successor, in Piccadilly, who supplied me with the *Edinb^gh and Quarterly Reviews,* I have had none sent me of the *former* since No. 117, Oct. 1833, nor of the *Quarterly* since *No.* 99, also *Oct.* 1833. If you can supply me with the Numbers that follow, viz. from 117—of the *Edinb^g* and from 99—of the *Quarterly* of both Publications, and continue to send them together with *Blackwood M.* at the times when they are published, I shall be glad to have them from you.

> I am, Sir,
> Your very Ob^t Ser^t,

MR. GEEVE,   NORTHWICH.
Bookseller, &c.

The intercourse between the two families appears to have been very lively, and young William must have derived more stimulus and inspiration from his uncle's bookshop than from his father's tailoring business.

In later years Sir William Crookes often told his astonished hearers that he clearly remembered learning to walk ! He said that one scene was indelibly impressed on his memory, and that it probably constituted his earliest recollection. Two gigantic figures, he said, appeared a long way off and held out their arms to him. (As they were both women, we may suppose they were his mother and his aunt Martha.) His back, at the time, was propped up against some hard object of great size. He tried hard to approach the giantesses across the immense intervening distance, and had almost succeeded in reaching them when they gave a shout and he promptly fell to the ground.

If Crookes's memory was not at fault, this recollection is very remarkable, as his age at the time cannot have been much more than a year.

Of William Crookes's childhood little has been recorded, and as his parents and most of his contemporaries are dead, the biographer has to glean a few facts from indirect sources. There is, however, ample evidence to show that the sphere in which young William was brought up was full of life and interest. The children of his father's first marriage were growing up, and were much about the house. " Lost 2 Habit Shirts, 1 Bonnet Cap, 3 Collars of Alfred " figures as an entry in Mary Crookes's diary, this " Alfred " being the second son of the first marriage, who eventually lived to celebrate his golden wedding. We also find that " Brother Richard came to town " and staid for four days—this being evidently Richard Scott, Mary's brother. About this time (1835) Mary Crookes " raised Elizabeth's wages from £10 10s. to £11 per annum "—in recognition, let us hope, of assiduous care devoted to the future discoverer of thallium.

Martha Geeves was a noted horsewoman, and her husband was in close touch with the literary and artistic world of the day. The memorable year of Queen Victoria's accession marked William Crookes's fifth year of life. Regent Street became a great centre of activity, and the Crookes family and its ramifications felt the full force of the stimulus. It was about this time that Joseph Crookes started to keep an autograph album. It is embellished with many portrait engravings, and contains some original contributions by prominent men of the day. Among these we find musicians like Franz Liszt, Sir Jules Benedict, S. Thalberg and Alexandre Olivier ; poets like Samuel Lover, Thomas Campbell, Samuel

Rogers, and Ebenezer Elliott (author of *Corn Law Rhymes*) ; Isaac D'Israeli, the author (father of Lord Beaconsfield) ; that adventurous improviser, Theodore Hook ; and Charles Mathews, the actor, who afterwards made a large fortune as a ventriloquist and entertainer. The collection also contains a letter to Joseph Crookes from Thomas Moore, the Irish poet, which I may quote here :

Dec. 9, 1834.

DEAR SIR,

The very day after I received your letter I returned an answer saying how much gratified I should feel by the honour which you intended me. What could have become of my note I cannot conceive. Regretting that you should have been, all this time, under the impression that I had neglected answering you,

I am,

Your faithful and obedient Servant,

THOMAS MOORE.

One of the most interesting letters of this time, however, is one addressed to Joseph Crookes by William Upcott. It reads as follows :

AUTOGRAPH COTTAGE, ISLINGTON,

May 20*th*, 1837.

DEAR SIR,

I thank you sincerely and I thank you heartily for the recovery of my long lost Manuscript—the sight of which instantly carries my recollections back to the days of my boyhood, and brings to my memory the forms and the features of the individuals therein noticed. Even at the present time, I can call to mind the delight experienced when engaged in the compilation and annotation of each Biographical Sketch, written in moments snatched from the hurry of a political Bookseller's business—so different to that of the publishers of the present time—and in the splendid days of Pitt and Fox, when political pamphlets were daily issuing from the press, between the important years of 1796 and 1800, when the last war was in full activity and the Union of Ireland was so warmly and generally discussed.

Some of the Articles contained in this MS. volume were selected and transcribed in Wrights' Shop facing Old Bond Street, Piccadilly—now pulled down—the leading Tory publisher of his time, and with whom I resided nearly four years, when I have seen sitting opposite my desk, around the cheerful fire—Dr. Joseph Warton, Editor of Pope's Works—George Steevens—William Gifford—George Canning—Wm. Combe, the Author of *Dr. Syntax, The Diaboliad, The Devil upon Two Sticks in England, The Royal Register, Letters of an Italian Nun,* and numerous other Anonymous Works—who contrived to procure occasionally a day rule, to stroll from the vicinity of the King's Bench, and mix with the distinguished characters that daily assembled at this political lounge. At other times have I there noticed the Marquis Wellesley, Lord Hawkesbury (afterwards Earl of Liverpool), John H. Frere, Geo. Canning, Wm. Gifford, Dr. Beeke, Edw. Nares—whose *Glossary* you are not unacquainted with—concocting materials for each succeeding Number of the Anti-Jacobin Newspaper, printed for Wright in 1798, and where I have been employed to copy and destroy the various papers on Finance—usually contributed by Mr. Pitt for that Journal, in his own handwriting, to avoid the discovery of the Author of these Articles in the printer's Office. In this Shop I have seen Mons.ʳ Mallet du Pan—Sir Francis D'Ivernois —Mons. Pettier, the political opponent of Bonaparte in his *L'Ambigu* —a periodical work that had a great Sale—and the French poet the Abbé De Lille—at that time nearly blind—all grouped together in some snug corner, debating on the news of the day or on literary subjects. It was during my residence in that situation I witnessed the affray between Peter Pindar and William Gifford —so much the conversation of the day, and the cause of several publications—one of which was entitled *The Battle of the Bards* —indeed, I assisted in turning Dr. Walcot from the shop into the street after he had received a violent blow on the forehead given by Gifford with Peter's own oaken stick which he had brought with him on purpose to chastise Mr. Gifford—for having written some very severe lines upon his character in his *Baviad and Moviad.* —I shall not easily forget the rage and violence of the Doctor on his being turned out—nor the remarks of the crowd that were gathered around him on this occasion. But nearly the whole of this distinguished group are now numbered with the silent dead !

EARLY DAYS

Once more accept my acknowledgements for the restoration of my small budget of Biographical Memoranda—and believe me to remain,

Dear Sir, your sincere Friend,

WILLIAM UPCOTT.

TO MR. JOSEPH CROOKES.

About this time Joseph Crookes bought Bute Cottage, situated near Regent's Park. Young William Crookes grew up to boyhood in this house, which was part of a village then called Park Village East. Of regular schooling he had very little, though he spent some years at a " College " in Weybridge, called Prospect House. A small trunk, the size of a suitcase, and covered with horse-hair, is still preserved as a family relic. It sufficed to contain the clothes which he took with him to school after the holidays.

In 1904, Crookes himself gave Mr. Francis Galton a candid opinion of himself and his relatives in a letter in which he wrote :

I can find no trace of such eminence in any of my ancestors as would bring them up to the standard you fix—a standard which is a perfectly reasonable one. With the exception of a younger brother, who died when he was twenty-one, and who was a clever mathematician and physicist, and who would certainly have made an eminent name had he lived,—with this exception I am what botanists would call a "sport." From my earliest recollections I was always trying experiments and reading any book of science I could find. A little older and I fitted up a cupboard as a sort of laboratory, and caused much annoyance and trouble in the house by generating smells and destroying furniture. I don't suppose any of my family even knew the meaning of the word "science," and I was always regarded as a bit of a fool, who would never get on. It is a great satisfaction to me to think that my parents lived long enough to change their opinion, and to feel proud of the reputation I was making.

16

The earliest letter which was written by Crookes and has escaped destruction is the following :

BUTE COTTAGE,

1841, *September 6th.*

MY DEAR MAMMA,

As you are in the country I think you would like to receive a letter from me to know if I am a good Boy and how I get on at school. I am not in disgrace more than once a week, which I hope you will not think too often for such a little Boy as I am. I hope you will find me improved in writing. I am already in Long Division in Arithmetic, which I do not think very difficult. I have had lately rather difficult lessons, particularly the last I learnt in Mangnall's Questions, it was a list of a few of the eminent men who flourished many years ago, and I was obliged to say them in regular rotation, which I could not very well do, their names being so very odd that I could not remember them. Father wrote to me on Friday to ask me to come home on Saturday by the two o'clock omnibus, but it was so very wet I was afraid Mrs. Keikhoffer would not allow me to come, but at last she sent some person into the School-room to say I was to go by the seven o'clock omnibus, you may guess, Mamma, how pleased I was. Jane has asked John Vinton to dine with me to-day, at which I am very glad, but it is so wet we shall not be able to take a walk this afternoon, which has rather disappointed me, but we must manage to amuse ourselves as we best can at home. I do not think I have any more to tell you, but I will write you another letter again soon, and do not forget to answer them. Give my love to Walter and Frank, uncle William, uncle John, and so good-bye,

My dear Mother,

Your affectionate Son,

WILLIAM CROOKES.

P.S.—I am com(e)ing home on good gooseday, I hope I shall have my brothers with me, &c., meaning you, Mamma.

MRS. CROOKES,
 at Mr. Scott's,
Souldern, near Deddington, Oxon.

There is another letter written five years later, when

Crookes was fourteen. It shows the typical development of the healthy boy of that age :

Prospect House,
November 8th/46.

Dear Father,

I hope you will excuse me for not writing sooner, but I thought it would be better to wait till after the 5th, as perhaps you would like to hear something about it ; I had laid in a stock of Squibs and Crackers, &c., two or three days before, and on Wednesday, when I was going away from College, I asked Mr. Wilson if he would give me a holiday the next day, as all the other boys were going to have one, but he, as I expected, refused. On Thursday, in the afternoon, I went out for a walk, and as I was returning I met some of the other boys with the manservant, who had gone to get a large tallow barrel, which had been promised us by a Gentleman a day or two before for our bonfire ; when we got home, although it was before tea-time, they had commenced letting off their fireworks, and I was just going to follow their example when the bell rang for the parlour boarders to get their tea ; a little time after the bonfire was lit, and we had some fine fun, but soon the fireworks were exhausted, and then I and a few others of the big boys went out in the town to see the fireworks there ; one of the streets named Regent Street was quite blocked up with people, it was as much as ever we could do to get through ; we expected to see the " town and gown " fight, as they generally do, but I do not think they did much this time, at least they did not while we were out. We did not stay long, and when we came home almost all the boys had gone to bed, so we had our supper and went to bed ourselves very tired. From what I have heard Mr. Wilson say, I expect Joe down at the beginning of next week, and Mr. Wilson says that he shall wait till then to take me out, and then we can all go together. I think I shall write to Aunt Geeves in a day or two. Ask Mamma to give my love to her when she sees her, and accept the same from

Your Affectionate Son,
William Crookes.

P.S.—I suppose I may expect at Christmas to see you comfortably settled at Brook Green.

He had a French drawing master, who delighted in the delicacy and accuracy of detail which William Crookes was able to impart to his drawings. Many of these drawings have been preserved. They comprise landscapes, figures, and insect life in pencil and indian ink, and, while somewhat devoid of imagination, show a wonderful steadiness of hand and correctness of outline.

When William Crookes was fourteen years of age, his father acquired Brook Green Farm, near Hammersmith, where he lived until his death thirty-eight years later. Thenceforth Brook Green constituted the country retreat of all the Crookes family, and for a generation its large garden nurtured in turn the childish and youthful imaginations of the younger generation of Crookes.

CHAPTER III

THE ROYAL COLLEGE OF CHEMISTRY

(1848–54)

THE MARRIAGE OF QUEEN VICTORIA to the Duke of Saxe-Coburg in 1840 had a far-reaching effect upon the relations between England and Germany. It led to a large influx of Germans into England, and to a Germanophile policy which lasted until the time of Bismarck and the foundation of the German Empire. The stirring events of 1848, during which the " Prince of Prussia " (afterwards Emperor William I) was stoned out of Berlin, were followed by a wave of emigration of the revolutionary elements to the *Hort der Freiheit* across the North Sea.

It is not surprising, therefore, to find that when it became a matter of public importance to appoint a head of a new College of Chemistry in London the appointment of a German chemist was taken for granted.

A proper provision for chemical training in the metropolis was necessitated by the industrialisation of England which was then in progress. As early as 1842 certain industrial leaders had proposed the foundation of a " Davy College of Industrial Science " in memory of Sir Humphry Davy. The formation of the Committee of the Privy Council on Education, the establishment of the Penny Post, the railway boom of 1844, all contributed to the strengthening of the demand for scientific training, and among the reasons given for contributing to the funds for such training, its probably favourable effect in view of the prevailing agricultural depression was much emphasised. This Royal College of Chemistry was opened in 1845 by the Prince Consort. It was temporarily housed in George Street, Hanover Square, but was subsequently transferred to more commodious premises at 16, Hanover Square. Its establishment, significantly enough, coincided with the repeal of the Corn Laws. The professorship had been offered to sundry distinguished German chemists, but declined by them. Eventually it was accepted by a young *Privatdozent* of Bonn University, aged twenty-seven, and named August Wilhelm Hofmann. This acceptance turned out to be a most fortunate one for the College, for not only did Hofmann gain the affection and stimulate the enthusiasm of the students, but he served the College in good and evil days with a devotion seldom equalled in academic affairs.

William Crookes entered the College in 1848, amid the turmoil of the Chartist riots. He travelled up daily from Brook Green by the omnibus, no doubt a vehicle much resembling the lumbering pirate described by Dickens ten years earlier.

Crookes threw himself into the work of the College with extraordinary enthusiasm. Bloxam, one of his seventy fellow-students, when recalling those days in after life, said that Crookes was remarkable for the extreme care and painstaking accuracy of his notes, especially the records of laboratory work. Mrs. Cowland, Crookes's surviving daughter, told me that one of the most remarkable things about her father was that his writing was always easily legible, and remained so up to his death. This quality, which is very evident from his extant correspondence, is all the more noteworthy when we consider the vast scale of his literary activity. It also indicates a fine balance of hand and eye such as has been pronounced the first essential of scientific ability.

Nor had the gifted student to wait long for his harvest. Within a year he gained the Ashburton Scholarship, which relieved his parents from the expense of his training for the next twelve months. By the end of that time Hofmann appointed Crookes as his Junior Assistant, and within another year Crookes became Senior Assistant in the place of J. S. Brazier. That was in 1851, when Crookes was nineteen years of age. It was the year of the Great Exhibition, held in the Crystal Palace, which was at that time erected in Hyde Park. The Exhibition was largely the work of the Prince Consort, and was probably the most signal service which that German prince rendered to his adopted country. The success of that Exhibition, and the funds derived from it, established the pre-eminence of South Kensington as the centre of science and art, and rendered possible the great and far-reaching pioneering work carried out since then under the auspices of the Science and Art Department (founded in 1853).

An idea of the amount of work accomplished by Crookes in the first two years of his training at the Royal College of Chemistry may be derived from the fact that at the end of that time he was in a position to publish his first original contribution to the advancement of science. It was a paper, " On the Selenocyanides," [1] read before the Chemical Society on June 20, 1851. Its scope is best indicated by the opening paragraphs :

The remarkable parallelism between sulphur and selenium, which has been traced in so many directions, left but little doubt

[1] Selenocyanides are analogous to the sulphocyanides, in which the metal is combined with sulphur (S) and cyanogen (CN). Ammonium sulphocyanide (AmSCN) is much used in photography.

WILLIAM CROOKES AS HOFMANN'S ASSISTANT, 1850.
ÆT. 18.

To face p. 22.

respecting the existence in the selenium series of a class of compounds corresponding to the sulphocyanides. In fact Berzelius (*Traité III*, 105) mentions that by fusing selenium with ferrocyanide of potassium a salt may be obtained which possesses the general characters of the sulphocyanide of potassium, but is less stable than the latter compound. This salt, however, appears to have hitherto escaped a closer examination. Berzelius makes no mention that it has ever been analysed, nor have any other selenocyanides been investigated. This circumstance induced me to re-prepare the potassium salt, in order to fix its composition by numbers, and to study at the same time several other selenocyanides, in order to establish more fully the character of this class of compounds.

He then goes on to describe the preparation, analysis, and properties of the selenocyanides of potassium, silver, lead, mercury, ammonium, barium, strontium, calcium, magnesium, zinc, iron, and copper, the greater number of which were compounds unknown to the chemistry of the day. He ends by acknowledging the "valuable advice and assistance" of Professor Hofmann, especially in placing at his disposal a large quantity of selenium.

Although Crookes was only nineteen when he wrote that paper, it bears not the slightest mark of immaturity. It is written in quite an adult style, and the most experienced chemist of to-day could hardly treat the subject with greater power or competence.

In view of the fact that William Crookes was trained as a chemist rather than a physicist, it is worthy of note that his first scientific paper was his last paper dealing with pure chemistry. This shows that his impulse was towards knowledge in the abstract rather than a narrow department of knowledge. It bears witness to his wide sympathies and free outlook. But it also accounts for the fact that an academic career was not destined to be his. And so his choice remained untrammelled, and he was free to advance in any direction in which the prospect appeared alluring. His family being, by this time, in easy circumstances, he might have drifted into the ranks of those numerous amateurs who follow any gleaming light, even though it be a will-o'-the-wisp, and waste their lives in vain speculation or in work which has long ago been completed by other and more competent investigators.

What saved Crookes from this fate was his consciousness of mastery. He felt fit to be in the front rank, he asserted his right to be there, he established a good title to his place, and maintained it against all competitors to the end of his life.

It is just possible that the state of chemical science in the fifties may not have been sufficiently inspiring to secure the whole-hearted allegiance of a young devotee of science. The doctrine of valency was just then being developed by Edward Frankland. The doctrine of " vital force " had already died with Wöhler's synthesis of urea, thus liberating organic chemistry from ancient and outworn shackles. But the " bonds " which Frankland assigned to each chemical atom, like tentacles capable of holding an equal number of tentacles outstretched by fellow-atoms, had not yet been generally accepted. Compound radicals like cyanogen, benzoyl, and cacodyl had, indeed, been fully recognised, but the fruitful conception of " types " had yet to be fully worked out by Gerhardt, Wurtz, Williamson, and Hofmann himself. Crookes attended Faraday's luminous lectures at the Royal Institution and took copious notes. And it is not surprising that Faraday's strong leaning towards physics should have communicated itself to his brilliant student. He was also attracted by the personality of Sir Charles Wheatstone, the Gloucester manufacturer of musical instruments, who became professor of " Natural Philosophy " at King's College, London (next to Somerset House), where the "Wheatstone Laboratory" is still proudly in evidence.[1]

The two sciences met on a common ground in photography, where light—a physical agent—was made to produce chemical effects. Crookes seems to have thrown himself with avidity into photographic work as soon as he had "made his bow" to pure chemistry in the company of his selenocyanides. In 1852 he tried to photograph the coloured rings shown by certain crystals between tourmaline plates in polarised light, but did not succeed until Wheatstone placed the "magnificent tourmalines and crystals" of King's College at his disposal. He then not only obtained the photographic records of those beautiful phenomena, using either calcspar or nitre, but also traced certain abnormal figures, due to rays beyond the visible spectrum, which had never been seen by eye at all.[2]

Meanwhile the Royal College of Chemistry was falling on evil days. The subscribers who had liberally contributed in the hope that the College would help to rescue agriculture from the prevailing depression saw no immediate realisation of their hopes. The students' fees, though substantial, did not cover expenses. Hofmann, with rare generosity, gave up first his proportion of students' fees, and then a portion of his salary, while abating nothing of his strenuous work. It was finally decided to hand over the College to the care of the Science and Art Department (1853). Some years later it was transferred to South Kensington and enlarged into the Normal School of Science, with which was associated the Royal School of Mines founded in 1851. The "Normal School" was,

---

[1] Wheatstone's best-known device is the "Wheatstone bridge," used in measurements of electrical resistance.

[2] See *Journal of the Photographic Society* for June 21, 1853 (read June 2nd): "On the application of photography to the study of certain phenomena of polarisation," by Mr. W. Crookes.

in 1891, re-named the Royal College of Science. At the present day this Royal College of Science is an integral part of the Imperial College of Science and Technology, rivalling the University of London itself in its importance as an educational centre.

Before Crookes left the Royal College of Chemistry (which he did in 1854), we find him engaged in a number of spectroscopic researches which have remained unpublished, but which have an important bearing on his claim to have anticipated to some extent the law of Kirchhoff and Bunsen, according to which a metallic vapour absorbs the rays of the wave-length of those which it emits ; in other words, that emission and absorption apply to the same wave-length. In an old notebook of " Original Experiments and Researches " we find the following under date of April 1854 :

EXAMINATION OF THE SPECTRUM PRODUCED BY COLOURED FLAMES

A fine vertical slit about 0·005 of an inch in diameter was illuminated with a spirit lamp, on the wick of which was placed some substance capable of imparting a colour to the flame. A spectrum was formed by an arrangement of prisms and a lens. The light being too feeble to show anything when projected on a screen, the spectrum was received direct into the eye by means of a positive eye-piece, and in the focus of the eye-lens a piece of glass, ruled with divisions 0·005 of an inch apart, was fixed. The yellow line which sodium gives was taken as the starting-point, and the other lines were measured and marked + and —, according as they were more or less refracted than this line.

Soda.—The spectrum given by sodium consists of a brilliant yellow line which, when observed with great care, was seen to be double, with no trace of any other visible rays. This line occurs in exactly the same place as the double black line D in the solar spectrum. Carbonate of soda was the salt used, as that was the nearest at hand, but any soda salt was found to answer, indeed the difficulty is to get a flame which is sufficiently pure from soda to be without the line. It is present with light of nearly all candles and lamps, and seems to reside chiefly in the exterior envelope of intensely heated air surrounding the flame. The following experiment confirmed the statement mentioned above :

The light from a white cloud was allowed to illuminate the above-mentioned slit. The spectrum obtained was viewed by means of an eye-piece, and when the line D was well in focus an assistant moved a spirit lamp with a soda flame backwards and forwards in front of the slit. Immediately a bright line darted along the vacant space of D ; and by regulating the brilliancy of the soda flame, the line was found to harmonise perfectly with the adjacent colours of the solar spectrum, and the (dark) line D was so completely obliterated that its former place could not be pointed out ; the other lines of the spectrum were entirely unaffected.

Though this account by the youthful spectroscopist contains nothing new except the experimental arrangement, which Crookes called his " spectrum camera," the experiment itself is an interesting variation of Kirchhoff's proof of the identity of the dark and bright sodium lines.

Moreover, spectroscopy was then in its infancy, and it required considerable enterprise and ingenuity to follow up the latest Continental researches and repeat them in an improved and simplified form. Indeed, Crookes kept himself up to date to such good purpose that he was soon able to outstrip all rivals and make a discovery of the first rank, as we shall see.

To this period of Crookes's life we must also assign the beginning of his friendship and collaboration with John Spiller, a young chemist employed at the Royal Arsenal, Woolwich. Spiller and Crookes were engaged for some time in the improvement of the collodion process of photography. In a joint contribution to the *Philosophical Magazine* of May 1854, the two young investigators discussed the use of glycerine for the purpose of prolonging the moist and sensitive state of a collodion film.

This friendship lasted for many years, until it was split on the hidden rock of " psychic force," as we shall see hereafter.

CHAPTER IV

PEREGRINATIONS

(1854–6)

THE YEARS 1854 AND 1855 are Crookes's *Wanderjahre*. In May 1854 he was appointed Superintendent of the Meteorological Department of the Astronomical Observatory founded at Oxford by the executors of John Radcliffe, the seventeenth-century physician. Crookes's connection with the Radcliffe Observatory lasted but a year, but his presence made an appreciable difference to the routine of the Meteorological Department. That routine required a great deal of personal attendance on the instruments indicating temperature, pressure, rainfall, the duration of sunlight, the direction and velocity of the wind, etc. Moreover, when readings have to be taken at regular intervals, say, of one hour each, it is necessary to watch the clock and time the observations carefully, as the resulting curves lose much of their usefulness without an accurate time record. It saves much personal inconvenience if the recording can be entrusted to an impersonal agent, and one may readily imagine that Crookes, conscious of his inventive ability, was somewhat impatient of old-fashioned and inconvenient methods. In a supplement to the Observatory records bearing the title, " Description of the Wax-Paper Process employed for the Photo-Meteorographic Registra-

28

tions at the Radcliffe Observatory, by William Crookes, Esq.," the desiderata of a process for making meteorographic records are set out as follows :

1st. The process adopted must be one combining sharpness of definition with extreme sensitiveness, in order to mark accurately the minute and oftentimes sudden variations of the instruments.

2nd. To avoid all hurry and confusion, it is of the utmost importance that the prepared paper or other medium be of a kind capable of retaining its sensitiveness for several days.

3rd. The contraction which paper undergoes during the numerous operations to which it is subject in most processes (in general rather an advantage than otherwise) is here a serious objection. For this reason, the experiment first tried, of transferring to paper the image received on collodion, preserved sensitive by the nitrate of magnesia process, was a failure.

4th. Strong contrast of light and shade, and absence of half tint, unfortunately so common amongst ordinary photographic pictures, is in this case no objection.

5th. It is essential to preserve the original results in an accessible form, and for this the Daguerreotype process is not suitable.

Crookes, after twelve months' unremitting attention to the problem, arrived at a modification of Le Gray's wax-paper process which was successfully applied to the barograph and thermograph already installed at the Observatory, to a " pluviograph " or rain-gauge designed and made by Crookes himself, and to a photometrograph or daylight recorder. The waxing of the paper was a device for preventing the dissolution of the size in the material of the paper by the successive baths and washings. (This was in the days before gelatine had been pressed into the service of photography.) This method of recording was carried out with the full approval and co-operation of

Mr. M. J. Johnson, the Observer, and with the help of Crookes's assistant, Mr. George Green. The following letters written by Green speak for themselves :

RADCLIFFE OBSERVATORY,
December 9, 1854.

DEAR SIR,

You will be glad to hear that during your absence I have carried the work on without an interruption. The only mistake I have made is misplacing a sheet which I took from the Barograph and put into the Photometrograph, but the images now developed are very plain and distinct.

I have not had time to make experiments with the Calotype process, but hope to make some on Monday.

I have the honour to remain,
Your obedient Servant,
G. GREEN.

W. CROOKES, Esq.

RADCLIFFE OBSERVATORY,
Dec. 19, 1854.

DEAR SIR,

For your satisfaction I write to inform you that all things are going on favourably. I get excellent pictures without dirt. We have forwarded our meteorological report to Le Verrier at Paris, so that your sheets will not be required. Mr. Johnson is suffering with Toothache.

I remain, Dear Sir,
Yours faithfully,
G. GREEN.

The third letter was written five years later, long after Crookes had left the Observatory. It is as follows :

RADCLIFFE OBSERVATORY,
March 29, 1859.

DEAR SIR,

I received your letter this morning and placed it before Mrs. Johnson, who desires to express her obligations to you both for your kind sympathy and for your kind promise of the negative, of which I will take the greatest care, and return it to you when the number of copies required are printed.

Your knowledge of Electricity would render great service to the Observatory could you be present for a short time, and when another director is appointed I will endeavour to avail myself of your instructions ; at present we can do nothing but to carry on the general work of the place without interruption. I continue to follow your method of preparing the wax paper, and can say I seldom fail to get a picture. I have practised other processes, of which I herewith send results. I have also been successful in some large wax-paper views.

I am, Dear Sir,
Yours very truly,
GEO. GREEN.

W. CROOKES, Esq.

In the summer of 1855 Crookes was appointed Teacher of Chemistry at the College of Science, Chester, a position which he held for one year. Beyond some appreciative letters of pupils written in later life we have no record of what happened there. But an interesting sidelight on the conditions under which Crookes taught is thrown by the following letter :

CHESTER,
21 *Nov.*, 1855.

DEAR SIR,
The leakage in the Heating apparatus is such as to prevent the use of a fire for the Laboratory. I much regret that you feel the cold so intense—the Pupils doubtless feel it also.

Under such circumstances it will probably be satisfactory to you to know that arrangements are made with Mr. Hutchinson and Mr. Beesley for your accommodation in the class rooms ; therefore you can henceforth give your lessons in the room in which you will find your classes.

Yours faithfully,
ARTHUR RIGG.

W. CROOKES, Esq.,
College.

The worthy Mr. Rigg apparently thought that chemical preparation and analysis could be carried on quite as "satisfactorily" in a class room as in a fully equipped chemical laboratory. Crookes's reply is not extant, but one can well imagine that the brilliant young chemist felt rather cramped in a small provincial college, and missed the stimulating mental atmosphere of the metropolis. It is certain that he paid frequent visits to Liverpool and Manchester, as well as the native county of his father, only seventy or eighty miles distant from Chester. Certain also that he was at this time courting a girl of nineteen, Ellen Humphrey by name, who lived at Darlington, an only child of a widowed mother, long acquainted with the Crookes family.

"My wife was considered very beautiful when she was young," Crookes wrote sixty-one years later, in a letter to an Italian painter. "Her eyes were of a blue-grey colour, and her complexion was a very good pink and white. Her hair was of a rather dark brown colour, and it kept its colour almost to the end."

The young chemist and his Yorkshire lass were married in London, at St. Pancras Church, Euston Road, on the 10th of April, 1856. The portrait of the bridegroom dated from this period shows a strong, determined face with deep-set eyes and rather a large mouth, the chin surrounded by the "Crimean fringe" and the hair brought forward in curly bunches over the ears after the manner of the time. The bride's face shows an oval outline in strong contrast with the breadth of Crookes's forehead. The eyes are profound, rather close together, and surmounted by heavy eyelids. The nose is well-shaped, and the mouth eloquent with a gentle appeal, though somewhat unsymmetrical. The hair is severely parted in the middle, and falls in ringlets over the shoulders.

WILLIAM CROOKES, 1856.
ÆT. 24.

To face p. 32.

The wedding ended Crookes's wanderings, and commenced a companionship the value of which is best shown by the inscription now adorning the " Nellie Cot " in the West London Hospital, Hammersmith :

THE NELLIE COT.

In perpetual remembrance of

ELLEN LADY CROOKES,

for sixty years the beloved Wife and faithful Companion and Friend of

Sir WILLIAM CROOKES,
O.M., F.R.S., D.Sc., LL.D.

CHAPTER V

THE BUSINESS OF LIFE

(1856–8)

WILLIAM CROOKES'S MARRIAGE took place at the end of the Crimean War and the beginning of the Indian Mutiny. His father had retired from business, and lived at Brook Green Farm, Hammersmith. His brothers, Joseph and Henry, were in the bookselling business, with a close connection with Bohn & Co. His brother Alfred had taken over the tailoring business at 143, Regent Street. All the children were grown up except Crookes's youngest brother, Philip, to whom he was devotedly attached.

Crookes's prospects at the time of his marriage were far from bright. It is true that his parents were well off, but there were many brothers and sisters, the result of Joseph Crookes's two marriages, and all of them were in honour bound (if not compelled by necessity) to make their way in the world.

Crookes had no regular income, no appointment, and no university or other distinctions to recommend him for one. It is true that he had had an excellent chemical training, but in those days a scientific training was no passport to a public appointment, and opportunities such as had come to Faraday and Wheatstone were rare.

Crookes had little opportunity of teaching chemistry,

and the choice of a profession must have met with formidable obstacles on every hand. He finally appears to have decided to throw his weight into the development of photography, in which, as we have seen, he was already an adept.

While still in the North, he took some photographs of the moon in collaboration with Mr. Hartnup, an amateur astronomer in Liverpool. The photographs were $1\frac{1}{4}$ inch in diameter, and were beautifully sharp. Through the intervention of Professor Wheatstone, he obtained from the Royal Society a grant of £20 towards the expenses of enlarging these photographs. A similar grant was made by the Science and Art Department towards the cost of experiments for finding a portable means of illuminating objects in dark places sufficiently to enable them to be photographed. On April 7, 1856 (three days before his wedding), Crookes wrote as follows:

To the Secretary of the Department of Science and Art,
Marlborough House.

18, STANHOPE STREET,
HAMPSTEAD ROAD,
April 7, 1856.
SIR,

I beg to acknowledge the receipt of your letter with the memorandum by Dr. Lyon Playfair. I enclose an appendix to my report, and if after its perusal the proposed demonstration before Mr. T. Thompson be thought at present necessary, I shall be happy to appoint a time for it.

As I am about changing my residence, may I request that letters, &c., be for the present addressed as above.

I am, Sir,
Your obedient Servant,
W. CROOKES.

The above letter shows no undue subservience to the Government Department addressed. It does not seem to have occurred to Crookes that the Department officials might possibly prefer to " appoint a time " themselves for testing his progress. But Crookes had good reason for maintaining his freedom of action for the next few weeks !

However, after a fortnight's honeymoon spent in Yorkshire, Crookes wrote a letter couched in terms more consonant with departmental dignity :

To Dr. Lyon Playfair.

18, Stanhope Street,
April 25, 1856.

Sir,

Having been out of town lately your letter of the 15th inst. has only just reached me. I take the earliest opportunity of acknowledging its receipt and saying that I shall be most happy to repeat any of my experiments before Mr. Thompson whenever it may be thought necessary for me so to do.

I am,
Your obedient Servant,
William Crookes.

In the course of the experiments referred to, Crookes discarded the voltaic arc between carbon poles " on account of the extreme labour and difficulty attending its use," though he acknowledged it to be the electric light *par excellence.* He also tried the " electric egg " and the mercury arc, but, in the absence of a cheap source of current, was stopped by the expense attending the use of batteries. He finally recommended an improved " limelight " consisting of two nozzles close together, one of them for blowing a stream of oxygen through a hydrogen jet issuing from the other, and the combined jet playing on a ball of magnesia supported on a pillar of platinum wire. He claimed that the use of this light was free from danger.

About this time Crookes compiled a 60-page *Handbook*

to the Waxed-Paper Process in Photography, which was published in 1857 by Chapman and Hall.

While laying the greatest stress on photography as a means of livelihood, Crookes lost no opportunity of extending his knowledge in order to qualify himself as a consulting chemist and pharmaceutical expert. Thus we find him studying such widely divergent subjects as food adulteration and the outcrops of English geological strata.

In May 1856 Crookes carried out some careful determinations of the specific gravities of eleven specimens of iron rails to be laid down in India, it having been suggested that there might be a connection between specific gravity and durability.

In November 1856 Crookes appears to have undertaken the editorship of the *Liverpool Photographic Journal,* published, among other periodicals, at Greenwood's " Steam Printing Works," 16, Canning Place, Liverpool. This engagement lasted until March 1857, and did not entail any necessity for leaving London. In March Crookes became Secretary of the London Photographic Society and Editor of its *Journal*—a position which for the first time gave ample scope for all his talent, and launched him on the editorial career which he followed for the rest of his life.

Another milestone in his life is a letter from Dr. J. H. Gladstone, a man slightly older than Crookes, who later made a great name as a spectroscopist.

21, Tavistock Square,

4 Feb., '57.

My dear Mr. Crookes,

I have sought in vain for any paper of yours in which you make known to the public what you told me about the soda flame. I suppose, however, I am at liberty to mention it (associated,

of course, with your name) on Friday evening—especially as I find
that through your notice of it at the Discussion Society, or somehow,
the fact is no secret.

I fancy you have tickets for the Friday evenings ; if not, and
you would like to come next Friday, please let me know.

I remain,

Yours very truly,
J. H. Gladstone.

This letter initiated a life-long friendship, which
extended to the families of the two men.

Crookes came to the Photographic Society with a
considerable reputation. His photographs of the moon
taken at Liverpool with the Hartnup telescope had been
exhibited at the Crystal Palace, then transferred to Syden-
ham. Their multitudinous detail permitted an enlargement of twenty diameters—a scale of enlargement which
would be difficult of attainment by means of a modern
gelatine negative.

Shortly after taking up the Secretaryship of the Photo-
graphic Society, Crookes was elected a Fellow of the
Chemical Society (F.C.S.), of which he lived to be the
President thirty years later.

At the last meeting of the Photographic Society before
the summer vacation of 1857, Crookes read a paper entitled
" The Albumen Process on Collodion." It embodied a
good deal of study of various expedients for obtaining
a perfectly clean negative. Glass is very prone to take up
impurities of all sorts, even when carefully handled.
Crookes therefore proposed in the first instance to coat
it with collodion, and then to use it as a basis for the albumen
process, which was already well known. Incidentally, he
invented a light-tight box of tinplate, such as is even
now extensively used, and a bottle for pouring albumen
free from froth.

An episode which belongs to this period is Crookes's

attempt to photograph a flying projectile. The story is best given by quoting some of the letters *in extenso* :

11, BRIDGE STREET,
WESTMINSTER, S.W.,
4 May, 1857.

DEAR SIR,

I am permitted by my friend Dr. Percy (School of Mines) to use his name as an introduction to you. I am anxious to have some conversation with you if you will permit me, on the subject of having photographs taken of a pair of gigantic wrought-iron Mortars to throw shells of a yard in diameter and of a ton and half weight—which are completed from my designs for Government.

Would you name any convenient hour to yourself after half-past 2 of to-morrow or after 12 on Wednesday, when I might have the pleasure of calling upon you for the purpose,

and oblige truly,
Your obedient Servant,
ROBERT MALLET.

WM. CROOKES, Esq.

11, BRIDGE STREET,
WESTMINSTER, S.W.,
5 Nov., 1857.

WILLIAM CROOKES, Esq.,
   Secretary, Photography Society,
      15, Stanley Street,
         Brompton.

MY DEAR SIR,

You will recollect my talking to you about Photographs of my large Mortars that have been recently tried at Woolwich, and will be again fired in a few days.

The Photographers of the Arsenal have greatly improved their results, so that it will probably be unnecessary to ask for the application of your ability to the original object of Photographing the Mortars and Shells. But another object much more worthy of your prowess now presents itself. My object is to know if you think you can effect it, and if so, I will take the necessary steps to get you professionally engaged by the War Department to perform it.

THE BUSINESS OF LIFE

It has become a matter of great interest in a *Gunnery* point of view to get a correct transcript of the trajectory of those large shells (a yard diameter) in their flight, which occupies 30 or 40 seconds in flying over a horizontal range of a mile and half, rising ¾ of a mile or so into the air. Now it has occurred to me that *possibly* you could upon an extremely sensitive surface obtain a trace of the flight of the black shell against the sky through which a regular curve could be afterwards drawn. I am informed that surfaces may be obtained so sensitive as to be darkened in a single second.

There will be no difficulty in so connecting a cover or spring flap to the camera that by a galvanic arrangement the camera shall be opened, and the surface exposed, at the instant that the shell quits the mouth of the mortar.

Your camera could be placed at one side, say ½ a mile off the plane of projection of the shell, thus:

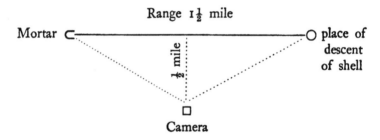

Range 1½ mile

Mortar ⊂————————————————O place of descent of shell

½ mile

□
Camera

or so that the *view* would subtend 90° or so, but if you deem it impossible to take in so extensive a field—we might commence with a *small range*, of, say, ¼ mile, the time of flight of which would be 8 to 10 seconds.

Our next day's firing will probably be in about 8 or 10 days—dependent, however, upon Lord Panmure, who will be present—so that if the thing be possible there is no time to lose in preparation.

I shall be here at my home, Delville, Glasnevin, Co. Dublin, for a week (and then in London), and shall be glad to hear from you as soon as possible, addressed either so, or to my Dublin Office.

Very truly yours,
ROBERT MALLET.

Crookes wrote as follows :

1, NEW COVENTRY STREET,
PICCADILLY, W.,
Dec. 22nd, 1857.

MY DEAR SIR,

According to your wish I attended at the trial of the 36-in. Mortar on Friday last, the 18th inst., with my photographic apparatus for the purpose of seeing whether it would be possible to obtain a sufficient photographic trace of the shell in any part of its flight to allow of its Trajectory being traced thereform. The weather was most unfavourable for such a purpose, as owing to the low temperature and dull light I could hardly hope to obtain the extraordinary rapidity necessary. On one occasion, however, during a gleam of bright sunlight, I did succeed in obtaining an impression of the shell as it was passing across a brilliant patch of sky, but so faint as to be hardly trustworthy. I am, however, inclined to think this a satisfactory result, as in brilliant summer weather I could without difficulty (even with my present imperfect apparatus) count upon at least a four-fold rapidity.

Having been unsuccessful in getting a satisfactory curve by means of Photography, I directed my attention to a plan which occurred to me when you first mentioned the subject : this was, tracing the path of the shell with a pencil, on a sheet of ground glass placed in the focus of the camera lens. After one or two attempts I found that although the *ordinary* camera and lens would not give an image sufficiently good to follow with the pencil point ; yet one which I had provided in case I were obliged to resort to this means answered very well, and enabled me to follow the path with the greatest accuracy ; the curve traced by the pencil being in no part of its course more than two diameters of the shell in error. I enclose two tracings of the paths of the 6th and 7th shells fired, together with an enlarged copy of the latter. These were taken from a point considerably behind the Mortar, and consequently the curve shown is a very oblique view of the true path described ; but of course there would be no difficulty whatever in projecting the real curve with the help of easily ascertained data.

I should feel much obliged if you would use your influence in obtaining for me permission to continue these experiments at future times when the Mortar is fired. Even supposing that *pure*

Photography should not eventually prove to be the readiest means of obtaining the Trajectories, and my plan of tracing the path on ground glass or other material should be found to give the best result, still I am convinced from my experience of Friday last that there are very few lenses capable of giving an image sufficiently intense, large, and distinct to be readily traced, and until all the conditions of success are discovered the skill and assistance of an experienced Photographer would be absolutely necessary, and in all cases a photograph of the mortar and horizon included in the shell's path should be taken before any change were made in the arrangement of the apparatus.

On a future occasion I should wish to plant the apparatus truly at right angles to the plane of the shell's flight so as to obtain the true curve direct : and I think it would also be very desirable to employ some means of registering the position of the shell at every second or half-second during its flight. This could easily be done, photographically, by fastening a screen to a pendulum which should vibrate in front of the lens and cross it at regular intervals : or, in the other case, by causing a momentary jerk, from a pendulum or other simple means, to be communicated to the ground glass at the desired intervals.

I remain, dear Sir,

Very truly yours,

WILLIAM CROOKES.

R. MALLET, Esq.

Mr. Mallet replied :

11, BRIDGE STREET,

WESTMINSTER, S.W.,

23 Dec., '57.

MY DEAR SIR,

I have forwarded your letter and curves to-day to Colonel Lefroy, R.A., Lord Panmure's Private Military Secretary, requesting him when he has read it to send it on to Col. Pickering, R.A., Sec. O. Select Committee, Woolwich ; and also recommending that you should be permitted to make further trials and possibly be on more complete success professionally engaged to complete the method of tracing trajectory.

Very truly yours,

RBT. MALLET.

WM. CROOKES, Esq.

The problem does not appear to have been pursued any farther at that time. The next great achievement in this direction, I believe, was when C. V. Boys succeeded in making a flying bullet photograph itself by means of an electric spark.

Crookes edited the fourth volume of the *Journal of the Photographic Society of London.* It extends over the period July 1857 to July 1858. The first and second volumes had been edited by Mr. A. Henfrey, F.R.S., and the third volume by the Rev. J. R. Major, M.A. It had become the custom to give the editor's name on the title-page.

Crookes retired from the editorship and secretaryship in 1858. His place was taken by Mr. Hugh W. Diamond, who, in his first editorial, rather unctuously announced that the *Journal* had outlived the necessity of appearing under the ægis of prominent men, and that it was no longer necessary to give the name of the editor. There is here some evidence of a current of jealousy, excited, perhaps, by the originality and brilliance of the editorship of Crookes. A quaint comment on Diamond's exordium is furnished by the fact that his name appeared regularly on the title-page of all the volumes of the *Journal,* until in 1869 the names became Diamond and Spiller.

I should like here to quote one more letter, written by Crookes to Mr. Rylander, who was the first to attempt artistic compositions made up of a number of different photographs. His most successful composition was entitled " The Two Ways of Life, or Hope in Repentance," and was built up from about forty negatives, each copied from living or real objects. As the following letter shows, Crookes was enthusiastic concerning the possibilities of this method :

THE BUSINESS OF LIFE

Dec. 8, '57.

MY DEAR SIR,

For the credit of Photographers I hope you will reconsider your determination of abandoning the glorious path which you have commenced, and if a few Hypercritical, Mock-Modest Scotchmen have no eyes or senses for anything beautiful and sublime, do not accuse the whole body of photographers of being insensible to all beauty in nature but sharpness.

If you knew the sensation which your glorious conception caused when exhibited at our meeting, and also the interest which your contributions caused at our Exhibition, you would, I am sure, feel amply repaid for the condemnation which you (in common, I suppose, with the old Masters) have received. I have mentioned to several gentlemen that you would, I thought, give us a paper on the subject, and I trust that you will not disappoint the general expectation of an interesting meeting because you have personally experienced the proverb of "throwing pearls before swine."

Your pictures of the "Blinds" are very curious and contain an interesting fact. I will give it more thought than I have had time to do at present—many thanks for them.

Believe me to remain,

Very truly yours,

WILLIAM CROOKES.

A. RYLANDER, Esq.

CHAPTER VI

THE CHEMICAL NEWS

(1858–60)

ON VACATING THE EDITORIAL CHAIR of the Photographic Society's *Journal*, Crookes made a two
years' agreement with Messrs. Petter and Galpin
for a supply of articles on photography to the *Photographic
News*, binding himself not to publish articles in any other
periodical.

He took a house at 20, Mornington Road, whither
he removed his wife and first child—a girl—as well as
his wife's mother, Mrs. Humphrey, who lived with them
until she died.

The house was fitted with a chemical laboratory, in
which Crookes planned to conduct chemical analyses
as well as research work.

One series of such analyses was made for the Coalbrook-
dale Co., of which his cousin, Charles Crookes, was manager.
The letter accompanying the results of the nine analyses
reads as follows :

Dec. 12, '57.

My dear Sir,

I have the pleasure of sending the results of my analyses
of the coals, iron ores, and pig-irons which you sent me. I did not
receive them till this day week, and as I understood you that you
wanted them as soon as possible, I have kept closely at them ever
since, and have finished the last this evening. I have made out

45

the bill at the same price as the previous analysis, but I must really ask for a little more for any further analyses with which you favour me. I have been obliged to leave other important things undone in order to get these done soon, and the amount of labour and thought required to do them properly is such that I cannot in future do them at the present low price. Do you think that it will be considered too much if I ask Two Guineas each instead of £1 8s. ? They are such as would be here charged £5 5s. for.

I sincerely hope that Charly is well by this time. Mrs. Crookes and myself are, I am glad to say, quite well, and unite in kind regards to yourself and all Friends at the Dale.

Believe me,

Very truly yours,

WILLIAM CROOKES.

C. CROOKES, Esq.

P.S.—A Merry Christmas when it comes.

The year 1858 made two notable additions to Crookes's list of correspondents. One of them was Howard (afterwards Sir Howard) Grubb, of Dublin, the maker of astronomical lenses and telescopes. The other was H. Fox Talbot, the foremost photographic discoverer of his time in England. A letter which sheds light, not only upon the history of photography, but also upon the public-spirited character of Talbot himself, is the following :

MR. CROOKES.

LACOCK,

Dec. 2, '58.

DEAR SIR,

I should be sorry you said anything so ungracious on my part, as to prohibit others from experimenting.

The fact is that in this, as in every other patented process, much must be left to the good feeling of the Public. It is impossible to draw a line of demarcation, but an Amateur can have no real difficulty in knowing whether he is only experimenting or whether he is really injuring a patentee. But a Patentee cannot safely declare that he permits all persons to exercise the invention for their amusement. The Lawyers would put the construction on it that he relinquished the patent. For that and similar reasons I must request you not to publish *anything* as from me, on the point whether Amateurs may or may not exercise the art. I should never think of troubling anybody who did me no harm. But I think that anybody who succeeded in making such perfect engravings as to wish to publish them, would probably at once communicate with me with a view to obtaining my license. At any rate he ought to do so.

The description of my process is correct. I have not as yet detected any error in it, and I am using it every day, just as described. The process of laying on the gelatine is quite easy, and good gelatine can be purchased of any Confectioner. If you publish an Almanac next year we will get up a much more perfect Chronology, at present there is not time. Herschel discovered the following " principles " : discovery of Hyposulphurous acid, discovery that hyposulphite of soda is a solvent of Chloride of Silver.

I myself was the discoverer of the broad principle that *positive* pictures could be made on paper by first making a *negative,* and then *fixing* it, and using it to obtain a second result, viz. a positive one. (Before that time all images or pictures on paper were *negatives,* which could not be fixed.)

In order to make out a list or statement of this kind, ample time should be given. Would it not be best to defer it till next year ? If, however, you publish now, I would remark that in 1853 I published my first method of engraving upon steel, and N. de St. Victor published his.

Yours very truly,
H. F. TALBOT.

In the course of the year 1859, Crookes matured a plan for an enterprise which, if successful, would automatically give him one of the most important and influential positions in the English chemical world. The project was to bring out a new chemical weekly under the name of the *Chemical News*. After consultation with other members of the family, Crookes decided to acquire the copyright and goodwill of an existing periodical called the *Chemical Gazette*, which was founded in 1843.

THE CHEMICAL NEWS

On account of his agreement with Messrs. Petter and Galpin, Crookes was precluded from making photography one of the features of the new periodical before 1862, but the following prospectus, which appeared in *The Athenæum*, shows that even with this limitation the scope of the new journal was quite sufficiently comprehensive :

A NEW SCIENTIFIC PERIODICAL.

On Saturday Next, *the 10th of December*, 1859,

Will be published, price Threepence, No. I of

THE CHEMICAL NEWS,

with which is incorporated

THE CHEMICAL GAZETTE :

A Weekly Journal

DEVOTED TO EVERY BRANCH OF

CHEMICAL SCIENCE,

AND AN ADVOCATE OF THE INTERESTS OF

THE CHEMIST AND CHEMICAL MANUFACTURER.

It is strange that in an enterprising kingdom like Great Britain, the all-important science of Chemistry, upon which so many of our arts and manufactures are based, should as yet be unprovided with a Weekly Record of its discoveries, improvements, and general progress. The object of the present undertaking is to supply this singular deficiency ; and it is confidently hoped that, by placing weekly before the Scientific Public the details of new discoveries, and the results of practical research in this and other countries, a very important and useful end will be attained.

To satisfy the requirements of the vast number of experimentalists in science, each publication will contain valuable Original Articles, contributed by authors of well-established reputation ; especial prominence will also be given to Critical Notices of new works, and to the consideration of recent Patents bearing upon chemical matters ; thus affording, both to the student and practical man, early and reliable information in all branches of Chemical Science.

48

Again, the large body of Pharmaceutical Chemists have long desired a medium through which their interests could be impartially represented, and which would convey to that important body the earliest intelligence of all discoveries or improvements connected with their branch of the profession, and at the same time be an organ of free discussion on all matters relating to their scientific progress, their legal and social status, and other questions affecting their present position and future advancement. To satisfy this requirement will therefore be a principal aim in the conduct of the *Chemical News* ; and in soliciting the support of the Chemists and Druggists of the United Kingdom, the Proprietors offer the strongest assurance that every exertion will be used to guard their interests.

To Chemical Manufacturers this Journal will render important service, by giving in detail all New Processes, British and Foreign, which may come within its province.

In order that every branch of the Scientific Public may be kept equally well informed in all departments of Continental research, arrangements have been entered into with some of the most eminent scientific men in France and Germany, through whose instrumentality will be furnished a Weekly Summary of all matters of a scientific or practical nature. The Readers of the *Chemical News* will thus be put in possession of the earliest intelligence concerning all branches of Continental Science.

The *Chemical News* will occupy a strictly independent position in Scientific Literature, and, being the organ of no party, will not hesitate to express an unbiassed opinion upon all current topics of interest ; at the same time its columns will be freely open to the discussion of all matters relating to Toxicology, Pharmacy, Agricultural Chemistry, and Abstract Science.

An important feature in the *Chemical News* will be the section devoted to Scientific Notes and Queries ; for it cannot fail to have been remarked how vast a number of interesting facts of daily occurrence in a laboratory of research are lost to the world, owing to the want of a readily accessible medium of intercommunication, in which such Laboratory Notes could find a convenient record. This department will, therefore, contain observations and inquiries relating to all branches of Chemical Science ; and by thus offering students and practical men facility for the interchange of their ideas and the communication of improvements and discoveries, the *Chemical*

E

THE CHEMICAL NEWS

News will necessarily become a most important stimulus to the Diffusion and Advancement of Scientific Knowledge.

As an Advertising medium, the *Chemical News* will be the best, since it cannot fail to be the most extensively circulated Journal in any way connected with Chemistry, Chemical Manufactures, Pharmacy, and general Scientific Information.

The *Chemical News* can be ordered through all Booksellers and Newsagents. Advertisements, Editorial and Business Communications are to be addressed to the Office, 12 and 13, Red Lion Court, Fleet Street, London, E.C.

I may here quote a leading article from the second number of the *Chemical News* (December 17, 1859). It is a good example of Crookes's vigorous style. It was written before iron shipbuilding became popular. The Admiralty were still under the spell of the experiments of 1845, which showed that the splintering effect of shot was greater in iron than in wood. The *Great Eastern* had just been built, and was about to demonstrate the superiority of iron over wood as regards lightness and strength. Here is the article in question :

WHY DO SHIPS ROT ?

The question is a very important one to a nation whose Government has paid more than half a million a year for the repairs of ships of war, not to speak of the enormous sums which must have been spent in maintaining its mercantile navy in good condition. Up to the present time chemists have busied themselves more in explaining the nature of the chemical changes which take place in the wood, and in devising means for preventing the disease, than in seeking for the efficient cause of the disorder. Our readers are familiar enough with the processes of Kyan, Burnett, and others, all of which are to a certain extent successful in preserving wood ; but in spite of all, the " timber plague " still continues its ravages, and a ship laid up in ordinary, when it is wanted for sea, is sometimes discovered to be in a condition which is quite extraordinary. What this " plague " is, chemists explain to us very clearly. They say it is a process of slow combustion, in which the elements of the wood under the influence of air and moisture become converted into carbonic acid and water ; and they give the process the fine-sounding name of Eremacausis. But what brings about this change ? No doubt there are many causes ; but a paper which we publish to-day reveals one which is perhaps the most important of all. It seems that when we build our ships we, with much effort and great noise, drive into the timber agents which are more destructive than the dreaded Black Sea worms. Hitherto it has been done in perfect innocence ; but for the future we shall know that, in the opinion of an eminent chemist, and an experienced dockyard officer, one cause of the rapid destruction of the hulls of ships is IRON NAILS !

How these act in promoting this destruction our readers will learn when they peruse M. Kuhlmann's paper, which has besides a practical bearing on other important matters. M. Kuhlmann believes that the energetic action of sesquioxide of iron is one cause of the spontaneous combustion which so frequently takes place in the waste of cotton and wool, and he thinks that the place where the iron (small fragments derived from the machinery) is deposited is the probable point of departure of the fire.

The *Chemical News* very soon got on a financially sound footing. There was, however, a certain amount of friction with the printers (Spottiswoode & Co.), and Crookes cast about for a more satisfactory arrangement for printing and publishing the journal. It was very natural, for family reasons, that he should look to publishers like Griffin, Bohn & Co. After some verbal communications, he sent them the following written proposal, which was also sent to Messrs. Reid and Pardon, printers, of Paternoster Row :

July 28, 1860.

C. GRIFFIN, Esq.,
10, Stationers' Hall Court.

DEAR SIR,
I enclose an abstract copied from the Books of the *Chemical News*, showing the estimated receipts for the last 4 numbers. I have also added the estimated expenses, by which it appears to

be bringing in, even with its present expensive management, a small profit.

I propose the following as the basis of arrangement between us :—The paper to be the joint property of Publisher, Printer and Crookes, each to have 1/3rd share. Publisher and Printer each to charge the property a fair and reasonable profit for what they respectively do. Crookes to pay writers and contributors as hitherto, and to give his own services as Editor, and to receive from Publisher and Printer 4 guineas per week. The profits to be divided into 3 parts, equally between Publisher, Printer and Crookes.

I will do myself the pleasure of calling on you on Tuesday at about 11 a.m. if convenient.

I remain,
Truly yours,
WILLIAM CROOKES.

ESTIMATED RECEIPTS FOR LAST 4 NUMBERS OF *CHEMICAL NEWS*.

	30 June 30			31 July 7			32 July 14			33 July 21		
By Sales of Current Number Trade	6	5	11½	6	3	9¾	6	12	0¾	6	9	3
„ „ „ „ „ Retail		16	11½		14	7½		14	5¼		19	3¾
„ „ „ „ „ Subscribers	2	14	0	2	14	3	2	15	0	2	15	0
„ „ „ „ „ Agents	1	0	8	1	0	2	1	0	6	1	0	6
„ „ „ Back numbers		12	9		18	4½	1	0	7½		14	3
	11	10	4	11	11	2¾	12	2	7½	11	18	3¾
„ „ „ Covers	1	8	0		9	4		17	6		4	8
„ „ „ Volumes		12	9	1	5	6		19	1½	1	5	6
„ „ „ Binding					2	0		9	0		2	0
„ Advertisements (less 20 % com.)	4	15	6	5	1	7	4	9	7	5	2	10
Total Receipts	£18	6	7	18	9	7¾	18	17	10	18	13	3¾

AVERAGE EXPENSES OF EACH NUMBER.

	£	s.	d.
Paper, 3 reams @ 1/1/–	3	3	–
Composition, working, &c.	8	–	–
Publishing expenses	2	2	–
Editorial, Writers & Contributors	3	3	–
Periodicals, &c. (say)		10	–
Shorthand writer (say)		10	–
Total Payments	£17	8	–

By October 1860 Crookes had made an agreement with Griffin, Bohn & Co., publishers, and Reid and Pardon, printers, for a joint proprietorship, to date from January 1861, by which the rights in the paper were shared equally between the three parties. The agreement was for one year, and thereafter terminable on three months' notice by any of the partners. Crookes was to receive $3\frac{1}{2}$ guineas per week in consideration of editorial services, to be increased to 4 guineas as soon as the half-yearly profits exceeded £50. Crookes was to be entirely responsible for the contents of the paper, excepting advertisements, and Griffin, Bohn & Co. were to manage the whole financial business. Provision was made for disposal of shares in case of death or bankruptcy, and a covenant was made against " any female person " holding a share ! History is silent as to which of the partners was responsible for this curious proviso. For the new year three columns were added to the letterpress, making 16 pages in all, and the price was increased to fourpence.

The names of the sureties against libel entered at Somerset House in connection with the *Chemical News* were :

(1) Joseph Crookes, Masbro' House, Brook Green, W., gentleman ;
(2) Alfred Crookes, 4, Clifton Terrace, Notting Hill, W., and 143, Regent Street, W., merchant tailor.

These are Crookes's father and brother respectively.

On November 15, 1860, Crookes wrote to Faraday at the Royal Institution to ask his permission to report for the *Chemical News* the great discoverer's Christmas Lecture to a juvenile audience :

DEAR SIR,
    Last Christmas I was kindly allowed the privilege of attending and reporting your course of Lectures in the *Chemical News*. I venture to hope that the manner in which I then acquitted myself of the responsibility of transferring your language to print will not cause you to withhold a similar permission on the occasion

of your forthcoming course " On the Chemical History of a Candle."

> I remain, Sir,
> Truly yours,
> WILLIAM CROOKES.

PROFESSOR FARADAY.

To this letter Faraday replied as follows :

[*Private.*]

> ROYAL INSTITUTION OF GREAT BRITAIN,
> 15 *Nov.,* 1860.

DEAR SIR,

I take your request as a great compliment, and as far as I can give it you have my full permission. My only fear is that I may not do credit to your good opinion. I know that my memory fails, and I know that the character of the lectures must fail with that. I had wished to cease lecturing altogether, but circumstances induced me to consent for this season. The risk on any account to withdraw is a trouble to me in some sense, for I really have had great pleasure in talking to the children. However, I will do my best.

> Ever truly yours,
> M. FARADAY.

W. CROOKES, Esq.,
&c., &c.

The reports of Faraday's lectures formed one of the chief attractions of the new volume. A notable communication to the same volume was made by John Spiller, who proposed the use of carbon in photography, as an absolutely indelible pigment. This is probably the first proposal of a process which has since been very extensively adopted.

The *Chemical News* was now fairly started on the prosperous career which still continues. A retrospect issued by Crookes in 1867 may be here anticipated, as it gives a masterly summary of the aims and achievements of the journal for the next few years :

FARADAY

For more than a Quarter of a Century the *Chemical News* (with its predecessor, the *Chemical Gazette*) has fully and faithfully represented the progress of Chemistry and cognate Sciences at home and abroad. It numbers amongst its regular or occasional contributors nearly every Chemist of note in Europe. In its pages some of the most important Chemical and Physical discoveries have for the first time been published, and investigators have frequently made use of its pages to secure priority of a discovery by a bare mention of facts and results before publishing their full papers. Its Correspondence columns possess, therefore, a lasting as well as an ephemeral value. Each new theory, and every fresh step in the changing phases of notation or nomenclature, which have occupied the attention of the Chemical world, have, during their tentative stages, been actively discussed in the pages of the *Chemical News* ; and to the opinions there elicited may be traced important alterations and improvements in existing theories.

The numerous Editorial Staff of the *Chemical News* comprises gentlemen in the first rank of science ; and the able and conscientious manner in which they have conducted the departments allotted to them, whether as Leader Writers, Reviewers, Foreign Correspondents, or Reporters, has made the *Chemical News* the first authority on all matters treated of in its pages.

Bound to no clique, connected with no institution, and representing no professional or trading firm, the Proprietor and Editor of the *Chemical News* can have no temptation to act or to write under any contingency otherwise than in a fearless and independent manner. No trade puffs have ever disgraced its columns, no unworthy book or patent has ever been commended in its pages ; and, based and conducted on these principles, the *Chemical News* has attained its influential position as the representative of scientific progress.

The reports of the Learned Societies have always been a distinguishing feature in the *Chemical News*. From the first number its readers have had presented to them every week a complete account of the proceedings of the Chemical, Royal, and Pharmaceutical Societies, and of the Royal Institution. These reports are given either *verbatim*, or in a form more or less condensed, according to the importance of the subject ; the reports in many cases being prepared by the speakers themselves.

Amongst the minor features of the *Chemical News* special

55

THE CHEMICAL NEWS

attention is directed to the " Contemporary Scientific Press " and to the " Chemical Notices from Foreign Sources." The former section has recently been introduced at the special request of many leading chemists. It purports to give as soon as possible after publication the title of every chemical paper in the world. To compile it every accessible scientific periodical is ransacked. The " Chemical Notices from Foreign Sources," giving a condensed account of every important paper as soon as published, is, to many readers, a section of still greater value. To give a detailed account of each paper would fill an octavo volume weekly. In this department, therefore, considerable judgment is required to decide what papers to omit altogether, and which to curtail, so as to allot to each subject its due prominence. The Editor is glad to say that this responsible office of selection and condensation is entrusted to a chemist thoroughly competent to carry out this design.

The reports of the meetings of the French Academy, and the graphic pictures of Foreign Science, sent regularly by the Paris correspondent of the *Chemical News*, one of the first scientific men in France, and a master of the gift of lucid scientific exposition— the Abbé Moigno—are features only lately introduced. We would, too, draw attention to the articles now appearing on the Paris Exhibition, from the pen of our special correspondent.

The *Chemical News* contains regularly twenty-four columns of matter, exclusive of advertisements, and whenever necessary this number is increased to twenty-eight or more columns. Arrangements are now in progress to permanently increase its size. With the next volume entirely new type will be employed throughout, and by diminishing the size of some of the type in which many of the articles are printed, still more information will be presented to our readers in each number. Other improvements are in contemplation, and will be carried out as soon as practicable.

The long period of existence of the *Chemical News*, and the high position which it has always taken, have gradually led to its introduction into all the public and private laboratories, the museums, institutions, and libraries on the Continent and in America. Its original and editorial articles are constantly reprinted in the Old and New World, and have been translated into German, French, Italian, Spanish, Russian, and other Continental languages. The attention of scientific investigators is earnestly directed to these considerations, for in the majority of cases the only rewards which the man of science can ever secure are fame and the appreciation of his labours by fellow-workers.

With the exception of the *Philosophical Magazine* (established in 1798) and the official Proceedings of the learned societies, no other English scientific journal enjoys this universal diffusion. The great number and conservative character of Continental institutions render it impossible for any journal, however ably conducted, to acquire such a position under many years of existence. The same conservative spirit which hinders the advancement of a young journal protects an old-established one. The *Chemical News* (with its predecessor) has been increasing in circulation and wide distribution for nearly a generation. In its youth many of the leading Continental Professors were at school or college. It has grown up side by side with them ; it has been their companion in their early days of laboratory work ; and it has now become an indispensable addition to their scientific libraries, and an oft-quoted record of the progress of chemical research.

The admission of an original paper into the pages of the *Chemical News*, therefore, secures its rapid diffusion over the whole world.

WILLIAM CROOKES.

CHAPTER VII

THE GREEN LINE

(1860–2)

IN SEPTEMBER 1860, Crookes applied for the vacant Professorship of Chemistry at the Royal Veterinary College, basing his claim, in the absence of any prominent teaching experience, upon "the high scientific and practical reputation now acquired by the *Chemical News*, of which paper I have been the sole manager and editor since its commencement." The application was not successful, and so Crookes was once more thrown back upon his literary and research career.

He continued dealing with photographic progress in the columns of the *Photographic News*, though his relations with Messrs. Petter and Galpin were sometimes anything but smooth. A letter which illustrates the situation is that sent by him to that firm on February 26, 1861. It says :

20, Mornington Road, N.W.,

Feby. 26, '61.

Messrs. Petter & Galpin.

Gentlemen,

I have received with some surprise a bill of 6/10 for *Family Paper*, "including Postage."

I beg to inform you that upwards of a year ago I wrote to you paying an account just rendered and requesting you that you would cease to send the *Family Paper* to me in future. This letter was received by you, as a receipt for the payment was returned.

For several months after that notice I ceased to receive the paper ; until when all our legal difficulties had been terminated,

58

and I had commenced to write " Scientific Gossip " and " Catechism " for the *Photo. News*, you of your own free will, and without any request on my part, commenced to send me the *Family Paper* again. Knowing that it was customary with all large publishers to have a certain list of presentation copies for their publications, I assumed that, as a delicate mark of appreciation of my writing in your paper, I had been placed on such a free list—and every week when the *Family Paper* has arrived it has called to my mind the magnanimity of your firm in continuing such politeness.

I feel sure you would not like the high opinion I have formed of your generosity to be lost for a few shillings. I will therefore assume that your clerk has sent me the account by mistake.

Your. obedient servant,
WILLIAM CROOKES.

Meanwhile the *Chemical News* was striding forward. Crookes added a list of chemical patents, which he aimed at making exhaustive. For this work he engaged the services of Mr. C. Greville Williams, who had written occasional reviews for him. A letter allocating one and a half columns weekly to Mr. Williams's contributions ends as follows : " I send herewith a book upon which I should like you to exercise *to the utmost* your castigatorial powers of criticism. A reviewer seldom gets so good an opportunity of speaking his mind strongly." One cannot repress a feeling of compassion for the unfortunate author thus ruthlessly handed over to the executioner.

It was to this same Greville Williams that Crookes made the first announcement of an epoch-making discovery. Crookes wrote to him on March 5, 1861, from 20, Mornington Road :

MY DEAR WILLIAMS,

Have you ever noticed a green spectrum line as far from Naα on the one side as Liα is on the other ? If not, I have got a new element. More particulars shortly. I will send you some of the compound which gives it, and it will be a nice thing for us to work it out together.

THE GREEN LINE

We ought to be first in the field with good descriptions and drawings with measurements of the hitherto undiscovered spectra of copper, BO_3, sulphur, selenium, tellurium. Can you get them plainly ? If you will give as good a written description of them as you can, I will prepare measurements. The so-called continuous spectrum of potassium wants mapping out. AsH_3, to my intense disgust, only gives a continuous spectrum. I was anticipating with pleasure some well-defined lines.

In haste, believe me,

Very sincerely yours,

WILLIAM CROOKES.

Three days later he was able to write more definitely :

March 8th, 1861.

MY DEAR WILLIAMS,

I fully believe that the substance I wrote about in my last is really a new element. I have found a mine of it ! a whole specimen tube of some impure selenium contains it in large quantities, so I have devoted all the morning to trying to isolate it. The selenium was powdered and submitted to numerous reagents, and at last I found the following plan of separating them. Boiling solution KCy dissolved selenium, but left a black insoluble residue. This residue when well washed gives the green line brilliantly, but I dare not burn too much of the body to see if there are any other lines at the blue end. I think there are not. . . . I fancy that the element existing in this residue (half of which I enclose) is a metallic (copper ?)-ide, for the following reasons. When gas, treated with HCl, a horribly stinking gas, rather like HS [$=H_2S$], but worse, is evolved, and then the line is altogether lost. Also when a portion of the black residue is put on to a platinum wire loop and moistened with HCl, immediately before it is introduced into the flame the green line is superb and quite temptingly measurable. Therefore I have sacrificed some in order to get some good measurements. I append the means of them, and they are very close. . . . I have again most carefully scrutinised sulphur, selenium, and tellurium, and I am satisfied that the line is not in either of them. It is quite different, as I think you will see when you examine what I have sent you. . . . It may possibly be a common metal existing as selenide, but I do not

60

think it is probable. I have tried all I can think of, copper, lead, bismuth, tin, etc., and they give nothing of the sort. . . . As soon as I have any more to tell you on the subject I will write. I wish I could isolate it. I think K Cy solution will separate it from selenium, but I cannot separate it from tellurium.

Believe me,

Very sincerely yours,

WILLIAM CROOKES.

P.S.—I should like us to announce it in next *C.N.* if possible. Oppenheim is working on the separation of sulphur, selenium, and tellurium, and he might come across it by chance.

In a subsequent letter, dated March 12th, Crookes gives Williams instructions how to obtain the element in sufficient quantity to show the green line :

"What can it be ? " he continues. " I have thought over the matter night and day, and really see no other possible conclusion to draw but that it is a true new element of the sulphur class. Unfortunately, I am so busy now with literary work that I have only a few minutes here and there to get into the laboratory, and lately I have only been able to work there so much as I have done by neglecting other work, which I must now stick to unremittingly for the next few days."

Some of this " other work " consisted in testing six soda-water machines for Messrs. R. C. Lepage & Co. These machines had been advertised as made " of pure tin and silver." Actually they were found to contain no trace of silver, but plenty of lead, so that their use was attended with danger.

The discovery of the new element was announced to the world in the *Chemical News* for March 30, 1861, and in the *Philosophical Magazine* for April 1861.

The latter article concludes as follows :

In order to remove any remaining doubt which there might be as to the green line being due to any of the elements mentioned in the above list, I have, moreover, specially examined the spectra produced by each of these bodies in detail, either in their elementary state, or in their most important compounds. Many of them give rise to spectra of great and characteristic beauty, but none give anything like the green line ; nor, in fact, is there any artificial spectrum, except that of sodium, which equals it in simplicity.

There still may be urged the possibility of its being a compound of two or more known elements, or an allotropic condition of one of them ; a moment's thought, however, will show that neither of these hypotheses is tenable. They would in reality prove what they are raised to oppose ; for nothing less could follow than a veritable transmutation of one body into another, and a consequent annihilation of all the groundwork upon which modern science is based. If an element can be so changed as to have totally different chemical reactions, and to have the spectrum of its incandescent vapour (which is, *par excellence*, an elementary property) altered to an appearance totally unlike that given by its former self, it must have been changed into something which it originally was not. This, in the present position of science, is an absurdity.

The method of exhaustion which I have adopted to prove the elementary character of the body which communicates this green line to the spectrum of the blue gas-flame,[1] may seem unnecessary as well as unchemical in the present state of the science ; I was obliged, however, to rely upon what I may call circumstantial evidence of its not being a known element, owing to the very small quantity of substance at my command (I believe I overestimate the amount which I have as yet obtained, at two grains), which precluded me from trying many reactions. The method of spectrum-analysis adopted to prove the same fact, although perfectly conclusive to my own mind, might not have been so to others, unsupported by chemical evidence.

The following diagram will serve to show the position in the spectrum which the new green line occupies with respect to the two lithium and the sodium lines.

Liα   Liβ Naα

New
Green Line.

---

[1] I need scarcely add that the line is quite distinct from either of the green or blue lines seen in a gas-flame which is undergoing complete combustion. It is moreover far more brilliant than these.

For confirmatory experiments on many of the observations mentioned in this paper, I am indebted to my friend Mr. C. Greville Williams. The detailed examination of the various spectra are at present being jointly pursued by us, and will be published as soon as completed.

Later researches showed that thallium was not an element of the sulphur group (sulphur, selenium, tellurium, ytterbium), but belonged to the triatomic aluminium group (aluminium, gallium, indium), and had its place in the neighbourhood of mercury, lead, and bismuth. This position could, however, not have been predicted on spectroscopic evidence alone.

The next problem was to find a name for the new element. In his letters to Greville Williams, Crookes sometimes called the substance simply " A." Choosing among a number of names indicating a green colour, such as viridium, malachium, etc., Crookes had the happy inspiration of deriving the word " thallium " from the Greek for a green twig. He announced it as follows in the *Chemical News* for May 18, 1861 :

FURTHER REMARKS ON THE SUPPOSED NEW METALLOID, BY WILLIAM CROOKES.

Assuming that further researches on this subject will confirm the correctness of the opinion which I expressed in a former paper,[1] and that the body there introduced to the notice of chemists will prove to be a new member of the large and increasing family of elementary bodies, I have thought it best to give in the present article a few additional observations which I have since made, and also to propose for it the provisional name of *Thallium,* from the Greek θαλλός, or Latin *thallus,* a budding twig,—a word which is frequently employed to express the beautiful green tint of young vegetation ; and which I have chosen as the green line which it communicates to the spectrum recalls with peculiar vividness the fresh colour of vegetation at the present time.

[1] On the Existence of a New Element, probably of the Sulphur Group. By William Crookes. *Chemical News,* vol. iii, p. 193, and *Philosophical Magazine* for April 1861.

THE GREEN LINE

I have not succeeded in finding this body in any selenium or tellurium ores which I have examined ; but two or three specimens of native sulphur, especially some from Lipari, have yielded it, the latter in such abundance that if the mineralogical specimen which I examined is a fair sample of what this sulphur is like on the large scale, it could easily be employed as a source of thallium in quantity. I have also found small traces of it in some crude sulphur sublimed from Spanish pyrites, for which I am indebted to the kindness of Mr. Thornthwaite.

The plan which I have found most effectual for separating thallium from the associated bodies is the following :

Finely powder the ore (selenium deposit or native sulphur, etc.) and mix it with its own weight of dried carbonate of soda and half its weight of nitrate of potash ; mix thoroughly, and project in small quantities at a time into a red-hot earthen crucible. When it has all been added, keep at a red heat until it fuses quietly to a liquid, when pour it out on to an iron plate. When cool, powder it and exhaust with boiling water until all the soluble portions are extracted. Filter, and add to the filtrate excess of NH_4O and NH_4S ; boil and filter. The precipitated sulphides are now to be boiled in a solution of one part commercial KCy to eight of water, until no more dissolves. Collect the residue on a filter, and wash well. Dissolve this residue, insoluble in KCy, in hot NO_4Cl ; dilute with water and filter if necessary. Add NH_4O in excess, then $NH_4O . O$ and boil. Filter, and add HS. Boil for some time, keeping the solution alkaline with NH_4O and smelling of HS. A dark brown precipitate will be produced, which gradually settles to a heavy black powder. Filter this off, and wash it well.

A portion of this, so minute as to be almost imperceptible to the naked eye, introduced into the blue gas-flame of the spectrum apparatus gives rise to a green line of extraordinary purity and intensity, a piece the size of a small pin's head being most dazzling, and quite equal to the yellow line of sodium in brilliancy. From the purity of this spectral appearance, I am inclined to think that this precipitate is thallium itself, uncombined with other bodies, and reduced by the HS from the state of oxide. This, however, is only conjecture. The quantity of the precipitate which I now have is far too small to settle the point by direct experiment. I am at present endeavouring to find a source of thallium which will yield it in quantity, and if, as I hope to be, successful, I propose giving further particulars at a future time.

Thus, at the early age of twenty-nine, Crookes had made his mark as a discoverer of the first class. His work on thallium secured him ample recognition from his contemporaries in science. But that sort of recognition is not always remunerative, and is sometimes the reverse, as it involves numerous obligations of the *noblesse oblige* kind. Crookes was, in any case, well advised to look for sources of additional income. Hearing of the forthcoming appearance of a monthly scientific review, he wrote to the editor, a well-known naturalist, as follows :

20, MORNINGTON ROAD, N.W.,
May 24, 1861.

SIR,

I have just received a prospectus of *The Review of Popular Science,* which will be edited by you. Such a work has long been wanted, and you have my best wishes for its success.

It will give me much pleasure to assist your undertaking to the best of my power, either by contributing scientific articles, or in any other way which might be thought desirable.

I remain, Sir,'
Truly yours,
WILLIAM CROOKES.

J. SAMUELSON, Esq.

Mr. Samuelson immediately asked Crookes to write an article for the first number, published in October 1861. The article was called " The Breath of Life," and dealt mainly with oxygen and its importance in the human organism. Incidentally, it gave a scathing exposure of the evils resulting from defective ventilation in offices, schools, law courts, and prisons. It concluded with a passage which indicates Crookes's outlook beyond imme-

diate scientific problems.   Referring to the conservation
of energy, he wrote :

> Just as mechanical motion is equally capable of being re-trans-
> formed into heat, light, electricity, or chemical action—just as every
> word we utter acting on the material atmosphere around us resolves
> itself into aerial waves of sound, which for ever afterwards vibrate
> with diminishing intensity, but expanding area, from one extremity
> of the atmosphere to the other, retaining always the same amount
> of energy as it did when the mechanical motion of the breath and
> lips gave it birth—so do the forces once born to activity when the
> candle is lighted live to the end of time undiminished in intensity,
> although changed in character.   When the flame is naturally
> extinguished these living forces do not die, but become absorbed
> in that vast reservoir of energy which is the source of all life and
> light upon this globe.
> And shall we then suppose that the soul of man is of less account
> than the flame of a candle ?   If philosophy can thus prove that the
> latter never dies, shall not faith accept the same proof that our
> own spiritual life is continued after the vital spark is extinguished ?

When the first number of the *Popular Science Review*
appeared it was well received, and Crookes wrote to
Samuelson on October 24, 1861 :

> I must congratulate you on its appearance.   I trust it is as much
> a commercial as it is a literary and scientific success.   As far as I
> have yet heard, opinions are entirely favourable, and for my own
> part I am highly flattered at the manner in which my small contribu-
> tion is spoken of, especially the conclusion—the part about which
> I was (like yourself) most doubtful.

We see here the first soaring of Crookes's inner mind
into the realms of the supernatural, his first excursion
into the " debatable land " between life and death which
he trod with such audacity—and such peril to his scientific
reputation—ten years later.

How careful Crookes was of his scientific reputation

at this time is shown by a letter addressed to Professor W. A. Miller, of King's College :

Septr. 14*th*, 1861.

MY DEAR SIR,
    I was much pleased to see from the report of the meeting of the British Association that you had turned your attention to the subject of photographing metallic and other spectra. You may perhaps be interested in knowing that I have been working in that direction during the last 10 years. I am just on the point of leaving home for a week or two, but on my return shall be most happy to show you my photographic negatives. They include solar spectra through complete quartz trains, far beyond the fixed lines, *t, u, v,* &c. ; indeed, I may say I have got out of the alphabet altogether : and also photographs of spectra of the electric spark between mercury, lead, tin, cadmium, bismuth, &c., poles, taken 8 years ago. I gave at the time photographic copies of some of these to Professor Wheatstone, to whom I was indebted for the loan of the electrical part of the apparatus, and to other scientific friends, and it was one of these copies which was placed in your hands at Manchester when detailing the results of your own experiments. These were taken with glass apparatus, but I have also taken the mercury and other spectra with a quartz train, by which the number of lines in the higher end is vastly increased.
    There has been of late so much misconception on these and collateral subjects that perhaps you will not object to make some allusion to my experiments in the event of an opportunity arising, as I see that my friends of the *London Review* intend shortly to give prominence to this subject.
                            Believe me,
                                My dear Sir,
                                    Very truly yours,
                                        WILLIAM CROOKES.
PROFESSOR W. A. MILLER,
    King's College.

The *Chemical News* triple proprietorship did not work smoothly, and Messrs. Griffin, Bohn & Co. retired from it in June 1861, leaving Crookes on the one hand

and Reid and Pardon on the other, as continuing proprietors, with the addition of Mr. F. C. Darley, who put
£500 into the business. Crookes at this time stated that
he had sunk £1000 in developing the venture, and claimed
£1 per week as interest on this sum from the joint proprietors. To this Mr. Darley demurred, and as shortly
afterwards the firm of Reid and Pardon ceased to exist,
the proprietorship was left " in the air." Finally, Crookes
decided to resume the entire proprietorship himself,
paying the successors of Reid and Pardon £200 for their
rights and refunding Mr. Darley. Some relief to all these
troubles was furnished by the effective co-operation of
one of the contributors, W. T. Fewtrell, who gradually
became Crookes's right-hand man in the editing of the
Chemical News. A letter which shows the intimate tone
of Crookes's correspondence towards this man is the
following :

20, MORNINGTON ROAD, N.W.,
Septr. 26, 1861.

MY DEAR SIR,
    I fancy we might be able to work the annual retrospect
of Chemistry together. At all events it is worth consideration,
whether if such be thought of it should not be published at our
office at the proprietor's own risk, for if Griffin took it he would
get all the plunder. It should be in the Liebig & Kopp style
—a scheme respecting which we have already had some talk, and
of which I have not given up the thoughts.
    I am just off for Brighton.

Believe me,
Very truly yours,
WILLIAM CROOKES.

W. T. FEWTRELL, Esq.

This project was, however, abandoned in favour of
another, by which a summary of chemical (and physical)
progress should be contributed to Samuelson's quarterly

review. The letter proposing this summary is characteristic of Crookes's self-confidence and enterprise. It also illustrates the fact that Crookes was at this time most anxious, for family reasons, to increase his income, and left no stone unturned in order to do so.

20, MORNINGTON ROAD, N.W.,

Novr. 1, 61.

MY DEAR SIR,

I think I can let you have a very perfect retrospect of Chemical Science. How many pages may I fill, as upon that depends the treatment each separate subject can receive. I should propose to take in as well as Chemistry and Photography, Physical Science, and to give a short abstract of *everything* of importance discovered since Septr. 30th, or shall I go further back ? I think I can make such a résumé a very interesting one to the public.

I append the headings of a few subjects to which I have given some attention, and which I could write a readable article upon. I have some others, but have mislaid my memoranda.

I ought to say, however, that I am so much in the habit of writing on *any* scientific topic which may arise, and have such excellent opportunities of getting the best scientific information, that I have no doubt I could please you on any subject connected with my department of science which you may suggest.

Believe me,

Truly yours,

WILLIAM CROOKES.

J. SAMUELSON, Esqr.

PROPOSED SUBJECTS.

Aluminium.
Preservation of stone.
Divisibility of Matter.
Electric & Lime light, &c.
The infinitely small ; in space, & time.
Spectrum Analysis.

&c., &c., &c.

Towards the end of 1861 Crookes was much preoccupied with the vicissitudes of the *Chemical News.* He asked Fewtrell's help in making the articles shorter and more numerous, so as to cope with the ever-increasing flood of chemical literature. He reported many public lectures himself. His shorthand, a copy of which is extant, was the Pitman system of that date, but he inserted plenty of vowels and made some errors in outlines. Nevertheless, his shorthand, like his longhand, is eminently legible.

At Christmas 1861 he wrote to Charles Crookes complaining that he had given him no analyses to carry out for some time. Crookes was, in fact, disappointed with the revenue from his laboratory work. His laboratory was a "small back room" at 20, Mornington Road, heated in the winter with gas. This was long before the modern gas "fuel" heater was invented, and so the conditions can hardly have been ideal. In any case, Crookes's revenue from this occupation did not exceed an average of £20 per annum, largely because he refused to attach his name to mere trade "puffs."

In 1862 the scientific world was stirred by the possibilities of the second great International Exhibition in London. In addition to numerous descriptions of the chemical exhibits published in the *Chemical News*, Crookes wrote an admirable article on the utilisation of waste products, some extracts of which are appended here as of both literary and historical interest :

When coal is distilled in close vessels for the purposes of the gas manufacturer, various other products are obtained at the same time. A large quantity of offensively-smelling water comes over ; the various sulphur compounds present in the coal yield up this element to the gaseous products, whilst a considerable bulk of tarry matter is also produced. Now the object of the manufacturer being to produce as much gas as possible in a practically pure state, all these accessory products were for many years looked upon as necessary evils, to be got rid of as quickly as possible. The gas-water and tar were thrown away into the nearest stream, where they killed the fish and poisoned the atmosphere for miles around, whilst the sulphur was removed, or at least supposed to be removed from the gas by means of lime, or, as more recently adopted, through a mixture of sawdust and oxide of iron, about which more will be said anon.

The first of these noxious products, the gas-water, has been utilised in the following manner. It owes its bad smell principally to the presence of ammonia and sulphur compounds, and it is only necessary to add some quicklime to this, for it to seize upon the acids with which the ammonia is in combination and liberate the alkali. This gas is conducted into chambers where it meets with carbonic acid, forming, after appropriate purification, the salt known in commerce as carbonate of ammonia, about 2,000 tons of which are made annually from this liquid. If, instead of distilling the gas-liquor with lime, a strong acid is added to it, hydrochloric for instance, itself a waste product, the compound known as sal-ammoniac is produced, which is of very great value in the arts, being the principal source of the more common salts of ammonia met with in commerce ; the *liquor ammoniæ* of pharmacy, or hartshorn, being made by distilling this purified sal-ammoniac with lime, and conducting the evolved gas into water. The uses of this ammonia are well known ; not only is it largely employed in medicine, but, when mixed with some aromatic substance, it is used in scent-bottles, thus affording a striking instance of the transmutations effected by scientific agency, the fetid liquid of the gasworks being transformed into a scent used by ladies as a cherished luxury. . . . One of the chief impurities in gas is sulphur. . . . Benzol consists of the elements carbon and hydrogen in the proportion of twelve of the former to six of the latter. The action of the nitric acid upon this is to remove one of the hydrogen atoms and put in its place an atom of peroxide of nitrogen. When this nitro-benzol is acted upon by certain chemical agents of the class called reducing (as for instance a mixture of iron-filings and acetic acid which is now generally used), the whole of the oxygen which the peroxide of nitrogen has brought into the nitro-benzol is removed, and two parts of hydrogen are added, so that the original benzol becomes transformed into a body containing twelve parts of carbon, seven parts of hydrogen, and one part of nitrogen. This is aniline, a substance which illustrates in a striking manner the effect that demand exerts upon supply. Some years ago all the laboratories in Europe did not contain a pound weight of it, whereas it is now manufactured by thousands of gallons at a time. There is still another stage to be passed before we get to the colouring matter, but here the change is by no means well understood, and the best processes are kept scrupulously secret. The action is, however, the reverse of the one just now described, being the addition instead of subtraction of oxygen, and it is by the employment of different oxidising agents that we get mauve, magenta, roseine, azuline, bleu de Paris, and other gorgeous dyes which have received arbitrary names. Mauve or aniline purple was first discovered by Mr. Perkin, and in his case at the Exhibition may be seen a very complete and beautiful collection showing the different stages of the manufacture, from the crude coal oil up to a gigantic block of the pure dye itself upwards of a cubic foot in bulk, and for the production of which the distilled products from 2,000 tons of coal were consumed. The tinctorial properties of this dye are very strong. Mr. Perkin illustrates this by exhibiting a gallon jar filled with a beautiful violet solution, the colour of which is communicated to the water by one grain only of the dye. To render this illustration more striking there is placed near it a similar sized jar filled with crude coal-tar, the whole of which would have to be employed to produce this single grain of colouring matter.

The gigantic scale upon which the manufacture of these colouring matters is carried on and the perfection to which it is brought are strikingly illustrated by Messrs. Simpson Maule, and Nicholson. This firm have succeeded in producing the beautiful colour known as magenta in a crystalline form, and one of the most striking objects in the whole department is the magnificent crown of acetate of rosaniline (the chemical name for magenta), which occupies so prominent a position in their case. One of the most curious points about this is the colour which it exhibits. The rich, deep, rose tint which the body communicates to silk, &c., instead of being concentrated and intensified until deepened almost to a black, here shines and glistens from the facets of the beautiful crystals with a rich metallic green lustre, sparkling in the sunshine like the plumage of tropical birds, or the wing cases of certain beetles. Crystals such as these are unattainable except when working on a manufacturing scale. Laboratory experimentalists had already ascertained the fact that magenta was capable of assuming a regular form ; but crystals such as compose this crown—the planes in some being nearly an inch across—can only be developed by manufacturers whose crystallising vats hold upwards of £2,000 worth of colouring matter.

Before leaving this subject, we may draw attention to the important branch of national industry which this manufacture is assuming, and pay our tribute of admiration to the skill and intelligence of the chemist who has succeeded in converting the most nauseous and repulsive by-products of gas manufacture into such lovely colouring agents. Through his exertions, England will cease to import colouring matters, and will become a dye-exporting country.

Another product of the distillation of coal now claims our attention. In 1841, Liebig said that it would certainly be one of the greatest discoveries of the age, if anyone should succeed in condensing coal-gas into a white, dry, odourless substance, portable, and capable of being placed upon a candlestick or burned in a lamp. Ten years afterwards, Mr. Young showed, in the Great Exhibition of 1851, a single candle made from paraffin, a waxy-looking solid, which was known to be obtained in small quantities from the distillation of peat, wood, or coal. Liebig's prediction was here fulfilled. Paraffin is absolutely identical in composition with the most luminiferous portion of coal-gas, and only differs from it in being in a more condensed state ; its percentage composition of carbon and hydrogen is, in fact, identical with that of olefiant gas. Another ten years, and the commercial manufacture of paraffin has assumed gigantic proportions, and Mr. Young's establishment for its production, at Bathgate, ranks among the largest chemical works in the world. The composition of paraffin, indeed, renders it pre-eminently adapted for the production of light. It is a beautiful wax, melting at about 130°, and when heated to a considerably higher temperature (as when burning in the wick of a candle) it decomposes into true olefiant gas, producing a beautiful white light. A paraffin candle amounts, therefore, to a perfectly-constructed and portable gasworks ; its flame is not like that of an ordinary candle, but is identical with that of the most perfect coal-gas, which is, indeed, self-produced as it is wanted, without any costly or complicated apparatus, and in a state of purity unattainable by ordinary means. . . . The manufacture of phosphorus—" that dark, unctuous, daubing mass "—which has now become so important and universal an agent of civilisation, is the last product from bones which claims our attention. It is prepared from the phosphoric acid contained in bones by heating it to a very high temperature with charcoal. The five equivalents of oxygen contained in the phosphoric acid are removed by the charcoal, and the phosphorus distils over. The crude product is afterwards purified by distillation and squeezing through chamois leather. One of the largest firms for the preparation of this element, Messrs. Albright and Wilson, of Oldbury, exhibit a most interesting case illustrating this manufacture, the beautiful semi-transparent, wax-like appearance and the large blocks in which it is produced are very accurately illustrated (phosphorus itself being inadmissible in the building).

The great consumption of phosphorus is of course in the manufacture of lucifer matches ; in this, two difficulties have to be contended with, phosphorus being the most inflammable body known, and also one of the most poisonous. Numerous had been the attempts to overcome these difficulties in the manufacture of this useful though humble commodity, but the liability to explosion, and the terrible disease to which the workmen who inhale the phosphuretted vapours are subject, seemed scarcely capable of being obviated, when Professor Schrötter, by one of the most remarkable discoveries in modern chemistry, effected an entire revolution in the manufacture. He discovered that when common phosphorus was heated for some time in a close vessel to a temperature of $470°$ F., it underwent a complete alteration in the whole of its physical characters. From a white, waxy, crystalline body, soft and flexible as lead, it became a deep red, amorphous, opaque mass, hard and brittle as glass. The white phosphorus quickly ignites by mere exposure to the air ; the red phosphorus will not ignite spontaneously, and may be packed up in boxes, in the dry state, without any danger. The white phosphorus is as poisonous as arsenic, and has a strong garlicky smell, whilst the red is without odour, and has no poisonous properties. The former is luminous in the dark, and melts at $108°$ F., whilst the latter is perfectly illuminous, and requires a temperature above $500°$ F. to melt it. Lastly, the white is freely soluble in various liquids, whilst scarcely any known solvent will touch the red modification. In spite of these striking differences, the red phosphorus answers quite as well for match-making as the common sort, and as the transformation may be effected with very little trouble, there is no doubt that the harmless variety will, in course of time, entirely supersede the dangerous variety.

Considering the frightful disease which attacks and destroys the jawbones of the workpeople employed in making common matches, and at the same time their highly poisonous properties, we think it the duty of every person to encourage to the utmost the manufacture of matches made from red or allotropic phosphorus.

CHAPTER VIII

SLINGS AND ARROWS

(1862-5)

ON APRIL 17, 1862, the Secretary of the International Exhibition at South Kensington wrote to Crookes inviting him to exhibit a specimen of thallium. Crookes exhibited a case containing (1) metallic thallium ; (2) thallium oxide ; (3) thallium sulphide. We may well imagine his astonishment and consternation when the list of awards of prizes at the Exhibition was published in July 1862 and a Belgian chemist named Lamy was found to have been awarded a medal for thallium, while his own name was not even mentioned !

Crookes immediately wrote a protest to the members of the jury in the following terms :

July 14, 1862.

DEAR SIR,

I find on reference to the awards of prizes at the International Exhibition that M. Lamy has had a medal given to him for thallium, whilst my name is not even mentioned in connection with the discovery.

Will you oblige me by saying whether it is intentional, or only by accident, that the credit of my discovery is thus given to someone else ?

I remain,
Truly yours,
WILLIAM CROOKES.

Through the intervention of Dr. Hofmann and Dr. Lyon Playfair, Crookes succeeded in having the award revised. A medal was awarded to him for the *discovery* of thallium, and another to M. Lamy for its production as a metallic ingot. It appeared that Lamy had observed the green line independently of, but later than, Crookes, had found a plentiful source of the element, and had exhibited an ingot of thallium at the exhibition early in July. His version was that Crookes had not recognised the metallic nature of thallium until he (Lamy) pointed it out to him. Crookes's version was given subsequently in the *Philosophical Magazine*. He says that he only exhibited three small preparations on account of the small amount of thallium at his disposal. These were displayed, at the opening of the Exhibition on May 1, 1862, in a case with the label " *Thallium, a new metallic element, discovered by means of spectrum analysis*," and with a card oh which was written " Chemical reactions of thallium, by which it is distinguished from every other known element. It appears to have the character of a heavy metal, forming compounds which are volatile below a red heat. It is reduced from its acid solutions by zinc in the form of a dense black powder, difficultly soluble in hydrochloric acid, readily soluble in nitric acid, etc."

On the 7th of June, being at the Exhibition, Crookes learnt from the Secretary that M. Lamy had just been to him, in company with M. Balard, and shown him an ingot of thallium. The Secretary, Mr. Quin, took them to Crookes's case and translated to them the labels, whereupon M. Lamy remarked that the substance exhibited as thallium was not the metal but its sulphide.

In a communication to the *Philosophical Magazine*, Crookes wrote :

. . . M. Lamy and his friends assume that . . . the black powder I exhibited as thallium was obtained by precipitating with hydrosulphuric acid, not the black powder described as precipitated by zinc. They assume that because in May 1861 I was, though doubtful as to the point, inclined to class thallium with the semi-metals, therefore in May 1862 I was ignorant that it possessed true metallic properties.

I trust the improbability, nay, the impossibility of this being the case, will be recognised by every chemist who has examined thallium. It is an element as easily reduced to and preserved in a metallic state as lead : can it, then, be imagined that I, who was so much interested in determining its characters—who had been for twelve months leaving no means untried to obtain a more copious source of thallium—who during that time had scarcely for a day relaxed working on this subject exclusively—is it likely, I say, that I should have been such an egregious blunderer as not to find out that it was a metal ? Why, as soon as I had obtained a dozen grains of one of its compounds fairly pure, I could scarcely try the simplest experiment without having the fact of its metallic character forced upon me in too positive a manner to admit of doubt. Consistent with this probability that I considered thallium to be a metal, is the fact that it was exhibited as being a metal.

The biographer may perhaps be allowed to insert here a list of dates which will help to clear up the various points in the controversy.

DATES IN THE DISCOVERY OF THALLIUM.

1861, March 5th : Crookes writes to C. Greville Williams announcing green line and probability of new element.

1861, March 8th : Letter to same, announcing certainty of new element.

1861, March 12th : Letter to same, giving method of isolating " new element of the sulphur class."

1861, March 30th : Announcement in *Chemical News* and *Philosophical Magazine* of " The Existence of a New Element, probably of the sulphur group."

1861, May 18th : Crookes published "Further Remarks on the supposed new metalloid" in the *Chemical News*, and proposes the name "thallium." Describes black precipitate obtained by H_2S, "probably thallium itself."

1862, March : Lamy (independently) observes green line.

1862, May 1st : Crookes exhibits metallic thallium as a heavy black powder, precipitated by zinc.

1862, May 16th : Lamy exhibits metallic thallium at the Imperial Society, Lille.

1862, April or May : Lamy learns of Crookes's anticipation and the name thallium. Obtains a large supply of mineral and prepares ingot of thallium.

1862, June 7th : Lamy shows ingot of thallium at Exhibition.

1862, June 9th : Lamy meets Crookes.

1862, June 19th : Crookes reads Paper before Royal Society on Thallium.

1862, June 23rd : Lamy reads Paper at the Académie des Sciences, Paris, on "The existence of a new metal, Thallium" (*Comptes Rendus*, 54, p. 1255).

1862, July : Lamy and Crookes awarded medals of International Exhibition for thallium.

From this chronology there can be no possible doubt that Crookes deserves all the credit of being (1) the first to discover the new element, (2) the first to determine its chief chemical properties, including its metallic nature.

On the other hand, it is practically certain that Lamy observed the new spectrum line independently, and was the first to produce a solid block of thallium. These claims were, in fact, very generously acknowledged by Crookes himself in a lecture delivered at the Royal Institution, so that, in spite of the glaring injustice at first perpetrated by the Exhibition authorities, the truth prevailed at last, and all ended happily.

Sir John Herschel, to whom Crookes sent a copy of his letter to the *Philosophical Magazine*, replied as follows :

SLINGS AND ARROWS

Aug. 22, 1863.

SIR,

I beg to thank you for your paper from *Phil. Mag.* in *re* Thallium. I am very glad to see it, and I trust it will prove final in settling a question which ought never to have been raised.

Yours truly,

J. F. W. HERSCHEL.

W. CROOKES, Esq.

But Crookes's troubles were by no means at an end. In October 1862 he had to expostulate with Mr. F. C. Darley for collecting money due to the *Chemical News* without authority. Also, on applying to Professor Frankland for renewed permission to report his Royal Institution Christmas lectures, he was met by a request for a sum for proof-reading, to which he replied as follows :

Decr. 2, 1862.

DEAR SIR,

I am much obliged for your kind permission to report your Xmas lectures in the *Chemical News*, and regret that I shall be unable to avail myself of your politeness. It would be most unreasonable of me to expect a gentleman in your position to work for so many hours a day for my benefit without adequate remuneration, and certainly the sum you name is a mere trifle compared to the very valuable lectures which it would secure to the *Chemical News*. I am, however, in this difficult position, that the expenses of reporting and getting up the lectures are already so heavy, that the circulation of the paper, large though it be, is insufficient to warrant such an outlay.

I should in the first place have to pay the shorthand writer 2 Guineas for taking down each lecture. The wood engravings with which the course should be illustrated (to make it uniform with the previous ones by Drs. Faraday and Tyndall) would not average less than 20/- each Lecture. If you corrected the proofs as in your former course I should have to pay the printer just double for composition, as they find it cheaper to reset the lectures in type from your corrected proof than to alter the original type   That was on the former occasion an additional expense of 30/- to 40/- each. Lastly, there would be your own honorarium, which for 4½ pages each lecture would be £14 5s. For the course of six I should therefore have to pay :

			£	s.	d.
Shorthand writer	...	...	12	12	0
Engraver	...	...	6	0	0
Extra composition	...	...	10	10	0
Dr. F.	...	...	14	5	0
			£43	7	0

which would be more than double the amount the three previous Xmas courses of Lectures have cost, each.

I trust therefore that you will not misunderstand the reason which will compel me to debar the readers of the *Chemical News* from the pleasure and advantage of reading your Juvenile Lectures this Christmas.

Believe me,
Truly yours,
WILLIAM CROOKES.

DR. FRANKLAND, F.R.S.

The years 1862 to 1864 were full of trouble. The vicissitudes of the *Chemical News*, and illness and anxiety at home, prevented Crookes from enjoying the fruits of his brilliant discovery, quite apart from the struggle for due recognition and fair play. His many lawsuits and his frequent appearance as an expert witness in courts of law had somehow gained him a reputation for undue pugnacity, though it is clear from his letters that he went very far to avoid a conflict, and often succeeded in renewing a friendship which appeared irretrievably broken.

A letter addressed to a member of the Council of the Pharmaceutical Society after his failure to obtain the Professorship at the Royal Veterinary College is illuminating :

SLINGS AND ARROWS

July 18th, 1862.

DEAR SIR,

I have just heard that someone, interested in my non-success, has circulated a report amongst the Council of the Pharmaceutical Society, that I am a very litigious and quarrelsome fellow, having, in fact, no less than three lawsuits connected with the *Chemical News* in hand at the present time.

I take the very earliest opportunity of denying this in the most emphatic and absolute manner. I am not in the remotest degree connected with any lawsuit whatever. I have the greatest horror of law, and would willingly put up with considerable sacrifice rather than adopt that means of redress. The changes which have taken place in connection with the paper I edit have been rendered necessary solely on financial grounds, the present printer and publisher having become part proprietors of the paper. I am happy to say that my relations with the present and former publishers, printers and proprietors have been uniformly of the most friendly description; and I have in fact employed the former printers of the *Chemical News* (Messrs. Spottiswoode & Co.) to get up the testimonials now before you.

It is also reported that my manner as a teacher is unpopular. My testimonials from those who have known me longest in that capacity afford the best evidence that this is not the case.

I am unable as yet to trace these reports to anything definite, and am consequently not in a position to give a more particular contradiction to them : but I am proud to say that there is no single incident in my career which will not bear the strictest scrutiny.

I remain, dear Sir,
Truly yours,
WILLIAM CROOKES.

P. SQUIRE, Esqr.

That Crookes was most willing and anxious to teach chemistry is shown by his arranging with Mr. John Yeats, headmaster of Peckham Schools, to give twenty experimentally illustrated lectures on chemistry for a guinea a lecture. At the end of the course Crookes set the following examination paper, from which we may gather the standard and range of his course of lectures :

82

1. Why does phosphorus burning in oxygen give a much more intense light than sulphur under the same circumstances?

2. If I burn a candle in such a manner that all the products of combustion are collected, will it become lighter or heavier? Explain why.

3. What are the products of combustion of a candle?

4. Explain accurately what takes place when sodium is thrown on water.

5. How is hydrogen prepared?

6. Explain the action of the oxy-hydrogen blowpipe.

7. Why does water, when added to quicklime, cause it to become hot?

8. How can water be frozen in front of a fire?

9. Suppose I boil water in a flask, and then cork it up tightly so as to have nothing but steam in the upper part of the flask, what takes place if I now dip the flask into cold water? Explain this.

10. How can the presence of carbon be shown in sugar?

11. What is produced when charcoal is burnt in oxygen?

12. Explain how it is that a small quantity of carbonic acid added to lime water produces a precipitate, whilst a large quantity of carbonic acid gives no precipitate?

13. How is this fact made use of in softening some kinds of hard water?

14. Is carbonic acid light or heavy? How can this be demonstrated?

15. What takes place when I breathe into lime water?

16. Explain the analogy between combustion and respiration.

17. What is the source of the warmth of our bodies?

18. In cold climates, the less clothing a person has the more food he requires. How is this?

19. What is the grand difference between the respiration of an animal and a plant?

20. Explain the process of the manufacture of gas. How is it purified?

WILLIAM CROOKES.

In January 1863 Crookes was much gratified by the receipt of an invitation to become a Fellow of the Royal Society:

SLINGS AND ARROWS

UNIVERSITY OF LONDON,
BURLINGTON HOUSE, W.,
16th Jan., 1863.

MY DEAR SIR,

I should be glad to see your name on the list of Fellows of the Royal Society, and if you have no objection to my doing so, would do myself the honour of proposing you for election into the Society. Could you spare a quarter of an hour on Monday afternoon to talk the matter over with me at University College, and oblige,

Yours very truly,
ALEX. W. WILLIAMSON.

Subsequently, Professor Williamson wrote : " I have much pleasure in congratulating you and ourselves on your being one of the fifteen selected by the Council of the Royal Society for election." There had been about fifty candidates. The election took place on June 4, 1863, and Crookes was formally admitted on November 19, 1863. The signatories of Crookes's " certificate " of proposal were the following :

General Knowledge.	*Personal Knowledge.*
J. P. GASSIOT.	A. W. WILLIAMSON.
BARTHOLOMEW PRICE.	M. FARADAY.
J. F. W. HERSCHEL.	JOHN TYNDALL.
E. W. BRAYLEY.	JOHN STENHOUSE.
	PHILIP YORKE.
	B. C. BRODIE.
	JOHN PHILLIPS.
	J. H. GLADSTONE.
	A. MATTHIESSEN.
	F. A. ABEL.
	J. H. GILBERT.
	B. STEWART.
	WILLIAM ODLING.
	D. T. ANSTED.

About this time Crookes had many consultations with Mr. Samuelson about a new departure of an editorial kind. Crookes's idea was to join Samuelson in the editorship of a new scientific quarterly, to be called the *Quarterly Journal of Science.* This would give Crookes ample scope for his wide scientific knowledge, and might also prove financially advantageous. He wrote to a number of prominent people for literary contributions. One of these, Mr. Nevil J. Maskelyne, replied as follows :

BRITISH MUSEUM,
Septr. 30, 1863.

MY DEAR SIR,

I have at this moment two or three heavy literary obligations unfulfilled to leading periodicals. I hardly see how I can promise you any immediate contribution. If in the course of my working off my present engagements—two of two years' standing !—I should have anything left to write about, I shall not forget your application to me now.

What is to be style and title of the new Quarterly ?

I am,
Yours truly,
NEVIL J. MASKELYNE.

The first number of the new periodical appeared on January 1, 1864.

Another move towards enlarging his income—there were already several children of Crookes's youthful marriage—was to take as an apprentice a pupil of Dr. Yeats's school, who showed considerable chemical ability. Crookes wrote to the boy's father as follows :

November 2, 1864.

MY DEAR SIR,

Before finally deciding to take your son as my apprentice, I thought it desirable to make a few inquiries from Dr. Yeats respecting him. It is a serious responsibility which Mrs. Crookes and myself will be undertaking ; the actual instruction in Chemistry will form but a small portion of our duties, and considering that Henry will form one of our household, and associate with our friends for so long a time, there are other things to take into consideration besides his intellectual capacity for acquiring chemistry.

I therefore went over to Peckham yesterday evening and had a long conversation with Dr. Yeats. You will be pleased to hear that his report is so far satisfactory as to remove any hesitation we might have felt in taking the responsibility of acting, *in loco parentis*, towards Henry during the next 3 years.

I shall therefore be happy to take your son as my apprentice for the term of 3 years. I will give him the best chemical education in my power and afford him facilities for acquiring an insight into manufacturing chemistry ; and when he has served his time I will either retain him as an assistant at a salary, or interest myself to obtain for him such an appointment as I may think suitable for him. He will board and lodge in my house, and shall be cared for and treated as a member of my family.

The premium I ask is £500, which will include Board, lodging and washing, in addition to Laboratory tuition for the three years.

I shall be glad to hear from you at your early convenience, as the addition of your son to my family will render necessary a larger house than I should otherwise have taken.

Believe me,

My dear Sir,

Very sincerely yours,

WILLIAM CROOKES.

B. SEWARD, Esqr.

All these months Crookes was seriously contemplating leaving London and moving to Manchester, in the hope of finding brighter prospects and more scope for his gifts in the industrial North. After visiting his friend Mr. P. Spence in Manchester, he wrote him the following letter :

October 31, 1864.

MY DEAR MR. SPENCE,

Since leaving your hospitable roof I have been constantly thinking of the proposed move. Inquiry amongst my friends and those able to advise has tended to confirm me in the opinion that such a change would be prudent, and I had begun to make up my mind seriously to the idea when a letter has arrived from Dr. Angus Smith (to whom I had written on the subject) which has acted somewhat as a damper.

He goes *seriatim* through the Manchester career of Dalton, Davis, Playfair, Bowman, Calvert, Frankland, Stone, Allan, Roscoe, and himself, and shows that analytical chemistry alone has not answered. He says : " No man has ever in Manchester made a decent living, or indeed any living at all by *analytical chemistry.*" Although he qualifies this by saying that " Several men, that is, about 5, have made a very decent addition to their living by *Consulting Chemistry.* This may be continued in Manchester, and probably will be increased as the value of chemistry becomes more known."

After giving the pros and cons on each side, he sums up as follows : " Taking now the world-view. If it be possible for me, I think a move to Manchester will bring you more rapidly into practice than staying in London, but the result will be less. However, it is often more agreeable to have our good rapidly, for life is short, and greatness often comes too late." The Dr.'s letter, which is very long, is on the whole rather favourable to my coming, but he takes a very gloomy view of the affair, and only predicts a very modest result. He has not shaken my opinion much with respect to the main fact. I am still inclined to come (I am only waiting for two letters before deciding), but I should like to know if you think there is any mental or constitutional peculiarity in Dr. Angus Smith which inclines him to be extra cautious and desponding. I fancy I have noticed it. Moreover, in his letter he says : " I always rejected little analyses which people sent for 5/- or 10/- and testings of soda and such unhappy things. . . . I was thrown on the larger work, and lost a great deal in that way, of course." He also leads me to infer that he has lost much by refusing to give evidence in law cases, or by not pleasing lawyers. Now I should especially seek to catch the little fishes as well as the great legal ones ; and, moreover, the £200 or £300 a year which the promised copper assays will bring in will be quite enough to encourage me in the face of a more desponding letter than Dr. Smith's. Still I should be very pleased to hear that he is likely to have presented the extra, double dyed-black side to my view. I have sent Calvert an offer which I don't think he will accept. I will write again as soon as I hear from him. Nelly joins me in kindest regards to Mrs. Spence, yourself, and your family circle.

Believe me,

My dear Mr. Spence,

Very sincerely yours,

WILLIAM CROOKES.

Crookes had been in correspondence for some time with Professor Calvert, of the Royal Institution of Manchester, who contemplated leaving that city, for the purchase of his laboratory and equipment. He now wrote to Dr. Angus Smith, a Home Office chemist, the following letter, which is given here almost entirely, as it sheds a most valuable light on Crookes's circumstances and state of mind about this time :

20, MORNINGTON ROAD, N.W.,

October 31, 1864.

MY DEAR DR. SMITH,

Your very friendly letter just received has set me thinking. I am grateful for the trouble you have taken in going so fully into the matter, and if I now proceed to argue upon the information you have given me, do not think that I wish to induce you to give me different advice, but only that I want you to look at the matter from my point of view, which may be different from yours. In the first place, I find " London " a *failure*. No doubt if I were to advertise constantly, and give puffing testimonials to tradesmen I could get a connection and make a decent living out of my Laboratory, but as for respectable work as consulting (in the wide sense you employ the term) or analytical chemist, I get next to none. I have made possibly £100 in six years at that work. My presence in a court of law, practically speaking, then my Laboratory and chemical education, are only of *money value* in so far as they enable me to exercise editorial supervision over the *Chemical News*. This being an amount of knowledge that any sharp person could get up in six months, nine-tenths of my " brain-force " is lying idle. This is the £ s. d. view of the case. There is another item. I was scientifically fortunate, but pecuniarily unfortunate some

88

years ago, in discovering thallium. This has brought me in abundance of reputation and glory, but it has rendered it necessary for me to spend £100 or £200 a year on its scientific investigation. Of course, it brings more than the value of this, *ultimately*, in reputation and position—but " whilst the grass grows, the horse starves," and it is the utter despair that I feel of ever making anything out of my London Laboratory that makes me anxious to leave. It is not that I neglect taking steps to make my wants known. I attend societies and converse with all the leading men, but . . . and the public all go in one or two grooves—their work is sent as a matter of course to the College of Chemistry, or to some of the well-known chemical schools attached to hospitals. There is plenty of work to be done, but there are so many eminent professors twice my age and standing that they absorb it all. I am aware that persons who have not lived in London are of a different opinion, but mine is the general experience of beginners here. When a great prize is to be competed for there are one hundred applicants from my class of chemists. One gets it, and is held up as an instance of the advantage of cultivating science and living in London, but what becomes of the 99 unsuccessful ones ? They starve in places of £150 a year, or are kept by their friends, or wait in the hopes of getting something better next time. . . . Within the last few years I have gone heart and soul into three competitions for a prize of £200 or £300 a year. I had a dozen rivals—everyone said I stood by far the best chance, but somehow or other I failed. I shall not try again.

I have thus given my reasons for not caring to stay in London. Now for the attractions of Manchester. Supposing I have moved there and find the Laboratory brings in *nothing at all*. I still have my present income scarcely diminished. But it won't be so bad as that. You refuse the small analyses. I can't afford to do so, and besides I feel I could go through the routine. *I am promised by Mr. Spence's new copper smelting company sufficient copper assays at* 10/- *each to bring in £200 or £300 a year*, and I have other trifles also promised which will give me a certainty of the latter figure. *Now all that is clear gain to me.* Moreover, there must be taken into consideration the other small analyses, which I shall lay myself out for and organise the laboratory for. Then there are the chances of pupils. Calvert tells me there are frequent applications, but he always refuses. I am accustomed to teaching, and like it.

There are also to be considered the chances of profit from entering more into Manufacturing Chemistry or dabbling in patents. Money has been made by yourself and most of those you name by these means. I have not had much experience at manufacturing, but I fancy my *forte* lies somewhat in the line which would be found most useful at Manchester.

Your letter has shown me that Manchester is not the place where a chemist is likely to make a fortune, but he can make a tolerable living by his profession if aided by a little private income. That is all I want. A stationary income will not do with an increasing family, and domestic necessities are apt to make scientific men very mercenary.

I have told Calvert that if he will guarantee by his books that his laboratory has been bringing in £500 or £600 a year, and can secure the laboratory to me for a long term of years, and will give me all his working apparatus (except balance and air pump, &c.), I will give him during the next 3 years such a share in the profits as will make up the £100.

I have decided to come, and *wish especially* to have his Laboratory. Will you therefore kindly see the secretary and see what can be done, for in case Calvert rejects my offer I should like to be certain of having his Laboratory before he interests himself for anyone else? He told me he could hand it over as a matter of course. Possibly you would be able to find out better than I could the time Calvert talks of leaving. I am thinking of moving soon after Xmas.

Respecting the article on the Report of the Mining Commission, I did not get your letter till Friday night owing to a delay of the post. I should very much like to have such a review. Can you not manage to write it yourself? As an article on the subject it would be an admirable topic for the *Quarterly Journal of Science*, and no one could handle the matter so well as yourself. If you would rather not do it, I would see if I could undertake it, but I would rather it were by you. Where can I see a copy of the report?

Excuse this long letter—you have brought the infliction on yourself by the very kind manner in which you have answered my former one.

Believe me,
My dear Dr. Smith,
Very truly yours,
WILLIAM CROOKES.

Realising that Dr. Calvert's business would be of little use to him without the use of his laboratory, the habitation of which belonged to the Manchester Royal Institution, Crookes applied for Dr. Calvert's Honorary Professorship in the event of his vacating it. He wrote at the same time to Dr. Calvert :

November 5, 1864.

MY DEAR DR. CALVERT,

Since writing to you on the 28th ult., circumstances have occurred which render it unadvisable for me to involve myself so heavily as it would be necessary if that proposition were carried out. I therefore beg to formally withdraw the offer contained in my letter above-named.

At the same time I may mention that I still contemplate coming to Manchester, and shall be happy to negotiate with you for the purchase of your goodwill in case you may be willing to part with it for a less amount than the one you have at first asked. I have delayed writing to you for some days, expecting to hear from you in answer to my offer. Please favour me with an answer at your early convenience, and believe me

Very truly yours,
WILLIAM CROOKES.

The above letter is a good sample of Crookes's *suaviter in modo* style, though in reality it might be considered a case of hard bargaining. In the event it failed of its object, for after some months of doubt Dr. Calvert made up his mind not to leave Manchester after all. It may be that the failure of Dr. Calvert's own London plans had more to do with this decision than Crookes's attitude. On March 29, 1865, Dr. Angus Smith wrote from Manchester :

MY DEAR MR. CROOKES,

Living as I have been in a small room and the dark, long winter occupied with my own work, and scarcely able to think enough of the doings of others to satisfy the demands of friendship, and emerging to-day from the house after such a prostration of intellect and such feeling as a cold gives one, I feel as if by coming out into the open world to-day I were entering on some bright land of promise. And yet when I consider the matter I have no bright words to say, and am rather compelled to give news unwillingly— the news is that Calvert has really decided (he told me himself) not to accept the London proposal. We cannot, then, have you at the Royal Institution. As to the copper works, I scarcely expected after last visit that you would go there, as most of the directors seemed to think—I speak from hearsay—that it was not a place good enough, and that they would be tempted to give more than they were willing to give. I think myself for the first year at least it would have been unsuited.

Another movement is on the wing ; they talk here of lifting the Royal Institution away and putting it down about the neighbour- hood of my Laboratory. This will be a very curious change in many ways, and it may be that it is fortunate that you did not come just now. It is especially fortunate that you did not give the money for the business as the whole of the value to be received was the *site*. A new Institution may have a very different arrangement, and I think I may say that I have not diminished your friends in that quarter. At the same time it may have no Laboratory at all, but all this is floating—still under active consideration. A better time will come for your movement here. . . .

We seem to get on with each other, and if so, may be useful to each other

I am,

Yours sincerely,

R. ANGUS SMITH.

The expectation of mutual utility was to be amply justified in the sequel.

While in Manchester, Crookes had heard rumours of an impending chemical appointment to the Woolwich Arsenal. He wrote to Spiller about it, who replied in an interesting letter which incidentally describes a great gunpowder explosion :

ERITH EXPLOSION

CHEMICAL DEPARTMENT,
October 17th, 1864.

MY DEAR WILLIAM,

I am glad to hear that you are spending such a pleasant time at Manchester. The weather here has been settled fair for a long while, I hope it has been so with you. Brown is at Waltham to-day, so that I must answer now for him, and will give him your letter to-night, for he lives at Woodland Terrace, Charlton. Try No. 12 on "spec."

I have seen Captain Scott, R.N., he does not know anything about a Chatham Appointment, but will ask the Lords when next he goes to the Admiralty. I have not seen him since. This reminds me that I last saw him on the day of the terrible explosion. Oh, what an event was that to us ! it shook our house so fearfully that I thought it was coming down upon us. Of course, at the moment I thought it was an earthquake, but on looking out of window (luckily none of our panes were broken) I soon saw a mighty cloud of smoke which has been well described as being like a mushroom. Henry and I dressed immediately, and just as we were starting for the Arsenal (thinking it was, of course, our Woolwich Magazine blown up) a shower of papers came down in our road, and some of these being gunpowder invoices, told me that it was Messrs. Hall & Co.'s premises at Erith. Our stone steps were quite covered with a black dust of gunpowder residue and blackened debris. Ada started up in bed and asked whether I had fallen out upon the floor. I assured her that I was not even Mr. Banting, but that another cause must be sought for the violent concussion we had just experienced. However, I found the Arsenal Gates besieged by a female multitude anxious to assure themselves of the safety of their better halves, and making my way in ascertained at once the precise locality which had just then been made known by telegraph. Henry and I started off immediately by train to Erith and saw a truly shocking sight. The dead and wounded being in the course of removal—two houses almost entirely knocked down, and a wide hollow basin, where once the magazines stood. The river bank destroyed or breached, and but twenty navvies at hand to repair it. We heard that messages had been sent to the Commandant, Woolwich, for troops to come at once, and to the Sewer Works near at hand for some or all of their navvies. The place smelt

93

horribly of sulphuretted compounds mixed with smouldering materials from one of the burning houses hard by.

* * * * *

P.S.—The Chief has gone with the Gun Cotton Committee to Allanheads. They are making experiments on blasting with gun-cotton.

The last number of *Chemical News* was very good.

This was the last of Crookes's futile efforts to obtain a public appointment. For two great plagues were already approaching the shores of Great Britain, and were about to create a situation requiring the services of just such a man as he was.

CHAPTER IX

THE CATTLE PLAGUE

(1865–7)

THE CATTLE PLAGUE OF 1865–7 was imported from Germany, and was sometimes referred to by its German name, *Rinderpest*. It commenced in Islington in July 1865, and the last case occurred in the Hackney Marshes in September 1867. While it raged over England it caused the destruction of more than 335,000 head of cattle, and the loss of three or four millions sterling.

Crookes does not seem to have concerned himself with the outbreak until October, when he happened to visit the Midlands to lecture at the Liverpool School of Science on " Explosions Viewed Commercially and Chemically." His notes for this lecture have been preserved, and they furnish evidence of careful preparation.

FRIDAY, *Oct.* 20, 1865.

It is perhaps necessary that I should explain the way in which I propose treating the subject I have chosen for this evening's lecture. Were I to impose no limitations on the subject I should have to take in nearly the whole science of Chemistry. For, explosions being, for the most part, the result of certain chemical actions exerting themselves in a very violent or rapid manner, it is quite possible, in an enormous number of chemical experiments, so to arrange the materials employed, that the reaction will take place with explosive violence. I have therefore to select, from

95

the very wide field open to me, a certain class of explosions, and I have chosen those which are most interesting on account of their present or probable future *utility* in the arts of peace or war, and illustrate a few of their points of agreement and difference, their special adaptation to the ends required, and the chemical phenomena involved in their ignition. I shall for the present leave out all reference to many explosions which are of a destructive or accidental character, such as explosions in coal mines, boiler explosions, and shall only treat of those which have been devised by man's ingenuity, and the forces in which are brought into play at any desired moment at the will of the operator.

What is an explosion ? It is due to a considerable and sudden expansion of matter, either gaseous, liquid, or solid. We shall have many opportunities this evening of becoming practically acquainted with explosions, and the special cause of each will best be explained when they happen. Three instances :

$$\text{I. } PCu_3 \text{ in HO.}$$
$$\text{II. } S+KOClO_3.$$
$$\text{III. } P+K_2ClO_3.$$

The last is Armstrong's shell exploder.

The first great explosive is gunpowder. This is too well known to need special description. Mixture of carbon, sulphur, and salt-petre. On burning, evolves large quantity of gas (800 times bulk). If this is allowed free vent no danger, but if confined it explodes. Mistake to suppose G.P. is so very inflammable. Not so inflammable as iron (Exp.).

Powder of different degrees of coarseness in plates to show rate of burning depends on size of particles. Large lump exploded.

Refer to Gale's plan of protecting (glass and plumbago). 1 G.P., 4 glass. *Very bulky.* How this acts, isolates particles. Water or oil do the same.

Next important explosive—the explosive of the future—Gun-cotton.

This evolves great amount of gas when burnt, and explosive force entirely depends upon the rapidity with which these come off. Show the gases, very slow combustion.

More rapid evolution of gas is not explosive. *Burn* the thread in air.

Burn thicker thread, nearly explosive. This gives an idea of how chemical action rises to explosive violence.

G.C. been before public many years. Schönbein. Dormant till General Leuk, Austria, took up subject. Our own War Department, Waltham Abbey, Mr. Brown, Scott Russell, Abel, &c., Messrs. Prentice, of Stowmarket.

Show rapid combustion passing into slow, by passing thro' hole in card.

Leave train to end.

1st form.—Fire train of G.C., 6 inches a sec   Explain action of hot gases.

2nd form.—Plaited cord, 6 feet a sec.

Comparison between G.C. and G.P.   1 lb. G.C. = 3 or 4 lbs. G.P. in guns.

To make gun-cotton formidable and destructive, squeeze it, compress it ; to render it gentle and manageable, give it room.

With G.P. it is just reverse. Give G.C. a light burden to lift and it will not. Invent a ruse, let it think it has a difficulty—put it in a strong box and it will overcome almost everything.

For instance, the G.C. yarn may be exploded on paper without danger, but if I want to make it do some hard work, blow up this box for instance, I must enclose it in a case. Explode glass tube under box. (Sit on box to keep it down.)

Explode indiarubber covered yarn in air. Sixty feet a sec.

This property makes G.C. valuable for mining and for shells. G.C. does not foul guns, makes no smoke, evolves less heat. G.C. not hurt by wetting, made and stored wet, perfectly harmless. Show this.

Nitroglycerine, or Glonoine, known 20 years, practical use thought to be impossible, difficult and dangerous to make. A. Nobel, of Hamburg. 3/- per lb.

Tho' not known much in England ; 10 tons weight have been used in Sweden in last 8 months.

Light yellow oil ; sp.g. 1·6, insoluble in HO, not volatile, very poisonous.

Not exploded by flame or spark ; put some on blotting-paper and heat on lamp.

Exploded by percussion ; will bear 212°, explodes at 360°.

H   97

1 Cubic inch yields 10,400 c. inches of gas, whilst G.P. only yields 800.

∴ 8 times power for equal weights.

13 times „ „ „ volumes.

Exhibit cartridges and Photograph. Agent, Mr. Cusel, 64, Wood Street, E.C.

Torpedoes, mines, &c. Difficulties in exploding mines, &c., at a distance.

In Crimean War the Russians defended Cronstadt by torpedoes across harbour. These ignited by glass tube of SO_3HO and sugar and K_3ClO_3.

Our Government had Select Committees to inquire into matter. As result, Wheatstone (one of the Committee) invented his magnetic exploder. Explain principle.

Explain new fuse (Beardslee's or Maury's).

Explode piece of G.C. over meal powder on plumbago fuse on table.

[Show action at a distance. Explode caps and quill tubes in different parts of the room.—Leave this till last.]

These exploders were consequently adopted by our Government, and instruments were made and used at Chatham, Woolwich, Gibraltar, Malta, Corfu, Belgium. They were not kept secret. One got to the Confederate States of America, and with this they blew up the ship *Tecumseh* in August 1863.

Immediately on hearing of this Captain Maury, who was then in England, had three exploders made on Prof. Wheatstone's principle, introducing some slight alterations of his own. (The plumbago fuse was his, I believe.) These were shipped in a blockade runner. This was captured and taken to New York. The cargo sold by auction. Mr. Beardslee bought these exploders, and thus became acquainted with their construction. He made some with slight alterations, but still adhering to Wheatstone's discovery, only making them in this way a copy of an old and less perfect invention of Prof. Wheatstone (patented, I believe, in 1848). Mr. McKay took the idea up. Came over to England, got introductions to the Admiralty. Had grand trials made at Chatham (see *Illustrated News*), and now asks our Government to give them £10,000 for an invention which was fully described in one of our Blue books nearly 10 years ago.

Show action at a distance. Explode caps and quill tubes in different parts of room.

Explode G.C. train.

In conclusion, beg younger hearers not to handle these things rashly. They are not playthings, but very dangerous. Safe in my hands because I know exactly how far to trust them, but terrible agents of destruction in hands of persons not acquainted with all their properties.

While Crookes was in the Midlands he visited Manchester and saw a Mr. Readwin in connection with a scheme for improving the methods of gold-mining. Crookes had taken out a patent for a method of extracting gold by amalgamation, and conducted about this time a considerable correspondence on the subject. But his visit brought him face to face with the serious menace of the cattle plague. He decided to try some new experiments with certain " tar-acids," particularly carbolic and cresylic acid. These were then just coming into use. They were extracted from coal-tar, the importance of which had not been realised until Faraday discovered benzene and Perkin extracted from coal-tar the first aniline dye, mauve. Hearing that the dreaded plague was approaching the farm of a Yorkshire acquaintance, Mrs. Carmichael, he wrote to her on December 7, 1865, advising her to use carbolic acid as a disinfectant, and sending a sample supply, with full instructions how to use it.

The experiment seems to have been very successful— so much so that Crookes thought of bringing the acid into use on a large scale and incidentally making some money out of his happy inspiration. He decided to take his friend, Dr. R. Angus Smith, into his confidence, more especially as the latter was coming into prominence as an expert on sanitary chemistry, and was already engaged in combating the disease. He wrote to him with a proposal to patent cresylic and carbolic acids as disinfectants, and offering to share profits. Dr. Smith, however, refused this, as he had been officially engaged by the Cattle Plague Commission.

The delicacy of his position was brought home to him by a question asked in the House of Commons on March 5th, which was reported in *The Times* as follows :

THE CATTLE PLAGUE.

Mr. W. MILLER asked the Secretary of State for the Home Department whether he is aware that the Cattle Plague Royal Commission, in their endeavours to discover the best disinfectant, referred the question only to a single individual—namely, Dr. Angus Smith, who reported in favour of carbolic acid, or Macdougall's powder, of which preparation he himself is co-partner with Mr. Macdougall, and which is generally believed by chymists to be no disinfectant at all.

Mr. BARING said that the question was one which ought to be put to the member for Calne (Mr. Lowe), who was a member of the Commission. Dr. Angus Smith was annoyed at the reference to himself, and he wished that the question should be deferred until the report he prepared had been made and presented to Parliament. He had, however, given them the following information on that part of the question :

" I never had any interest, profit, or advantage from the sale or manufacture of Macdougall's powder, or of any other substance made by him, or by anybody else." (Hear.) Dr. A. Smith added :

" I do not recommend Macdougall's powder as the best disinfectant." (" Hear, hear," and a laugh.) And upon the third point he said :

" Carbolic acid is not Macdougall's powder, but a liquid not manufactured by Macdougall." (A laugh, and " Hear, hear.")

With regard to the second branch of the question he had received the following from Mr. Montagu Bernard, the secretary to the Commission :

" Dr. Angus Smith was the person employed by the Commissioners to report on disinfection and disinfectants. He is an eminent chymist, as everyone knows, and had previously turned his attention to the subject. After a long series of experiments on a great number of substances he reported in favour of chlorine, muriatic acid, sulphurous acid, and the two tar acids (otherwise called carbolic and cresylic acids). On a consideration of his report, the tar acids were deemed by the Commissioners most likely to be efficacious and best suited for general use. They then instructed a younger, but distinguished chymist, Mr. W. Crookes, F.R.S., to go to a district where the disease was raging to test in several ways the efficacy of the selected substances, and to ascertain by personal experience the best and simplest modes of using them. Mr. Crookes has been for some time at work and the accounts received from him are very satisfactory. Macdougall's powder is a preparation containing carbolic acid, with sulphites of magnesia and lime. It was among the many substances tested by Dr. Angus Smith, and he recommended it as useful in some ways, making no secret of the fact that it had been first produced by himself, together with Mr. Macdougall, ten years ago." (Hear, hear.)

In answer to the latter part of the question, he had to state that the Home Office had no means of forming an opinion in regard to disinfectants, but when the Cattle Diseases Act was passed the Secretary of State for the Home Department directed a letter to be written to the Royal Commissioners, asking them to furnish him with the best plan for disinfecting premises, etc., from the contagion of the cattle plague. That information was supplied, and it had been circulated throughout the country. From the constitution of the Commission, and the manner in which they had directed the experiments to be made it was impossible to suggest any body of persons better qualified to come to a proper conclusion on the matter. (Hear.)

The patent for Macdougall's powder was granted to Robert Angus Smith and Alexander Macdougall on January 26, 1854, and in 1867 Alexander Macdougall applied for an extension of the patent. Dr. Smith wrote in one of his letters to Crookes : " I hear Condy wanted to use non-volatile liquids in washing the air of the Houses of Parliament, and also he set in motion that *vile* question put as to my connection with Macd.'s powder. . . . There are some very fame-greedy persons at work."

It is certain that Crookes derived no financial advantage

from his labours in combating the cattle plague. Even if he had had an interest in the sale of carbolic acid—there is no evidence that he had—its use never became general. People complained of its corrosive action on the skin, and Crookes had considerable difficulty in answering this objection.

A great part of Crookes's difficulty consisted in the active or passive resistance of the inspectors appointed by local authorities under the " Cattle Diseases Prevention Act." Their only weapon of war against the disease was the pole-axe. If any of them believed in disinfectants, they confined themselves to the old-fashioned chloride of lime bleaching powder, which acted at all events as a deodoriser. This attitude finds a parallel in the maxim frequently enumerated by the cook-private in the Great War : " If the meat smells good, it *is* good."

Crookes found it necessary to flourish his official powers on some occasions, as he did in ordering Dr. W. H. Ryott, one of the inspectors, not to slaughter the infected animals on Barroby's farm at Dishforth, but to reserve them for investigations.

Crookes's report " On the Application of Disinfectants in Arresting the Spread of the Cattle Plague " was embodied in the Third Report of the Cattle Plague Commission. It is a model of lucidity and close scientific reasoning. The science of bacteriology was still in its infancy. Cohn had only begun his work ; Pasteur was still a chemist rather than a physiologist, and Koch did not isolate the anthrax bacillus till ten years afterwards. Yet Crookes was quite clear that the disease must be due to *germs*. A few extracts will suffice to show this :

There are weighty reasons for deciding that the infecting matter is neither a gas nor even a volatile liquid. . . . The infection is capable of being carried considerable distances in clothing or running water. . . It seems clear that the disease must be communicated by the agency of solid, non-volatile particles. . . . The specific disease-producing particles must, moreover, be organised, and possess vitality ; they must partake of the nature of *virus* rather than of poison. . . . Many considerations tend to show that the virus of cattle plague is a body similar to vaccine lymph, and consists of germinal matter, or living cells, possessing physiological individuality. . . . The foregoing view differs from the prevalent notion that the virus of contagion consists of decomposing organic matter, declining from a complex to a more simple chemical constitution, and during its degradation inducing decomposition in the neighbouring particles of matter. . . . According to this view the ferment would be constantly diminishing, whereas in reality it constantly increases in bulk. The hypothesis is therefore insufficient to explain the prodigious procreative power of the original particle. This power belongs only to the nature of an organised germ, capable of producing multiples of itself by a process of nutrition and subdivision.

The Commissioners were sufficiently impressed with Crookes's report to recommend his method of disinfection. Among their " Recommendations for Disinfection," issued on February 23, 1866, are the following :

Wash the woodwork of the sheds everywhere with boiling water, containing in each gallon a wineglassful of carbolic acid. Then limewash the walls and roof of the shed with good, freshly-burnt lime, adding to each pailful of whitewash one pint of carbolic acid.

Cleanse the floors thoroughly with hot water, and then sprinkle freely with undiluted carbolic acid.

Lastly, close all the doors and openings, and burn sulphur in the shed, taking care that neither men nor cattle remain in the shed while the burning is going on.

But Crookes was fighting an almost hopeless battle against human inertia and conservatism. His last letter to Professor Bernard shows that he was willing to try other expedients, such as dosing with chlorine, which met with less prejudiced opposition.

Crookes's engagement by the Cattle Plague Commission was for two months, and came to an end on April 6, 1866. His " memorandum of expenses " is interesting :

THE CATTLE PLAGUE

Memorandum of Expenses paid by William Crookes on Account of Cattle Plague Experiments, from Feby. 7th to April 6th, 1866.

	£	s.	d.
Railway expenses	13	12	6
Dog Carts, Driver, Ostler, &c.	4	10	6
Cabs, &c.	1	8	0
Hotel Bills, Servants, &c.	12	12	6
Expense of Cattle, Keep, &c.	22	10	6
Professional assistance in experiments, men, &c. ...	4	3	0
Carbolic and Cresylic Acid	12	4	0
Sundry Chemicals	2	7	3
Instruments for injection, Thermometers, aspirator, &c.	7	8	6
Carriage of Parcels, Goods, &c.	1	10	4
Telegrams	1	4	0
Postage, Books and Sundries	0	18	2
	£84	**9**	**3**

Crookes was subsequently engaged for a month by the Earl of Lichfield to disinfect his Staffordshire estate. He charged the moderate fee of two guineas a day, as is evident from the following account :

Rt. Honble. the Earl of Lichfield to W. Crookes, F.R.S.
June 1866.

For expenses during visit to Stafford in May and June on Account of the Cattle Plague, as under :

	£	s.	d.
Railway	13	9	0
Cabs	1	5	6
Carbolic Acid	3	0	6
Thermometers	5	18	6
Altering pump, &c.	1	0	0
Telegrams	0	8	6
Lunch, Servants, Men and Sundries	1	16	9
	£26	18	9

For Professional attendance from May 9th to June 7th,
25 days, at 2/2/0 .. .. 52 10 0

£79 8 9

After six months of arduous and unpleasant work, Crookes found himself but moderately successful and ill requited. His Manchester plans having come to nought, and his gold-extraction schemes being delayed, he thought of applying for an appointment as a municipal gas chemist. On consulting his friend, Angus Smith, he received the following characteristic letter :

MANCHESTER,
June 5, 1866.

MY DEAR MR. CROOKES,

I fear I cannot tell anything regarding the appointment of the gas chemists. I think if I were you I would apply to the City authorities, straight to the Mansion House. There will be a person there whose business it will be. Dr. Letheby will know most—if you know him.

Nobody knows a perfect way of trying the amount of sulphur in gas, any more than you.

I have a paper on water which has been largely circulated in India by the Government. I left a copy for you and for Mr. Fewtrell. I shall tell it to be sent.

Yes, I think the gas chemists would be appointed from their position, not caring about the fact of their having been busy with gas, as it is a problem.

Mr. Ward is here.

I would not dispute the matter with Mr. Stone.

I am dead tired, Mr. Ward having kept me up till daylight. I think Miss May Spence has paid me the greatest compliment I ever received, and it is also decidedly original. It is in saying that she will send no special message as she knows I do not require it. Such a message is good for a man's spirit, he endeavours to act up to the character for which men and women give him credit, and even if it were not true, to have said it would make it true.

No, I cannot come. I am utterly tired. May you all be happy

Yours sincerely,
R. ANGUS SMITH.

THE CATTLE PLAGUE

During the winter of 1866–7 the Cattle Plague abated somewhat, and hopes were entertained of its disappearance. In December, Crookes warned the public against undue optimism in the columns of the *British Medical Journal* (December 22, 1866), saying that if the stringent measures of protection were relaxed there would inevitably be a new outbreak in the spring.

The following letter, written to Dr. Calvert, illustrates the scrupulous correctness of Crookes's attitude in matters involving his professional opinion. It should be remembered that Dr. Calvert manufactured and supplied the bulk of the carbolic and cresylic acids issued in fighting the rinderpest.

20, MORNINGTON ROAD, N.W.,
Jany. 2, 1867.

MY DEAR DR. CALVERT,

I shall be very glad to carry on any examination you wish respecting the antiseptic value of your acid and powder, and from what I already know of them I could make a report that would be considered satisfactory ; but I would rather you did not ask me to undertake an examination of Macdougall's or any other disinfectant at the same time.

I have repeatedly been applied to by manufacturers, and have invariably refused, to give reports in which their products are compared with those of rival makers. On one occasion this was urged so strongly on me that I took the advice of one of the leading Professors of Chemistry in London, and he showed me so clearly the injury it would do me to give such a report, that I made up my mind never to be tempted.

Besides, I really do not think that such a comparative report as you wish for would be of much service. It would be so evidently a one-sided document that no one would pay much attention to it, in spite of my name being attached to it ; and Macdougall would immediately get some one else to examine his preparations against yours, and thus the little good the comparison might otherwise do you would be entirely neutralised.

Were you, however, to consent to your products alone being

examined the investigation could be as full as you like, and there could be no suspicion of partisanship.

Mrs. Crookes joins me in kind regards to Mrs. Calvert and yourself, wishing you all the pleasant compliments of the season.

Believe me,
Very truly yours,
WILLIAM CROOKES.

The recrudescence of rinderpest predicted by Crookes set in as the summer of 1867 approached, and the neglect of the precautions recommended by him was vigorously criticised in several quarters. The *Medical Times and Gazette* (June 1, 1867) wrote :

CATTLE PLAGUE AND DISINFECTANTS.

If Mr. Crookes were not one of the most modest and retiring of men, we should long ago have had a question put in Parliament as to the steps which the Privy Council had adopted, beyond the slaughtering of beasts and arrest of transmission, for the purpose of checking the progress of this new outbreak of cattle plague. The question would have been asked whether any directions or recommendations had been issued to dairymen, inspectors, and stock owners as to the best method of procedure when the disease breaks out in a farm or other establishment of cattle in their neighbourhood. The neglect of issuing such recommendation leaves each inspector at liberty to follow his own devices. We have heard of a shoemaker being appointed by a local authority to such a post, and an appointment of the kind does not necessarily bring with it the requisite knowledge for the proper performance of the duties of the office. In London everybody knows that the committees locally entrusted with the arrangements for checking the outbreak are not constituted of men very learned in preventive medicine, so that, as things stand now, they are very much in the position of blind men being led by the blind. Besides, the absence of any official instruction leaves the general public to believe that nothing further is known as to the method of prevention than that the sick and the healthy should be indiscriminately slaughtered. We are not quarrelling with the procedure as far as it goes ; all we say is, that the preventive measures should not stop here. The slaying of cattle, indeed, necessarily puts a stop to any further progress of the disease in the particular herd affected, and so far lessens the chance of attack in neighbouring herds ; but still the latter are not safe so long as the virus is undestroyed. And the virus is not destroyed by mere slaughtering. It hangs about the shed ; it is concentrated in the manure dropped upon the floor of it or in the fields ; it is blown about in dust ; it is carried about on the boots of herdsmen, and, in fifty indefinable ways, endangers the whole neighbourhood of an infected place for miles round. So far as the issue of instructions to stock owners goes, the neighbourhood of infected places may continue to be endangered. One would almost think that the Lords of Her Majesty's Privy Council were without all scientific assistance. Certainly, for any practical use they have been to those in authority, Mr. Crookes might never have made the accurate and convincing experiments which he did make in respect of the several methods suggested for the destruction of cattle plague virus. As their lordships and Dr. Williams appear so innocent of all knowledge upon this subject, it will not be much amiss for us to enlighten them a little.

First, then, Mr. Crookes has shown that the favourite disinfectant—namely, chloride of lime—is about the least efficient of any of those substances which are reputed to possess disinfectant qualities. Chlorine itself is very little better. True, it will destroy the virus in time if one uses enough of it ; but as it acts by way of oxidation, and as living virus resists oxidation longer than dead oxidisable matter, it follows that before chlorine gas can attack a virus everything else that it can oxidise will be oxidised first—everything, in short, such as sulphuretted hydrogen, which has no more to do with the spread of cattle plague than their lordships' black hats. And if chlorine is so extravagant and questionable a disinfectant, what can possibly be expected from a sprinkling of chloride of lime, perhaps one of the most adulterated disinfectants known —adulterated in the sense that, as commonly sold, it contains an excesss of carbonate of lime and chloride of calcium, and not more than about 20 per cent. of chlorine in a form that is available for disinfection.

Secondly, Mr. Crookes has shown that there exist in sulphurous acid and carbolic acid substances which are absolutely destructive of every kind of living thing of low organisation, such as cattle plague virus is supposed to be—that these substances not only destroy the virus, but attack it at once, and, moreover, arrest all tendency to putrefactive decomposition in animal matters with which they are mixed. But even this is not all that his observations have brought to light, for the deductions of the philosopher have been thoroughly borne out by the observation of the practical sanitarian. Mr. Crookes is a scientific chemist, but this is no reason why his practical observations are to be ignored.

Crookes himself wrote a letter to the President of the Privy Council, of which the following are some extracts :

DISINFECTION.

Extract of a letter sent to His Grace the Duke of Marlborough, Lord President of the Council, June 14/67 :

" Much of the so-called disinfection which has been adopted in London and elsewhere is of no value whatever. I was one of the scientific reporters appointed by the Cattle Plague Commission to investigate the subject of disinfection. After a long and searching inquiry, it was ascertained that the best results were obtained by combined disinfection with the tar-acids and sulphur fumigation. These results were adopted by the Commission in their official recommendations for disinfection, which were distributed over the country. But the agents recommended to be used were unfamiliar, few would take the trouble to inquire into the reasons for which they were preferred, and as their use was not made imperative, they did not meet with a fair trial. At present they are almost unused, and chloride of lime—a deodoriser of very limited use, and much inferior to sulphur and the tar-acids—is trusted to almost entirely. I believe the present renewed outbreak of cattle plague, when not a re-importation, has occurred from the use of imperfectly disinfected premises. Too much reliance is placed on mere visible cleanliness, and too little faith is put in bodies whose efficacy is only apparent to the reason. It cannot be too strongly urged that *deodorisation* is only half-way towards *disinfection*. But in ordinary attempts at disinfection the only test of efficiency which a workman would employ is the sense of smell, and many cases have occurred in which a *deodorised* shed, to all outward appearances *disinfected*, was still in reality full of infection. It happens that the stinking vapours of decomposing animal or vegetable matter are of comparatively little danger in the atmosphere, whilst the deadly virus-cells of infection are scarcely appreciable to the sense of smell.

" These are no mere theories which I have ventured to bring before your notice. By acting in accordance with the views here expressed, I have succeeded in protecting healthy farms in the most plague-smitten districts of England from the march of the closely investing plague. I have arrested the progress of the disease when it has commenced ravaging a farm with virulence. I have put them to a still severer test, and have prevented the communication of the plague to healthy animals when kept by themselves in highly infected sheds ; and have preserved a healthy animal from the plague when tied in a close shed by the side of a diseased and dying animal. Lastly they have enabled me to counteract the virus of the plague when actually present in the system of the animal, and thereby, for the first time, effect a rational cure for the cattle plague.

*   *   *   *   *

" In making these remarks I wish to state that I am not a veterinary surgeon, nor have I the slightest direct or indirect interest, or any commercial relation whatever, with the subject. I write solely from a scientific standpoint, my object being to show how waste of life, property, and food may be avoided, and the results of scientific investigation made available for economic purposes.

" W. C."

The cholera epidemic which followed close upon the rinderpest, and became particularly prevalent in England in the summer of 1866, moved Crookes to offer his free services to the Government for the purpose of carrying out researches. He submitted a programme, and, considering the extreme risk to the investigator involved, we may, while admiring Crookes's courage, rejoice that he was not authorised to carry out his public-spirited intentions.

His part of the fight against cholera was confined to publicity work and service on local cholera committees in London, which he discharged faithfully until the epidemic abated.

CHAPTER X

GOLD AMALGAMATION

(1865-7)

IT WAS IN 1865, at the age of thirty-three, that Crookes, for the first and last time in his life, had a vision of amassing a vast fortune. The manner in which he pursued this vision, and the novel expedients and methods he adopted to find his way into what was to him an unknown region of activity, enable us to form a new estimate of his character. He reveals himself as self-confident, intensely optimistic, resourceful, masterful, sensitive of his honour and dignity, but easily appeased and always ready to shake hands after a stand-up fight. In his legal and financial transactions he showed an acumen and ability somewhat rare in academic circles, and though his dream was not realised, he probably made as much of his opportunities as a highly specialised business man could have done in the circumstances.

In this whole period Crookes does not appear so much as a scientist as he does as a chemical engineer. The purely scientific pose of the student days is gone. He is now the practical man of affairs, bent upon turning an invention to the utmost advantage. The invention itself is nothing but the adaptation of ideas already current. Crookes himself embodied in his notebooks of the period an extract from Gmelin's *Handbook of Chemistry*, vol. i, p. 382 (1851):

Traces of Bi, Sn, Pb, Cu, and Ag dissolved in Hg separate on the addition of K amalgam and HO, because an evolution of H gas takes place on the Bi, etc.

Here we have practically a note on the metallurgy of silver, dissolved in mercury, and precipitated by potassium amalgam and water.[1]

Another significant extract which probably was immediately seized upon by Crookes's active and enterprising mind, was from the *Chemisches Centralblatt*, subsequently published in the *Chemical News* :

Number 17 contains an original article by Bunge " On the Action of Sodium Amalgam on some Metallic Salts." A tolerably strong aqueous solution of ferric chloride, acidulated with hydrochloric acid, when treated with sodium amalgam becomes clearer and clearer, and at last colourless, the whole of the iron being reduced to protoxide. By the further action of the amalgam all the iron is withdrawn from the fluid, and is found as iron amalgam. The author is of opinion that the reduction of ferric to ferrous chloride, and of the latter to metallic iron, proceeds simultaneously.

Chromic chloride suspended in acidulated water undergoes a similar reduction. The amalgam of chromium was fluid, and rather unstable. On standing, its surface became covered with a dark grey powder of metallic chromium.

Mercuric and gold chlorides are similarly reduced. In the former case calomel is first produced.

The chloride, iodide, and bromide of silver are very quickly reduced. (The author made all these experiments in the dark.) This reaction, it is thought, may be usefully applied in the analysis of the haloid compounds of silver, instead of fusing these compounds off with an alkaline carbonate. It is only necessary to place a small quantity of the powder in a test glass with a little water, and add thereto some pieces of sodium amalgam, and the chlorine, iodine, or bromine may be detected in the solution in the ordinary way.

Crookes appears to have determined to investigate the

[1] Now denoted by the formula H_2O, since O was found to be bivalent.

action of sodium amalgam, not upon metals in solution, but upon native metals, disseminated in ores, and to study the difference between pure mercury and mercury combined with sodium. These investigations were secretly made about Christmas 1864, and by February 1865 Crookes had lodged his first patent application in London. It was dated February 11, 1865, and claimed as original " the employment of an amalgam of sodium, or such other alkali metal as aforesaid, in treating ores or substances containing gold or silver for the extraction and separation therefrom of the precious metals." This specification was not published till January 1866.

What Crookes did not know at the time, nor for several months afterwards, was that a German-American chemist, Professor Henry Wurtz, at one time Chemical Examiner in the U.S.A. Patent Office, had patented the sodium-amalgam process for gold extraction in 1864, probably inspired by a chemical investigation abroad just as Crookes had been. Be that as it may, Crookes lost no time before interesting his friends in the new process. He discussed it with several Manchester chemists while on his tours connected with the Cattle Plague. He met with some scepticism, notably from his friend, Dr. P. Spence. Crookes wrote to him to point out that his objections were without foundation ; that sodium acted rather more mechanically than chemically, and that, when mixed with mercury, it did not readily oxidise under water. On the same day, Crookes wrote to Mr. Readwin, the proprietor of the Gwynfynydd mine in the Welsh goldfield, asking him for facilities to test his process with several tons of ore. By April he had carried out some tests in London, which were very satisfactory, so that he felt ready to operate on ten or twenty tons of ore in Wales. He obtained permission to try the process in three Welsh gold mines.

To encourage Mr. Spence, sen., he offered to hand over to him half his American patent rights.

The papers for the American patent were sent from England on May 17, 1865. The application was dated June 13, 1865.

The patent was not granted, on the ground that the use of sodium amalgam was no longer novel.

Crookes was undismayed. Having, in the course of his experiments, found that several additional operations sometimes became necessary, he embodied these in a new specification, and applied for a fresh patent, which was dated August 30, 1865, in London. The gist of the specification is contained in the following passages :

There are several causes of loss in extracting gold by amalgamation. (*a*) When certain sulphurets or compounds of arsenic, antimony, bismuth, tellurium, or other elements are present along with the gold (and especially if the latter occurs in " pyrites ") the mercury in which the mineral is triturated or ground becomes " floured " or powdered, i.e. it becomes subdivided into excessively minute globules, which, owing to the film of tarnish they have contracted refuse to unite and are consequently washed away, as it is almost impossible to effect their separation from the heavier portions of ore by any known process of washing in water. (*b*) The presence of some of the above-mentioned minerals affects the mercury in another way, viz., by "sickening" it. I believe the "sickening" and "flouring" are frequently due to the same cause, the former being the preliminary stage to the latter. "Sick" mercury has lost its fluidity, and will not flow with a bright surface, neither will it touch gold except with great difficulty. The ill effects of "sickening" are not so great as those of "flouring," for "sick" mercury can generally be restored by distillation, in which case there is not much loss, the chief objection being that it will not take up gold from the ore ; but "floured" mercury is not only entirely lost, but it carries away with it all the gold which it may already have taken up. (*c*) Another very serious cause of loss is this—even when the mercury preserves its bright metallic condition, and is in the most active state ever met with in commerce, it will seldom take up more than half or two-thirds of the gold present in the ore, owing to the precious metals being naturally tarnished on its surface, and resisting the action of the mercury, except when they are ground together for a longer time than is usually practicable.

A partial remedy for these evils has been provided by an Invention for which Letters Patent have been granted to me, bearing date the Eleventh day of February, One thousand eight hundred and sixty-five, and numbered 391, and which Invention mainly consists in the addition of sodium, either in the form of an amalgam, or in the pure metallic state, to the mercury employed for the purpose of amalgamation.

The present Invention consists in the addition of certain other metallic substances to the mercury used for amalgamation, that is to say, zinc and tin either separately or in combination, and either in the form of an amalgam with mercury, or directly in the pure metallic state ; or instead of so combining those metals or either of them with the mercury in the first instance, they can be reduced electro-chemically from compounds so as to unite with mercury in the act of separating ; or ores of or compounds containing the same may be united with the ores or substances containing the precious metal, and the whole ground together and submitted to the action of mercury, in which a metal has been dissolved or reduced, which is capable of reducing to the bright metallic state the metal which it is desired so to introduce in order to assist and facilitate the process of extracting and separating the precious metal.

In either of these modes of combining the before-mentioned metals or either of them with the mercury used for amalgamation, the result will be to more or less protect the mercury from chemical action, and also to prevent the " flouring " or " powdering " or " sickening " of the latter, and also to increase its tendency to unite together into larger globules.

In order to facilitate the separation of gold and silver from the substances containing those metals, or with which they are mixed, when the surfaces of the precious metals are in a tarnished or soiled or greasy state, as when they occur with pyrites, sulphurets, chlorides, or minerals containing arsenic, antimony, tellurium, or bismuth, it is advisable to introduce sodium into the mercury, in accordance with my said former Invention, together with one or both of the above-mentioned metals ; or either of them may with advantage be used for the purposes of amalgamation in combination with sodium, as employed in my said former Invention.

To provide for the various varieties of ore, Crookes specified three distinct amalgams, prepared as follows :

PREPARATION OF AMALGAM A.

This is a simple mixture of sodium and mercury. A convenient strength is made of three parts of sodium and ninety-seven parts of mercury. The preparation of this amalgam is mentioned in the Specification of my said former Invention above referred to, and upon which the present Invention is an improvement. I have not considered it necessary to enter into minute details respecting the precautions necessary to be taken against accident in the preparation of this amalgam, as these are well understood by manufacturing chemists, and are always adopted by them when preparing it. Moreover, sodium amalgam is an article of commerce, and its preparation is described in numerous books on chemistry which have been published long before the date of either of my Patents. In illustration of this statement, I may refer to the *Chemical News*, Number 151, October 25, 1862 (vol. vi, p. 216), where its preparation is fully described by myself ; also to Gmelin's *Chemistry*, English edition, 1851, 1852, vol. vi, p. 103.

PREPARATION OF AMALGAM B.

Weigh out seventy-seven parts of mercury, three parts of sodium, and twenty parts of zinc ; unite half of the mercury with the sodium with the customary precautions. Then melt the zinc, remove the crucible from the fire, and just before the metal is going to solidify, pour in the remainder of the mercury with constant stirring ; then add the mixture of the sodium with the first half of the mercury, and heat the whole together till the two amalgams are both melted ; stir with an iron rod, and cast the mixture in an ingot mould into any desirable form.

PREPARATION OF AMALGAM C.

Weigh out seventy-seven parts of mercury, three parts of sodium, ten parts of zinc, and ten parts of tin ; proceed exactly as in making amalgam B, only adding the tin to the zinc in the first instance.

Of these, amalgams B and C were useful in preventing the mercury from " sickening " or " flouring " (B being used in the case of silver ores), while amalgam A was powerful in increasing the mutual affinity between mercury and gold or silver, and in bringing back " sick " or " floured " mercury to the bright liquid state.

The " claim " in this specification was " the employment of zinc and tin and such other metals as hereinbefore mentioned as being applicable to the purposes of this invention, and also of such several processes as aforesaid, for the extraction and separation of gold and silver from the ores or substances containing them, and for the treatment of mercury employed for such purposes."

About the same time Crookes met Major F. T. Rickard, Inspector-General of Mines of the Argentine Republic, and talked the matter over with him. The impression made upon this gentleman by the project must have been favourable, for in November, when patent matters were progressing favourably, Crookes wrote to him giving him power of attorney for the Argentine and Chile, and asking him to secure first patents of both processes, which, Crookes believed, gave them " complete mastery " of gold and silver ores. He also told him that a large company was in process of formation in England to work both patents.

The " large company " was by this time in active process of formation. He obtained favourable opinions of his process from Professors Hofmann, Miller, Frankland, and Odling, and from Mr. Robert Hunt, F.R.S., Keeper of the Mining Records, Museum of Geology, London. He also took the advice of Professor Wheatstone, but the most active person in forming the company was Mr. F. O. Ward, of Manchester. Crookes's letters to Mr. Ward occasionally furnish the liveliest reading, the two men often coming into sharp disagreement. On all important decisions Crookes, as usual, consulted his father, and it was at Brook Green that most of the weighty business arrangements were decided upon.

On November 23, 1865, Crookes wrote to Ward to express his dissatisfaction at the outcome of their conversations : " Why waste more time," he wrote, " in the attempt to mix oil and water ? After four hours' conversational shaking we *appear* to become homogeneous, but the next morning the two layers are as distinctly separated as ever. Still, if it pleases you, I am ready for another shake."

The next " shake " resulted in a provisional agreement by which Crookes was to receive from the proposed company the sum of £11,000. But another misunderstanding arose as to whether that sum was to be paid in cash or shares. Crookes wrote somewhat sharply to Ward, saying :

If you talk about being bound only by the "strict letter of our agreement " ; when you accuse me of "playing at chess " with you in this matter ; of writing letters "pour prendre date " ; of bringing a "witness " and endeavouring to entrap you into admissions which I could afterwards use against you ; of trying to get all the work I can out of you and then turn you off with no reward whatever ; when I see that every word of yours, written or spoken, is dictated in the most narrow-minded diplomacy ; when I who have been accustomed to meet gentlemen with outspoken sincerity and perfect candour find that I am treated as if I were a Machiavelli or a Talleyrand ; is it surprising that I should at last take a leaf out of your book and refuse to be bound by your version of a desultory after-dinner conversation when it is at variance with my own clear recollection of the same, and also with my written letters on the subject.

Why will you not agree to meet as gentlemen and men of business ? We have each too great a stake to risk losing it by quarrelling.

* * * * *

But your view seems to be that I am to find the coach and horses and you are to be coachman, whilst I may stand up behind as footman and look ornamental. Now that does not suit me. I am accustomed to *work*, and if you can't find employment for me in the concern I must find it for myself. It is an absolute impossibility for me to sit dreaming away my time, idling.

*   *   *   *   *

Well, my pugilistic lessons have at all events taught me to receive " punishment with a smile, to come up to time and then to shake hands heartily and be good friends when the spar is over."

The relations between the two men were strained for a time, but an apology by Ward was cordially accepted by Crookes, and energetic steps were taken to work the patents in various countries. The manufacture of sodium amalgam was undertaken by Messrs. Johnson, Matthey & Co., of Hatton Garden, and a Captain Hitchins was deputed to negotiate for the sale and working of the patents in the United States. Crookes also entered into negotiations with Mr. W. A. Henry, Attorney-General of Nova Scotia, and proposed that the latter should see to the patenting of the amalgamation process in Nova Scotia, whereupon Crookes would assign to him half the rights thus acquired.

The Attorney-General, after some correspondence, let the matter drop, but a new opening appeared in Bolivia, where Sr. Aramayo, a correspondent of Professor Wheatstone, conducted successful experiments with silver ores containing bismuth and other valuable substances.

Wheatstone disdained all desire to take steps in these matters himself, and handed them over to Crookes, who wrote on December 31st to Sr. Aramayo, giving full advice on the sale of the metals and particulars of the sodium amalgam process. He wrote again a month later to warn him against sending crude bismuth as " bismuth " pure and simple, on account of the disastrous affect it would have on the market.

Crookes's letters to Ward towards the end of 1866

show evidence of much worry and anxiety, probably due in the main to the rivalry of Wurtz in America, but also, no doubt, affected by the trade depression of that time. The only hope in South America seemed to be based upon prospects in Brazil, where Captain Burton had taken charge of Crookes's interests. On Captain Burton's advice, Crookes took out a third patent, specifying the use of a small admixture of silver instead of zinc in the amalgam. He had little hope, however, of any success in a country " where law and justice are unknown, and intrigue and political influence carry the day."

Meanwhile, Wurtz himself had taken the first step towards a *rapprochement* by giving Crookes an account of his invention. As a consequence, Crookes acquired for Wurtz a feeling of respect which was not annulled by a certain amount of public controversy. The following letter addressed to Wurtz is extremely interesting in this respect :

LONDON,
Novr. 26th, 1866.

PROFESSOR H. WURTZ.

DEAR SIR,

I beg to acknowledge the receipt of your favour of the 6th inst. with the enclosures. I read your paper with great interest —you will see that the principal portion of it has already been printed in the *Chemical News* ; and I shall always be pleased to lay before the public in a similar way facts and observations respecting the action of Sodium in amalgamation with which you may favour me.

I must however take exception to one paragraph in your valuable paper—that is the part in which you refer to myself in the question of priority of discovery. Whilst I feel in the highest degree flattered at the manner in which you have spoken of me, I beg you will not place a wrong interpretation upon my conduct. The priority of discovery of any scientific truth is a fact which must be settled for one or for another by an appeal to facts ; and it is as much out of the power of either of us to make concession of this point to the other, as it is opposed to my wish to inaugurate or prolong an utterly fruitless argument on the subject.

The history of invention abounds with instances of simultaneous discovery ; and I am very willing to believe that the discovery of the practical value of Sodium in Gold and silver amalgamation was a *bonâ fide* independent discovery on your part, as I hope you will believe it to have been on mine. I have been silent on the question of priority of invention partly because it was a subject in which I really felt little interest—partly because I have a great objection to a *paper discussion*, and mainly because my time has been so fully occupied about other matters that I have had no time to go into the question. You will see a few remarks on the subject in No. 333 of the *Chemical News*, written by Mr. Vaughan, my patent agent. Whether the paragraph in your paper will render it necessary for me to take public notice of this question I must take time to consider.

But setting aside these questions, which are devoid of interest: Is it not possible for us to unite our forces, and acquire that strength which the union of two such original and experienced investigators in this subject cannot fail to give ? We each possess information which would be useful to the other,—we each possess patents which the other would like to have—and we each possess considerable power of annoying or assisting the other.

I think I may assume that your object in taking a patent out was to gain money by the invention, and I confess I was mainly actuated by the same motives. Now, putting on one side barren discussion as to the exact date when the idea first entered our brains, can we not unite our forces, and as men of business see that it is in friendly collaboration, more than in passive antagonism, that the road to wealth lies ?

Captain Hitchin was authorised by me, and if he calls again you can treat with him, but I do not think he will follow the matter up.

I will send you the file of *Chemical News* you require. The price of advertisements varies as to size. If you will send me the space you wish to occupy I will get the publisher to give you an estimate.

GOLD AMALGAMATION

I write this letter in a friendly spirit, but I wish you to be good enough to consider it *private*, and without prejudice to any plan I may subsequently adopt.

Believe me to remain,
Very truly yours,
WILLIAM CROOKES.

The " paragraph " referred to was a passage in Professor Wurtz's paper on " The Utilisation of Sodium in Gold and Silver Amalgamation " read before the American Association for the Advancement of Science, 1866.

It ran as follows :

I shall but refer briefly to the conflicting claim to priority of discovery which was entered by Mr. Wm. Crookes, one of the most learned, industrious, and successful of the English scientists. The graceful concession of this point, which Mr. Crookes is considered of late, both by scientists and jurists, to have made, by his own silence, and by publications in the journals ostensibly authorised by him, it is thought will not prove any appreciable detraction from the laurels so well and worthily won by him in the field of chemical discovery.

Before making a public statement concerning priority, Crookes wrote to Ward to enlist his " fluent pen and clear habits of expression." The statement eventually appeared in the *Mining Journal* of December 15, 1866. It acknowledged the probability of simultaneous discovery, but made it clear that Crookes's solution of the problem was entirely original.

The exchange of amicable messages between Wurtz and Crookes took the sting out of the controversy and averted a legal conflict which had just been started by an injunction served on Messrs. Taylor and Sons, of San Francisco (Crookes's agents) by the Wurtz Amalgamation Co. against the further sale of sodium amalgam, 80 lbs. of which had been sent to California by Messrs. Johnson,

Matthey. This injunction was withdrawn later, so that in June 1867 we find Taylor and Sons selling Matthey amalgam containing 25 per cent. sodium freely in San Francisco in packages of from one to twenty pounds.

The close of 1867 was marred by a family bereavement which affected Crookes deeply. The following year saw a gradual decline in Crookes's interest in gold. Potassium cyanide, a true solvent of gold, was coming into use, and although the sale of sodium amalgam continued into the seventies, the new agent was destined to supersede both Crookes's and Wurtz's discoveries, and automatically relieve them from any further efforts to capture the gold-winning industry. The *auri sacra fames* came to an end, and Crookes, after his dip into the turbulent life of the metal speculator and company promoter, became once more a scientific chemist, ready to turn his talents on to any new chemical problem which might arise.

CHAPTER XI

THE ECLIPSE EXPEDITION

(1870)

O N SEPTEMBER 22, 1867, an event happened which profoundly affected Crookes's outlook on life. It was the death of his youngest brother, Philip, aged twenty-one, while engaged on a cable-laying expedition to Havana, from yellow fever.

Crookes, who was very much attached to his brother, felt the blow keenly. His sorrow was mixed with anger that such a thing should have been allowed to happen, and when the commander of the expedition, Mr. F. C. Webb, gave out that Philip had contracted yellow fever when on a private expedition into the town of Havana, his indignation knew no bounds. He immediately set up in type and circulated a letter to the cable-laying company. This letter was as follows :

[*Confidential.*]

To the Directors of the India Rubber, Gutta Percha, and Telegraph Works Company.

GENTLEMEN,
      In justice to the memory of my brother, Philip Crookes, whose life, with that of thirteen others, has been lost in the disastrous Cuba and Florida Cable Expedition, and on behalf of those who have to mourn the loss, I beg to call upon you to institute a full inquiry into the causes which have led to such lamentable results. All the evidence which has yet come before me points steadily to

the fact that a grave responsibility for this awful loss of human life rests on Mr. F. C. Webb, and justice and humanity both demand that the conduct of this gentleman, who commanded the expedition, should be fully investigated.

I would urge that Mr. Webb's conduct ought to be rigorously challenged on the following points :

I. Assuming that the exigencies of the contract made it necessary for the cable expedition to carry on operations at so deadly a season, why was so much valuable time wasted on arriving at Florida ? My brother writes, under date September 8th :

" We first reached Havana on the 26th July ; we got to Key West on the 27th ; *stopped there a week fitting up* (which we might just as well have done in London before starting), and made a beginning on Saturday, August 3rd."

Mr. Brand, your excellent Secretary, informs me that Mr. Webb's instructions were to do this fitting up on the passage. Had the Company's orders been obeyed, the week thus gained might have saved the lives of most of those who fell victims, and certainly that of my brother, who was the last to die.

II. What object had Mr. Webb, when laying the cable from Havana to the shore end at Key West, in insisting upon going out of his course ? My brother gives me a drawing of the ship's path on September 6th–7th, which agrees with the one in your possession ; he writes that Mr. Webb—

" Contrary to the advice of the pilot and captain, had the ship steered to the eastward ; the current we afterwards found was also taking us east, and so at 8 a.m. [September 7th] when we stopped, we were in the position you see, having contrived to get 20 miles out of our course during a run of 70 miles."

Mr. Brand also tells me that the American frigate, which led the way straight to the buoy, repeatedly signalled that the *Narva* was going wrong ; and it was well known that the Gulf stream was running rapidly to the east, and required a westward course to counteract its influence. It is impossible to conceive for what reason this advice and these signals were disregarded ; and from an extract from the *New York Herald* for September 6th, which Mr. Brand has shown me, it would appear that a repetition of such apparent perversity was only prevented by the captain refusing to steer a second time out of his course.

The extract in question states that—

" Captain Dowell took the course south, half-east, and *despite all remonstrances*, kept on clear through. It was the same course taken by the *Lenapee*, which acted as our pilot ship."

Such extraordinary proceedings demand the closest inquiry.

III. Such an important expedition should have been provided with an ample staff of practical electricians. We were led to believe that such was the case. It appears, however, from my brother's letters, that the expedition was *starved* in this respect ; absolutely the whole of the electrical work on board the *Narva* devolved upon him, a mere youth of twenty-one ; and had he broken down under this severe strain sooner than he did, there is little doubt but that your cable would have been entirely lost. My brother's letters, written when he was in full health and strength, give a very painful idea of the mental and bodily strain in which he was kept the whole time, and leave no doubt that it was the exhaustion consequent on sheer overwork which caused him at last to succumb. He speaks of having to keep up the tests every minute, for hour after hour, with no one to relieve him, and being obliged to ask one of the cable hands, Roe, who happened to know something of telegraphy, to assist him in sending messages for five or ten minutes. On another occasion he had to take the same tests for *sixteen hours uninterruptedly*, at a crisis when the least inattention would have risked the safety of the cable. Is it surprising that he speaks of dropping off to sleep in the act of sending a message ? At another time he was kept at this work for more than *twenty-four hours* at a stretch : he was, however, unable to keep up this continuous strain, and for a few hours during the night he was forced to obtain the services of a man to watch the indications of the instrument, and wake him at every hour, when he got up and did ten minutes' work.

Then follows a description of days of weary dredging in the rain, and of nights spent out on an open barge, till he was thoroughly worn out, and he relates how one day he sat down in his cabin immediately after dinner, fell asleep, and did not wake till four in the morning.

Not one word in all his minute account of the proceedings does he say of any other electrician but Mr. Donovan, himself, and the cable hand, Roe, who once assisted him in sending messages. I ask what were the other electricians doing on board, that the whole of this responsible work devolved on my brother?

IV. When at last the cable was successfully laid, and yellow fever was attacking one man after the other, why was the *Narva* kept waiting in that deadly climate for ten days, and, I am informed, in defiance of the orders from home to leave instantly? On September 12th my brother wrote announcing that Mr. Webb and the captain had at last decided to leave. He said:

" There can be no question about the wisdom of their decision ; we have suffered terribly, and now that our work is done, the fact of our having guaranteed the cable ought not to be permitted to outweigh considerations for men's lives ; no one on board, I may safely say, would have counselled running away before both cables were completed ; but they were finished last Saturday [September 7th] ; and if I had had charge of the expedition, by Sunday morning we should have been on our way to New York."

Instead of that, the coal which might have carried them there, was burnt up whilst they were waiting for the next ten days ; hence the necessity of going to Havana to coal.

V. After the lamentable death of my brother, Mr. Webb ought, in common humanity, to have communicated the sad fact to his relatives with as little delay as possible. Not until three days after he had arrived at New York—four days after my brother's death —did he send a telegram to London announcing " *all well* " on board ; and it was not till the day following that he telegraphed the mournful news to your Secretary.

My brother's account of his attack is the following :

" To-day [September 13th] I have been ashore, running about after some telegraph instruments we left here, and *am rather tired.*"

Mr. Webb's version is :

" He had been ashore at Havana, and got wet in a shower of rain when coming off, and felt unwell, but went ashore again the next day, and then got worse. *He was not ashore on business.*"

Gentlemen, I appeal to you in the name of my lamented brother, and of those who have died with him, not to allow Mr. Webb's conduct to pass unchallenged. I pray you to cause a full and searching inquiry to be made into all the questions I have here raised. Mr. Webb may possibly be prepared with satisfactory explanations. In that case, no one will be more willing than myself to withdraw all that I have urged against him ; but until then, I, and all the other members of our family, have no alternative but to look upon Mr. Webb as in some degree answerable for my brother's life.

You will be kind enough to receive this letter as confidential.

I have the honour to be, Gentlemen,

Your obedient Servant,

WILLIAM CROOKES, F.R.S.

20, MORNINGTON ROAD, LONDON,
October 15, 1867.

Thirty copies of this letter were printed. One was sent to each of the nine directors. Messrs. Heugh, Henderson, Wardrop, Beattie, Hancock, Silver, Child, Jones, and Bannatyne. The ship's officers received six (including Webb), Crookes's father three copies, and copies were given to " Mother, Alfred, Jane, Walter, Frank, and Mr. Geeves."

The *Narva* left England on June 27, 1867. Philip came of age on August 29th. On September 10th he sent his last letter to Crookes. On September 13th he wrote his last letter to his father. On September 17th Mr. Medley and six others were dead of yellow fever, and a telegram to that effect was sent to London, which naturally created alarm in the Crookes family. The ship sailed that day for New York. Philip was then already ill. A report of his case was given by Dr. Andrew Dunlop, surgeon to the expedition, in the *Lancet* of February 15, 1868. It is as follows :

The following five cases of yellow fever have been selected from amongst those which occurred on board the *Narva*, as they exemplify the different forms which the disease assumed.

P. C——, aged twenty-one, electrical engineer. Attacked on the 16th of September. Suffered from dyspepsia for about ten days before the attack. On September 13th he went ashore at Havana for about an hour, and returned to the ship at about 5 p.m., getting thoroughly wet on the way off. On coming aboard he complained of feeling sick, but took a good dinner afterwards. The following day he was ashore from 9 a.m. till 4 p.m, walking about nearly the whole of the time. In the evening he complained of severe headache, which passed off after a good night's rest. On the 15th he made no complaint. But on the morning of the 16th he had a rigor ; and when I saw him at 6 a.m. his pulse was 100, his skin was hot and dry, and his tongue was thickly covered with a yellowish-white fur. There was no headache. He had nausea, but no vomiting, nor was there any epigastric pain or tenderness. Bowels constipated, pain in the loins, and aching pains in the limbs. Ordered two compound rhubarb pills immediately, an effervescing saline mixture, and tepid sponging.

September 17th.—The pills acted twice. Perspired a good deal during the latter part of yesterday. Great disgust at all sorts of food, the smallest quantity causing nausea. Complained of pains in the loins this morning. A sinapism was applied, which was followed by relief. Passes his urine in large amount. It is of a smoky colour, and has a strong, peculiar odour. Other symptoms unaltered. To have three ounces of brandy.

18th.—Complete defervescence. Skin cool, but dry ; pulse 60, irregular and intermittent ; tongue and lips covered with aphthæ. He vomited once this morning after taking some tea. Frequent scanty mucous stools, accompanied with tenesmus. Occasional sighing respiration. Champagne was ordered, but he could not take it ; it was "too sweet." To have four grains of sulphate of quinine every four hours.

19th.—Pulse 76, intermittent. Very restless. Other symptoms unaltered.

20th.—Slight delirium during last night. Great restlessness, constantly getting out of bed if not watched. Intellect obscured. Pulse 76, regular but weak.—At 9 p.m. he complained of severe pain in the epigastrium, and an hour afterwards vomited about two pints of dark-brown liquid. Slight epistaxis during the day. Bowels confined ; urine scanty.

Delirious during last night ; urine suppressed ; œdema of lower eyelids ; pulse 100, thready ; skin hot and dry ; conjunctivæ and whole surface of a yellow tinge ; no vomiting to-day ; became delirious towards evening.

22nd.—Black vomit at 2.20 a.m., again at 3 a.m. Died at 3.10. The *Narva* left Greenhithe on the 27th of June, having on board five engineers connected with the cable, five officers, four engineers, nine seamen, nine firemen, and nineteen cable hands, besides the carpenter, cook, steward, etc., making a total of sixty souls. She arrived at Havana on the 26th of July, with all on board well ; and as the yellow fever was unusually prevalent, she was not taken into harbour, nor was anyone allowed to land ; but after taking on board some gentlemen connected with the enterprise, we sailed for Key West, where we arrived early the following morning. At the time of our arrival in the latter place there was, I believe, little fever there ; but the cases kept increasing in number, until at last the inhabitants admitted that the disease was more severe this summer than it had been for several years past. We left Key West, and commenced to lay the Cuba cable on the 3rd of August, and it was about this time that sickness first began to appear amongst us, principally in the form of short, but sometimes severe, attacks of febricula. About the middle of the month nearly every-one of the cable hands and one or two of the crew suffered from diarrhœa. On investigation, the cause of the outbreak was found to be twofold—first, the water in the cable tanks, which were immediately beneath and opening into the sleeping place of the cable hands, was found to be foul, and giving off sulphuretted hydrogen; and secondly, the beer served out to these men was becoming sour. Fresh water was pumped into the tanks, grog was served out instead of beer, and the epidemic disappeared. On the 11th of August we returned to Key West, and lay alongside the wharf until the 14th, when we went to sea again. On the 19th we returned once more, and remained until the 21st, when we left to lay the Florida cable. We arrived at Punta Rossa, where our first death occurred on the 24th, and returned to Key West on September 1st. We stayed there until the 12th, when we ran over to Havana, and took in a hundred tons of coal outside the harbour. We came back on the 15th, and finally sailed for New York on the 17th.

From the 23rd of August until the 23rd of September we had twenty-three cases of yellow fever ; of these, fourteen died. Eleven cases were sent ashore at Key West to be placed under treatment at the hospital or elsewhere, or were taken ill while ashore ; of these eight died. Eleven cases were treated entirely on board ship ; of these, five died. Besides these cases of yellow fever, nearly everyone on board suffered from febricula, or some disorder of the digestive system, due to the influence of the climate.

The introduction of the fever amongst us could not be traced to any infectious origin ; it evidently arose from our coming under the local or endemic influence of the disease. It was chiefly those who were most called upon to be ashore in the island who suffered, and many of the cases were distinctly traceable to evening or night-work ashore, or day-work on the beach, or in shallow water while laying the shore ends. The greatest mortality was consequently amongst the cable hands and cable engineers. Of the nineteen cable hands, fourteen had yellow fever, and ten died. Of the five electrical engineers, three were ill (two with yellow fever, and one whose case was doubtful), and two died. Only two of the crew had yellow fever, and one of them died. It is curious that with the exception of the second (ship's) engineer, the disease attacked every one in the steerage—viz , the chief officer, the chief engineer (of the ship), the second mate, and the third and fourth engineers.

The chief engineer was the only one of them who died   It was several times suggested that the cable hands were more susceptible to the disease on account of their being less accustomed to work in tropical climates than the crew.   But this could not have been the case, as they had, almost without exception, either been sailors or previously engaged in similar work.

Whether yellow fever is contagious or non-contagious has been much disputed ; and although a belief in its *personally* non-contagious nature is gaining ground, the contrary opinion is still widespread. Certainly, as it appeared aboard the *Narva*, the disease was not contagious.

Crookes paid a touching tribute to his brother's memory by setting up in type, with his own hand, Philip's letters written on board the *Narva*.

The Commander of the expedition took proceedings for libel against Crookes. These began by the following letter :

159, Fenchurch Street,
London, E.C.,
6 *Novr.*, 1867.

Sir,

Mr. F. C. Webb has consulted us with reference to the printed statements issued and circulated by yourself containing a grave charge against him, besides many cruel insinuations with imputations as to his professional capacity and conduct in connexion with the laying of the Cuba and Florida Cables and he has instructed us to obtain for him redress at your hands. We have consequently to require from yourself a retractation of the charge and a withdrawal of and apology for the various statements and insinuations made affecting him to be printed and circulated in like manner with the charge and statements made by yourself.

Mr. Webb whilst willing to make every allowance for excited feelings occasioned by the loss of a Brother yet he cannot permit your charge to remain unchallenged or your statements in other respects uncontradicted and which as they appear to us are wholly without justification. To your charge against him Mr. Webb gives (through us) his most distinct and emphatic denial and to your other statements and insinuations he instructs us to say that whilst they are all exaggerated and inaccurate they are for the most part simply untrue.

Mr. Webb has no desire to occasion yourself unnecessary pain under your present loss and affliction but in justification to himself he has no other alternative than to require retractation and apology from you and this he is willing to afford you an opportunity of giving before any further steps be taken by ourselves in the matter.

We are, Sir,

Yours most obediently,

Tamplin & Taylor

William Crookes, Esq ,
   20, Mornington Road,
     Regent's Park, N.W.

Crookes refused to apologise or retract, and the case was tried at the Court of Queen's Bench on December 12, 1868. It was reported in the *Observer* of December 13th as follows :

COURT OF QUEEN'S BENCH.—Sat., Dec. 12th.

[Sittings at Nisi Prius, at Guildhall, before Lord Chief Justice Cockburn and Special Juries.]

Webb *v.* Crookes.—This was an action for libel. The defendant pleaded not guilty.

Mr. Giffard and Mr. M. Griffiths appeared for the plaintiff, and the Solicitor-General and Mr W. H. Holt for the defendant.

It appeared from the plaintiff's case that he was an engineer, and had been engaged in superintending on behalf of the Gutta Percha Company the laying of two submarine cables, one from Havana to Key West, on the coast of Florida, and the other from Key West to Punta Rossa. After the cables had been successfully laid, but, unfortunately, while the vessel conveying the cables was lying at Havana, many of the officers and crew took the yellow fever, which was then raging at the port, and, among others, the defendant's brother died of the disease. On returning to England the plaintiff was informed that the defendant had published a pamphlet in the shape of a letter, addressed to the directors of the Gutta Percha Company, in which he charged the plaintiff with having occasioned the delay of the vessel at Havana by his neglect, and called for a thorough investigation of all the circumstances. The letter concluded with the words, " Be kind enough to receive this letter as confidential." The defendant had given the letter complained of to a tailor in Bond Street. The plaintiff in bringing the action only sought to obtain a retractation of the charges contained in the alleged libel.

The Solicitor-General said that an offer had been made on behalf of the defendant to retract the charges by letter, but that the plaintiff had demanded that the retractation should be published in all the morning papers. The publication alleged was denied to have been by the defendant.

At the conclusion of the plaintiff's case an apology was made on the part of the defendant, and a verdict was entered by consent for the plaintiff for 40s.

The death of Philip Crookes was a turning-point in Crookes's life. It brought him into close touch with Mr. Cromwell F Varley, F.R.S., the telegraph electrician, who was interested in spiritualism, and who persuaded Crookes to try and get into communication with his dead brother by spiritualist methods. In this, as we shall see, he believed himself to be successful, and he eventually embarked upon his famous attempt to bring the scientific world to a formal approval of the "occult" phenomena he had been studying. A fuller account of this attempt must be reserved for a special chapter.

The years 1868 and 1869 marked a lull in Crookes's scientific career. For part of 1868 he suffered from ill health. He gave up every attempt to develop the sodium amalgam process. The long struggle and the final disappointment discouraged all financial and large commercial operations. He stuck grimly to the *Chemical News*, defending his "child" from plagiarism in America and from the threatening rivalry of *The Laboratory* at home. He even set up a printing office of his own, and practised the art of type setting. He contributed articles to the *Quarterly Journal of Science* on photometry, on the Melbourne Observatory, and on the great solar eclipse of 1868. With the help of Dr. Ernst Röhrig, he brought out an adaptation of Kerl's *Metallurgy* (Longmans, 1869). He began the writing of books on aniline colours and on beetroot sugar. He also designed a binocular spectrum microscope, and showed it at the Royal Society. He gave a lengthy opinion on the professional education of medical students to the General Medical Council, in the course of which he proposed the total abolition of lectures on chemistry, since "the only value of a lecture consists in the demonstration of experiments, and in affording a supplement to the textbook of the class," while no lectures can impart manipulative dexterity essential to the successful performance of the simplest analytical operation. He also advocated the inclusion of physics as essential to surgery and ophthalmology.

As long ago as 1860, Crookes had taken photographs of a partial solar eclipse seen at Woolwich, in co-operation with his friend John Spiller. The important spectroscopic

Proceedings, No. 112, 1869.

observations made in the United States during the total solar eclipse of August 7, 1869, were fully described by Crookes in the *Quarterly Journal of Science* for January 1870. The Royal Astronomical Society urged the British Government to arrange for several eclipse expeditions for December 1870, when the track of totality would cross Spain and North Africa. William Huggins, V.P.R.S., the wealthy brewer who became a highly distinguished amateur astronomer, with a private observatory at Tulse Hill, where he specialised in stellar spectroscopy, invited Crookes to join one of these expeditions. Crookes had become acquainted with him through spiritualism, in which Huggins was keenly interested.

The Government Eclipse Expedition was divided into two parties, one for Spain and Algiers, the other for Sicily. The former was divided into three detachments, in charge respectively of the Rev. S. J. Perry (Cadiz), Captain Parsons (Gibraltar), and Dr. Huggins (Oran). The Sicilian party was in charge of Mr. J. Norman Lockyer, and included Mrs. Lockyer, Professor Roscoe, Herr Vogel, Mr. Vignolles, and Professor Thorpe. Dr. Huggins's party consisted of Admiral Ommaney, Lieutenant Ommaney, Professor Tyndall, Rev. F. Howlett, Mr. Carpenter, Captain Noble, Dr. Gladstone, and Crookes, and therefore comprised the greatest proportion of scientific luminaries. All observers were to sail on a transport, H.M.S. *Urgent*, on December 6, 1870.

The chief object of the expedition was the study of the corona, or white halo surrounding the sun and extending in wisps and streamers to sometimes more than the length of the sun's diameter. Its relation to the " chromosphere " or region of coloured prominences was also to be investigated, especially as the method devised by Janssen and Lockyer of observing the chromosphere without an eclipse was not applicable to the corona. In the official instructions to observers, the objects were stated as follows :

NOTE.—The objects to be obtained are :—

1. To determine the actual height of the chromosphere as seen with an eclipsed sun ; that is, when the atmospheric illumination, the effect of which is doubtless only partially got rid of by the Janssen-Lockyer method, is removed. If the method were totally effective, the C line, the line of high temperature, should hardly increase in height ; but there can be little doubt that the method is not totally effective, so the increase in height should be carefully noted.

2. To determine if there exists cooler hydrogen above and around the vividly incandescent layers and prominences. To do this the band of the spectrum just above the stratum which gives the hydrogen lines before totality and during totality, should be carefully examined, to notice (*a*) if any traces of the hydrogen spectrum exist above the region which before totality gave the hydrogen lines, and (*b*) what lines extend outside the hydrogen spectrum, and whether they also exist with it in the lower strata.

3. To test the American observations of last year as to the existence of a line at 1474 in the corona spectrum, by seeing if it be visible above the region which gives the hydrogen spectrum.

4. To determine whether any other gases or vapours are ordinarily mixed up with hydrogen, but remain invisible with the uneclipsed sun in consequence of the absence of saliently brilliant lines in their spectra.

The *Daily Telegraph* of December 12th contained a well-written leading article on the Eclipse Expedition, of which we may quote the following :

Favouring winds and bright weather may well be invoked for the good ship *Urgent*, now bound from England to the Mediterranean with a cargo of philosophers. Horace would have written another ode, like that beginning *sic te Diva potens*, to propitiate the Deities

of air and water in behalf of such a precious craft. The *fratres
Helenæ lucida sidera* might indeed be very naturally addressed ;
for these philosophers go a sky-gazing ; they are merchants in a
grand commodity of aerial fact, chapmen of celestial science, sun-
worshippers, whose only gain will be made from the gold of the
solar fires and the silver of the empyreal light. One of the earliest
works of fiction which marked the rise of German literature was
the celebrated *Stultifera Navis*, or "Shippe of Fooles." Here is
a "shippe of wise men"—and we may, indeed, add of women—
who are bound through the wintry seas, giving up Christmas, and
daring the perils of the grim Bay of Biscay, to add, if possible, to
our stock of knowledge about that great orb which governs the
system and the nature of his radiance. The Bay of Biscay is seldom
a very quiet place ; but in winter it is generally as wild and turbulent
as winds and waves can make it. If, therefore, Neptune be there
or thereabouts just now—unless in classic phrase he "hears more
favourably," as Davy Jones—we would gladly vow to him, say, a
Naval Constructor or a Tory member or two, as an offering to
induce him to pass the word for fair winds and fine weather while
the *Urgent* speeds along. His quarrel with Phœbus-Apollo we know
was settled at the end of the *Iliad*, so there is no reason why he should
not bid "sleek Panope upon the waters play," and "rule the waves"
straight for Britannia's solar-eclipse expedition. If we have heretofore offended the stern God of the Sea, by stretching telegraphic
cables along his sands and oozes—if he is angry with Dr. Carpenter
and the crews of the *Lightning* and *Porcupine*, for dredging his very
deepest secrets up—let him not remember all that now. It takes
so much to make a wise man that Nature is always parsimonious
of them ; fools are plentiful, and good honest "lords and squires,
and men of each degree" ; but the *Urgent* carries a cargo of brain
which would be all but irreplaceable for civilisation. If we ran
through the entire list of the scientific company on board that
vessel, each name would be a good plea for favour from Neptune.
Only to select specimens : there are Professor Tyndall and Dr.
Huggins ; there are Captain Noble, Mr. Crookes, and Mr. Carpenter ;
there are the Rev. F. Howlett, Mr. Ladd, and Admiral Ommaney,
with others, whose loss altogether would be far worse than the
sinking of a whole armada of Spanish galleons, loaded to the gun-
wales with silver. We cannot underwrite such a ship and cargo ;

for who is going to put a price upon Professor Tyndall, who to estimate the discoveries of Dr. Huggins during the rest of his natural life, or to put a market value upon the next metal identified by the discoverer of thallium ? We must trust them to winds and water " as they sail with the gale through the Bay of Biscay, oh ! " But if the " cherub that sits up aloft," and other nautical divinities, do not keep a bright look-out for such a company as this, they ought to be discharged from the service outright. Perhaps it is too much to ask that eclipses should in future take place in more convenient spots than beyond the sea. For a corona, as for a crown, much can be endured ; and all we ask of the Marine Deities is to take our *savants* safely to Gibraltar, Cadiz, and Oran, and return them to us again well and sound.

Alas ! if spectroscope and polariscope could reveal to the *Urgent's* company those scenes which blot that fair France down whose coast they are sailing, what a contrast to their own mission it would furnish ! How stupid and wicked appear those battlefields before Paris and along the Loire, compared with the nobler battles of Man against Ignorance, to fight one of which the *Urgent* takes out her scientific troops ! Or if by chance the Angel of the Sun could at one glance witness the poor dead French and Germans festering in the potato fields at Champigny, and this Ship of Wisdom steaming down to study the phases and phenomena of his eclipse, what an odd breed we should seem ! How could the angelic meditator, if he were not very far-sighted indeed, comprehend a race which at the same time commits wholesale mutual murders, and yet knows enough to establish a chemistry and geography of the central orb ? Such folly side by side with such insight and skill would surely appear as incongruous as though the mastodon and dinotherium were grazing upon Norfolk pastures and cannibals of the Flint period were mixed with modern society.

The importance of the event was enhanced by the fact that the next occasion on which an eclipse would take place within easy reach of England would not occur till seventeen years afterwards.

From the diary kept by Crookes during the expedition we may take the following glimpses into his luggage :

Contents of Wicker Case.

Battery	Files and Screwdriver
Vacuum tube	Battery wires
Eclipse glasses	Ink
Flame spectrum apparatus	Ammonia
Tubes containing Tl, Li, Na	Salt
Mercury for zinc plates	Tacks
	Cartridges

Contents of Dispatch Box, exclusive of Stationery.

Kreosote	Telescope	MS. Sptl. books
Glycerine	Spectroscope	String
Nightlights	Sticking-plaster	Paper piercer
Eau de Cologne	Needle and thread	Scissors
Seidlitz powders	Magnesium wire	Matches
Sulphate of zinc	Water brush	Screwdriver
Carbolic acid	Travelling lamp	Stop-watch
Alum	Lieberg	Lead pencils
Chlorodyne	Aneroid	Pocket mirror
Chloral	Rules and scale	Plain cards
Gum	Tape measure	Soap
£5	Chloroform	Letter scale
Copying ink	Cigars	

Contents of Hat Case.

Best neckties, gloves, handkerchiefs
Hat
Fez
Mittens,
Linen cuffs,
£5

Contents of Leather Bag.

Boots and shoes
Books :
   Tyndall, *Imagination*
   Roscoe, *Spectrum*
   Footfalls
   De Quincey, Vols. I, II, III
Thick coat

THE ECLIPSE EXPEDITION

Dressing Case.

Brandy, O.D.V., and orange
Slippers
Eclipse pictures
Ebonite cup
Fur cap

Opera glass
£15
Dusters
Thin coat
Chocolate

Contents of Large Trunk.

Frock coat
Best dress suit
Best morning trousers
Pistol
Brush case
6 White shirts
4 Night shirts
6 Vests
6 Towels
Common dress suit
Black waistcoat
2 Flannel shirts

Books :
 American Eclipse
 Canadian Eclipse
Copying book
Quarterly Journal of Science, No.28
Journal of the Franklin Institute (4 numbers)
Miscellaneous papers
Collars
Socks
Shirt fronts
Cigars

After these important preliminaries, the diary itself proceeds :

ECLIPSE EXPEDITION.

December 5, 1870, Left Waterloo 11.30 a.m.

Arrived at ship *Urgent.*

Mr. Huggins. Mounting equatorial on tree-stump cut off at an angle. Remarkable result of grafting an equatorial on a living tree in a tropical climate : budding spectroscopes and baby prisms.

Remedies for sea-sickness. Carpenter and berth. Changes with Oxford men.

Captain Noble, the jocular correspondent of the *D.N.*, etc. Carpenter and *D.T.* A council as to sending notices to papers.

Anecdotes. The effect on bucolic mind of photographer changing a plate in middle of a field, having enveloped his head and waist in a large yellow bag. Appearance that of a gigantic yellow puff ball on a black stalk. The curate and the horse which had to take sea baths.

Tyndall and coronal beams.　Cold wind at moment of totality.
Theories.　Air diathermanous.　Query, aqueous vapour.
Bed.　Difficulties of undressing.　Worse in morning.

December 6th.—After breakfast walked on deck.　A host of
young Oxford men, some nearly boys, pressed into service as observers with polariscope.　" What is a polariscope ? "　Favouritism
and jobbery.

Urgent a mere shell a week ago.　Sudden orders to fit her out.
Everything done since Tuesday last.　Interior fittings, curtains,
beds, sheets, china, glass, stores, etc., all got in red-hot haste.　All
spoons and forks newly plated.　Not cost Government less than
£10,000, half of which might have been saved if more time had
been given.　Most of men and officers come from *Duke of Wellington.*
Everyone strange to ship except Captain Hamilton.　This expense,
however, will not appear in Estimates.　Red tapeism—parsimony
in treatment of observers and lavish expenditure of money in other
ways.

Professor Newcomb and Mrs. Newcomb (the only lady on
board).　Conversation as to American eclipse.　Two parties in
U.S., the Naval Board and the Observatory.　The Naval Board
report not printed yet.　The Observatory report out first.　Dr.
Coffin the head of the observers.　His name not mentioned in my
article in *Quarterly Journal of Science* by accidental omission.

Swinging ship for compass deviation.　A long job, some hours.
Queen's ships do not adjust by permanent magnets as in the Mercantile Marine, but by table of errors and deviations.

Telegrams sent to *Echo* and Nelly.　Also long letter to Nelly,
posted and sent by officer who superintended swinging of ship.

Carpenter and Noble sent to other papers.

At last moment last night a telegram arrived from Admiralty
(in answer to urgent appeal from Huggins) ordering the ship to
take the Oran party on to Oran from Gibraltar.　This is good
news, for I shall be with the *Urgent* all the time, and shall not have
to leave the ship for a smaller one.

Bread and cheese lunch at 12.30.　Too early for appetite.

Tyndall came up and told me he had seen Coleman [1] a few days
ago in a Turkish bath, both naked.　Coleman commenced spiritualism at once.　Tyndall then asked me whether I was going on.

---

[1] Mr. Benjamin Coleman—E. E. F.

Yes, and as he seemed evidently wanting to talk on the subject, I spoke as strongly as I could, telling him of many of the phenomena I had seen, but confining myself to the physical movements only. He was at first inclined to ridicule and explain away, but Huggins then came up, and the combination was too strong for him. He said he was very anxious to see these things for himself, and reminded me that I had promised to invite him. I said I would, provided he would undertake to come at least six times, even supposing the first five times were failures. Also he must impose no conditions at all, but just witness what happened. He was very willing to do so, and would like to meet Home. He said that at first he felt he ought to have accepted any invitation to meet Home on the same conditions as Faraday imposed, but now he would ask for nothing at all except to witness the phenomena. He would come in an earnest truth-seeking spirit, and really try to investigate. I feel that to have got Tyndall into such a mode of thinking is a great triumph. Whilst we were discussing the subject on the poop others came up, and the conversation became general. At dinner the conversation soon turned on the subject, and Tyndall happened to sit next to a gentleman (Captain Collins) who has seen Home many times, and thoroughly believes in his genuineness. He told him many things which corroborated what I had said. Captain Noble attacked Home as an impostor and charlatan, but Huggins and I then joined in, and the combination was too strong. Tyndall and several others present spoke very fairly as to the necessity of investigating the phenomena, and I see no difficulty as soon as I return to London (with Mrs. St. Claire's help) to satisfy these sceptics. I sat next to Revd. Mr. Howlett at dinner, and we had some conversation on the higher phenomena. He will soon believe it all if he has any opportunities.

After dinner I was introduced to Home's friend, Captain Collins. He knows Lords Adare and Lindsay, and was present at many of the *séances* described in Adare's book. The conversation then turned on ghosts and spirit seeing. One of the Lieutenants (Mr. Smith) told us several things which had happened to him. On one occasion he was abroad and saw his grandmother's face suddenly appear before him. By the next mail he heard that his grandmother had died that very time. He told us several things of a similar sort, and then others began to draw upon their experiences. Nearly every one present had told us something supernatural which had come under their experience. The whole thing reminded me strongly of a Xmas number of *All the Year Round*, and if a good editor had been present to weave the whole together a capital number could have been made of it.

The sea was beautifully calm all day, and we rapidly passed the Isle of Wight (along the Undercliff), and then lost sight of land. The ship was going very steadily between nine and ten miles an hour, and the officers say we can reckon upon that rate of speed being kept up night and day till we reach our destination. One or two passengers were ill. I felt nothing whatever of sea-sickness. Went to bed at about eleven. The motion of the ship was increasing. In the night it got worse, and at 3 a.m. I was rolling about from side to side, and could get very little sleep after. The noise of the screw was just like a gigantic shuttle going thump, thump at the rate of 45 times a minute. I heard crockery smashing, and things falling down frequently.

Wednesday, December 7th.—The ship decidedly livelier (she is said to be a very lively ship). Mr. Carpenter looked very ill, but stuck to it that there was nothing the matter with him ; as, however, he did not get up to breakfast I suspect it was something more than laziness which kept him in bed. I had difficulty in dressing, as the sides of the cabin kept running up against me, and the floor kept bobbing up. After a time, however, I managed to get my things on and went on deck. A beautiful morning, the wind freshening, and the sun peeping out occasionally behind clouds. The ship rolling about in a very decided manner, and the walking up and down being very difficult.

The rest of the voyage as far as Cadiz is best quoted from an account written by Crookes for *The Times* :

On the 6th of December last Her Majesty's ship *Urgent*, a screw transport of about 2,000 tons, commanded by Captain Henderson, left Portsmouth with the Cadiz, Gibraltar, and Oran detachments of those who had been selected to make observations on the total phase of the solar eclipse on the 22nd of December. The next day the peculiar propensity of the *Urgent* began to make itself manifest in a degree which, although thought trifling by sailors, proved highly unpleasant to some of the passengers on board.

THE ECLIPSE EXPEDITION

Whatever good qualities the ship may possess—and that she has several has been amply proved on this voyage—she enjoys the peculiarity of rolling violently. The day after we started the wind freshened, the sea rose, and about noon, when the Bay of Biscay was entered, the motion of the ship was such as to send below all who were not good sailors, while the clinometer showed an angular movement of 12° each way. The fillets of wood, technically known as fiddles, which divide the dining-table in the saloon into longitudinal compartments, were now fitted on before lunch, but even with this protection it was difficult to keep the dishes and their contents from jumping over ; as it was, one joint took a flying-leap over all obstacles and deposited itself on the floor. When we got fairly into the Bay, the movement increased considerably, and our dinner was eaten under somewhat amusing circumstances. The glasses and bottles had to be supported on the hanging shelf above the table, the plates we were using would suddenly shoot out from under our hands, while any attempt to arrest their movement was followed by the contents jumping either into our laps or over our opposite neighbour.

During the night the wind became squally, and breakfast on the morning of the 8th found only a very select party to partake of it. By noon we were about the middle of the Bay of Biscay, and there were ominous signs of wind and bad weather. Several of the party, having found that the incessant rolling from side to side all night long interfered materially with their rest, gave up their fixed berths and determined to try hammocks and swing cots ; they accordingly enjoyed a somewhat quiet night in spite of the liveliness of the ship. So long as the motion is confined to rolling, the fact of sleeping in a cot almost entirely neutralises the annoyance, for being suspended from above by a strong rope at the head and foot, a movement of the vessel parallel to the line of suspension is not communicated, but when the vessel pitches fore and aft a swing cot or hammock is of very little use.

On Friday morning, the 9th of December, Cape Finisterre was in sight, and there was a general sigh of relief at seeing the southern extremity of the much dreaded Bay of Biscay, forgetting that the swell from the Atlantic, urged forward by a gale of wind, was as liable to strike the ship on one side of Cape Finisterre as on the other. After dinner, the wind, which had been increasing in violence all day, veered round till it was right in our teeth, and the roll of the ship somewhat changing to a pitch assumed a screw-like movement, which sent below many who had hitherto bravely resisted the terrors of the Bay. About midnight, when I turned into my cot, it was blowing half a gale, with every prospect of an increase. My cot was suspended a little on one side of the centre of the main deck, so that for some time I amused myself by watching the angular movement of the ceiling, as it bent down to me first on one side, then on the other, and when the pitching movement was superadded I tried to picture to my mind the actual movement of the ship as it would be seen on deck. Suddenly I was nearly knocked out of the cot by a blow against the side of the cabin, and then it slowly receded, and I seemed to go up and up on the other side until the edge of my cot was, as near as I could estimate, not more than six inches from the ceiling. Back I went again, and, simultaneous with the thump against the cabin side came a smash as if all the crockery on board had broken loose, then a most horrible crash at the stern and a dashing of water made everyone rush out to discover the extent of the damage. One gentleman found the ship's carpenter, and was not reassured when he heard it reported that the stern-post had been carried away, and the ship was filling with water. Sleep being out of the question, I dressed and went on deck to observe the extent and appearance of the storm. There the sight was the grandest I ever beheld. The wind was blowing a stiff gale ; the great Atlantic rollers came toppling in as if they would overwhelm us. Every wave was crested with white foam which was severed by the wind from its parent wave and driven wildly across the ship. We were wallowing in the trough of the sea, and each wave caused us to rock from side to side in a helpless manner. The men at the wheel, who should have been actively employed in turning the head of the vessel to meet each wave, were standing idle, and on looking at the compass I saw that we were turned quite out of our course—drifting, in fact, before the storm. Just then the lieutenant of the watch asked the steersman if she would answer the helm yet. " No, sir ; not in the least ! " " Quartermaster, send another man to the tiller ropes, and go yourself. We must get out of the trough soon, or ——" I could not catch the alternative for the noise of the wind. On inquiry I was told that the tiller ropes, which were quite new and made of hide, had stretched considerably, and until they were tightened the rudder would not act.

Notwithstanding our critical position, I must confess that the spirit of investigation overcame all other feelings, and I determined to make the best of the opportunity now before me of seeing the various phases of a thorough Atlantic storm in midwinter. The bumping which aroused me and the heavy lurches which followed were caused by the ship falling off into the trough of the sea, while the terrific crash was occasioned by a gigantic wave crumpling up the stern boat and carrying away a baulk of timber which stretched across the screw well. The rolling was so excessive that even practised sailors could with difficulty keep their feet, and it was only by a judicious course of hanging on to ropes and rails that one could drag oneself to a suitable post of observation.

I wanted, if possible, to measure more or less roughly (1) the angular roll of the ship ; (2) the height of the waves ; and (3) their distance from crest to crest. The oscillation of the ship was not quite synchronous with the advance of the waves, and, consequently, the roll of the ship increased and diminished regularly after a certain number of vibrations. We were swinging somewhat like a top-weighted pendulum, or a metronome, set to vibrate about 15 strokes a minute, but the motive power gave us additional impetus at somewhat less frequent intervals. The moon was tolerably clear overhead, being obscured only by a slight scud ; by its light were visible the great waves coming broadside on, as if they would overwhelm everything before them. As the wall of water approached, the vessel heeled over away from it, and then rose to the top, and after quivering for an instant on the crest the ship seemed as if it would turn on its side, and it literally did tumble down at an angle of 30° into the deep valley beneath. Here another wave caught it and turned it on the opposite side, and as it struck the vessel near the time of its return oscillation a few more degrees were added to our roll. Again the ship heeled over, and again the times of vibration and impulse appeared coincident. At each roll her arc of vibration increased until it became an effort even to hold on by a rope. The ship, however, was very dry, and so long as solid green water could be kept out there was no immediate danger ; but her bulwarks were dipping deeper and deeper each time, and I watched with some interest the advent of a gigantic wave which could be seen in the distance toppling far above its fellows. The previous wave had brought us nearly to the water's edge, but we were rather losing time in the rival oscillation of ship *versus* wave. On it came — the ship gave a shudder as it struck her, turned wearily over, and sank her bulwarks lower and lower until the gangway fairly dipped under water. The sea in tons poured in, everything movable fell over to the side, one of the sailors was thrown completely across the deck, receiving some injury from striking against a heavy gun, and for a moment the ship was stationary, appearing to hesitate whether she could recover herself or whether she should not complete the revolution, and join the ill-fated *Captain* at the bottom of the Atlantic, at a depth of about two miles from our present position. The want of synchronism saved us. This wave on turning slightly helped the ship to recover herself, and for the next few rolls the arc of oscillation gradually decreased. I watched this alternate ebb and flow of vibration, of *maximum* and *minimum* of waves, for some time, but only on one occasion did I see the two *maxima* coincide, and only once was a green sea shipped.

During some of the severest rolls I tried to get an idea of the angle through which the ship moved. Eye estimates showed frequently an angle of about 30°. Among the numerous ropes which stretched about the rigging at all inclinations I fixed my attention on one, which, whenever the ship made rather a heavy roll on the port side, became nearly perpendicular to the horizon. Occasionally a right angle was just reached, while about every twelfth or fourteenth roll this angle was exceeded by several degrees. The next day, on measuring the amount which this rope declined from the perpendicular, I found it to be 35° ; so that the ship occasionally must have rolled about 40° over on one side. That this estimate is not excessive is proved by the fact that the clinometer, which was only graduated to 35°, was found during the night jammed against its bearings, thus showing that 35° had been exceeded. On subsequently measuring the angle through which my cot must have moved to bring it to within six inches of the ceiling, it showed an angle of 61° ; but from this angle must be subtracted an unknown amount to allow for the swing given to the cot by the thump it had just received at the opposite side. The amount to be deducted could not, however, be more than 12° or 15°, leaving a residual arc of 46° to 49°, through which the ship must have rolled on that side.

I next attempted to judge of the height of the waves. After taking a firm position amidships, at a part where the top of the bulwarks was 20 feet from the normal water-line, I brought my eye level to them, and noticed when the deck was horizontal how high the advancing waves appeared. When the ship was in the trough most of the waves overtopped the bulwarks, while the giants, which, fortunately, only came at rare intervals, towered at least 10 feet above us. The inclination of the vessel was changing so rapidly, and the circumstances were so opposed to exact observation, that I cannot be certain to within a few feet. There is no doubt, however, that some of the waves were quite 30 feet from trough to crest.

The deck and saloon through which I passed to get to the stern cabin, where the tiller ropes were being tightened, presented a most chaotic appearance They are surrounded by cabins, from each of which issued at every roll the noise of trunks, boxes, lamps, washstands, &c., grinding against one another, and sliding about from side to side on the floor. This was varied by the ejaculations of alarm or groans of despair from the prostrate occupants. One of the most eminent of the expedition was devoting the whole of his attention to the saving of two valuable chronometers from the universal wreck. With one in each hand he received, with outspread elbows and extended knees, the alternate shocks with which he dashed against the sides, regardless of the bombardment which his shins were receiving from the smash of furniture in his cabin. Another philosopher, who evidently looked upon portable property under such circumstances as dangerous as a live shell, was busily engaged in throwing everything movable out of his cabin into the saloon, where his chattels speedily got mixed up with the wreck of tumblers, moderator lamps, legs of chairs, coal-scuttle, and cinders from the stove which had broken loose and were surging about from side to side. Another of the party was trying to collect together his valuables, which had originally been safely deposited on a table. In some incomprehensible manner his trunk had been thrown on the top of the table, and had wrenched off three of its legs. A fourth observer, who had persistently kept his bed from the first day, was in a state of deshabille, hanging on like grim death to a corner of the table.

In the stern cabin the quartermaster and five or six stalwart seamen worked steadily and silently, as men doing their duty in critical circumstances, with ropes and cord securing the ends of the hide ropes to the tiller, the rudder being jammed mechanically hard round on the port side. At every roll in one direction, when the strain was to a certain extent taken off the ropes for a moment, a tug was given by all the men with a will, and when the lurch was on the opposite side they simply held on, with clinched teeth and rigid muscles, so as to keep what they had just gained. In this manner, slowly and painfully, a little of the slack was gradually got in, sometimes a quarter of an inch being gained, and sometimes not a tenth. A learned professor was doing not the least important service he has rendered to science by holding a tallow candle to light these men of muscle in their work ; and it was with a profound sense of relief that at last we heard the announcement that the ropes were quite taut, and that the vessel would once more answer to her helm. The monotonous roll soon assumed the screw-like character produced by its combination with a pitch, the vessel's head was steadily brought round to windward, and the *Urgent* was at last rescued from her perilous position in the trough of the sea.

On returning to deck the scene was somewhat changed. The wind, if anything, was fiercer than ever, the moon obscured by rainclouds, and the waves meeting the ship diagonally, while at every pitch the screw rose out of the water, making the ship quiver from end to end by the terrific hammering which it occasioned on its re-immersion. At the bows the scene was fearfully wild. The figure-head dipped into nearly every wave, as, owing to the length of the vessel, she was not able to rise readily, while not unfrequently the crest was dashed against the starboard bow and deluged me in salt water Two men, muffled in countless tarpaulins, were on the look-out. One of them said that he had been on board one of the Channel Fleet the night the *Captain* went down, and that gale was nothing compared with the one we were in now. "It was not above five miles from here that the *Captain* turned bottom upwards, and she didn't roll half as much as this ship," he screamed in my ear. Conversation, indeed, was only to be carried on in shrieks and screams, for the noise of the wind, which was now blowing a hurricane, was loud enough to overpower all other sounds. In the early part of the night it appeared to howl through the rigging, and it had been ascending the gamut of unearthly sounds, until now it could only be compared to a long-sustained shrill whistle. Its force was such that, although worked by engines of 400 horse-power, our screw was unable to hold its own against the wind, for we drifted sixty miles out of our course that night. The masts bent before the storm, and moving from one end of the deck to the other could only be accomplished by a series of gymnastic efforts. The rain, which began to fall about 2 a.m., stung one's face like sharp needles. Looking forward from the bows one saw a solid black wall of water towering above the vessel, and as the figure-head actually dipped into the hill it seemed impossible to believe that we were not going straight through to the other side. Up the vessel turned, however, though inclined at such an angle that on looking back towards the stern the horizon appeared high up at the top of a rugged range of mountains, while as we pitched down-wards the stern was lifted high up into the air.

I could get no satisfactory estimate of the breadth from crest to crest of the waves. Taking into consideration their position with respect to those portions of the ship to which they were opposite at the same time, when we were crossing them nearly at right angles, I should consider their width to be from 60 to 80 feet.

At eight the next morning, Saturday, the 10th, the wind almost suddenly sank, and the barometer, which had been steadily sinking all night, remained stationary, and then began to rise again. Soon the wind blew from the opposite quarter, but this being in our favour was not so objectionable. We had evidently passed over the centre of a cyclone ; its energy, however, was pretty well exhausted, and towards midday the severest storm which, according to Captain Henderson, the *Urgent* had ever been in might be considered at an end.

The remainder of this day and Sunday were uneventful ; our adventures of the Friday night formed almost the sole topic of conversation, and after a small adventure on Monday, which delayed our arrival at Cadiz for another twelve hours, we steamed into Cadiz Harbour on the morning of Tuesday, the 13th, just seven days from the time of leaving Portsmouth.

Let Crookes's eclipse diary tell the rest of the story :

In the morning we were off Lisbon, but saw nothing of it as it is up the Tagus.

Monday, December 12*th.*—At 2 this morning we passed
Cape St. Vincent, and then bowled along well, the wind for almost
the first time being of some use. We made this morning 11
knots an hour. In the afternoon we began to look out for Cadiz.
Soon white houses and a tall white lighthouse commenced to appear
above the waves. "There's Cadiz," everyone said, and the ship's
course was altered direct to the lighthouse. As we neared it the houses
got higher until we could see a small town, on a low sandy shore
appearing. Then a pilot boat put off to us, and a man was seen in
it waving his hat violently to attract our attention. "There's
Lord Lindsay," cried Huggins, who was looking through his aluminium telescope. The word went round, and the ship was stopped.
The man came alongside, when, instead of Lord Lindsay, he turned
out to be a seedy-looking pilot who could not speak English. We
mustered sufficient Spanish, however, to find out that the place
was not Cadiz, but that he would take us there. This was a thorough
sell, so we gave him a sovereign and bundled him back, and steamed
away a little further south. The lighthouse (a new one not on our
chart) had misled the master, and the village it seems was Chipiona.
As it got dark the lighthouse of Cadiz appeared, but the navigation
being difficult, it was thought better to lay to all night at sea. So
here we are, some miles from shore, very little wind, and no steam
up, rolling about in a helpless manner. We expect to be in Cadiz
to-morrow morning by breakfast-time.

Tuesday, December 13*th.*—At daybreak this morning we steamed
towards Cadiz, and saw an ugly reef of rocks ahead, which quite
accounted for the Captain's caution last night. After a little delay
a pilot came on board, and away we went into Cadiz Harbour.

Cadiz stands very beautifully. It is built on a small peninsula,
and the houses being of white stone and of some architectural beauty,
the town looks very pretty. A boat came off from shore, and told
us that there was some difficulty about quarantine. We ought
to have got a clean bill of health in England viséd by the Spanish
Consul. For the present we must hoist a yellow flag, and no one
must land on any pretence whatever. They had had a telegram
from Lord Granville, who had asked them to give us every facility
for landing instruments and passengers without delay at Custom
House. It was suggested that if these were *facilities*, it would be
better for a few obstacles to be thrown in our way next time. After

a wearisome delay of some hours the same person came alongside with the joyful news that we could haul down our yellow flag, and a few minutes after several boats with gorgeously dressed officials came up, and an immense deal of bowing and gesticulating was done, cigar in mouth. One man was particularly splendid, his coat was covered with medals to such an extent that it was difficult to see the colour of the cloth beneath. Captain Noble suggested that he had never been vaccinated for medals, and had had a very severe attack of the disease in consequence, and had broken out all over. After lunch the Cadiz party left, and whilst the Doctor was going to the shore and back (to sign some paper) two bum-boats came up and purchases were made. Figs, oranges, cigars, &c., were indulged in. I bought nothing, for I thought it better to defer purchases till the return journey.

Just as we were about to steam off to Gib. a boat came alongside and a somewhat seedy individual came up with a tale that he was a distressed British subject who was desirous to go back to Gib He had no paper from the Consul, and his account of himself did not seem very plausible, but our Captain, who is the best-natured fellow in the world, let him stay. The cool impudence of an auctioneer's clerk (as he said he was) stopping a Queen's ship with pennant flying and saying he wanted a passage to Gib. beats anything I ever heard of before.

We got a little later French news at lunch. I gave Mr. Moulton a letter to Nelly to post at Cadiz.

After dark we got the first sight of Africa, and soon the land on each side narrowed till we got to the Gut of Gibraltar, where the strait is only nine miles wide. Just at this part the moon rose right ahead of us, and the appearance of the high mountainous land on each side and the bright, almost full, moon shining in front along the sea was beautiful in the extreme. The height of the mountains on each side narrowed the apparent width very much.

The grand rock of Gibraltar now rose from the waves and rapidly became the most prominent object in front; the straits now widened to about 20 miles. The "ape's head" mountains on the African coast being a bold and not unworthy opponent to the rock of Gibraltar. We anchored in the bay near the mole.

Wednesday, December 14th.—At about 10 I left the ship with a gentleman from the town known familiarly as Sacconi's devil.

He showed me the Post Office and other principal places. Mr. Carpenter and I then walked about the town, went through the markets, and looked at the shops. The number of Algerines in their national costume was remarkable, and from the variety of Moorish curiosities in the shops it is evident that any amount of presents can be procured. The variety of out-of-the-way fish and fruit in the market was curious. We then went up the rock, and finding a sergeant who appeared willing to show us round, went over some of the fortifications and galleries, and had a general view of the place. We then went again into the town, and I bought a few things, and then went back to the ship to finish a letter to Nelly. Just before dinner Admiral Ommaney came up with an invitation for me to dine with Sir Frederick Williams (of Kars), the Governor of Gibraltar, at his official residence, "The Convent." After dressing, the party consisting of Admiral Ommaney, Tyndall, Huggins, and myself, started, and on reaching our destination met Captain Noble. The dinner was a very excellent one, served up in good style.

Thursday, December 15th.—I am greatly disappointed to find that there were no letters for me, and that we shall leave Gibraltar this morning before the P. & O. steamer, which is expected to-day with mails from London of last Saturday, comes in. At about 9 a.m. the gunboat which is to take the Estepona party started, and in about an hour we followed on our way to Oran.

The Mediterranean was as calm and smooth as a pond, scarcely a ripple to be seen, and there was no wind. The appearance of the rock of Gib. is singularly grand viewed from the Mediterranean side, resembling a lion couchant, the head towards Spain and the tail towards Africa. Soon the African coast disappeared, and we skirted the Spanish shore nearly all day. The little wind which now blew was rather chilly, coming as it did from the Sierra Nevada range of mountains, which could be seen in the distance, their tops covered with snow. The day passed without any event at all. My head ached rather badly all day (the result of the dinner last night—or perhaps the penny cigars !), but towards night, after a nap, it got better. The phosphorescence of the sea was very beautiful, the track of the vessel was left in a sheet of silvery flame, and looking down the screw well the whole body of water seemed a mass of light, which illuminated the surrounding objects. I tried to get

153

a spectrum of this light, but could only see that there was little or no red in it.

Friday, December 16th.—We passed close along the African coast all the morning. It is extremely bold and picturesque, high mountains alternating with beautiful green valleys. Not a tree, however, was to be seen anywhere, and the heights were perfectly bare of vegetation. After some delay we at last anchored in Oran bay, and had the usual officials in gold lace on a visit of inspection. Huggins and Admiral Ommaney went ashore, and as the captain thought it better for no one else to leave till they had come back to say it was all right, we were kept prisoners for nearly 4 hours —much to our disgust. At about 4 p.m. we left the ship, and for the first time put our foot on African ground. I was disappointed at the appearance of Oran. It is an inferior edition of a dirty French town, and has all the vices and inconveniences of a low garrison town without much redeeming points of Oriental life. Moors and Arabs and darker gentry there are in abundance, and the quaintness of their costumes, in spite of the dirt and filth about them, is very picturesque. Still there was quite as much to be seen at Gibraltar. The streets of Oran are wide, and there are many good shops. It is quite as large a place as Boulogne, but of course vastly inferior as far as the French life is concerned. From a comparison of the photographs I should say it greatly resembled Scarborough in outward aspect and scenery. On the high ground around are perched forts, and one tremendous hill close at the side of Oran has a large fort on it. Every other man one meets is a Zouave, or Chasseur d'Afrique, and the place is entirely under military rule.

We returned to the ship to dinner, and afterwards went out in a party to see " life " in Oran, which consisted in going into a *café chantant* of the lowest description, sitting beside the biggest blackguards I ever saw (together with some decentish people), drinking a villainous mixture of coffee and curacoa, and seeing some highly disgusting dancing, terminating with the can-can. On returning to the ship we had a committee meeting. Little Huggins's bumptiousness is most amusing. He appears to be so puffed up with his own importance as to be blind to the very offensive manner in which he dictates to the gentlemen who are co-operating with him, whilst the fulsome manner in which he toadies to Tyndall must be as offensive to him (Tyndall) as it is disgusting to all who witness it. I half fancy there will be a mutiny against his officiousness. Wrote to Nelly.

Saturday, December 17th —Huggins and a party went to see some appropriate sites for our eclipse observatory, and Captain Noble and I waited behind writing letters, and then we went into Oran and walked about. In the afternoon purchased a Burnous of an Arab maker for 20 francs. Saw the site for our observatory. They have found close to the railway station, where they had decided to pitch their tents, a house and grounds belonging to a Scotchman named   He has been in Oran for some years, and appeared glad to render his countrymen a service. He has placed a field at the disposal of us all. Walked through the Arab quarter of Oran. This is very curious, and purely Oriental. The women are walking about in all manner of costumes, according to whether they are Mohamedans or Jewesses, and the men are dressed in Burnouses, &c. Saw an Arab wedding, but could not see the bride, or in fact distinguish which she was, for all the women were dressed alike in long table covers, coming to a point at the top and reaching to their heels, a small aperture for one eye to peep through being in front of their faces.

Sunday, December 18th.—Cloudy and rainy. We had service on the main deck. Mr. Howlett preached, the text being from Amos viii. v. 9 : " And it shall come to pass in that day, saith the Lord God, that I will cause the sun to go down at noon, and I will darken the earth in the clear day." The sermon was a very excellent one. In the afternoon walked up to the observatory, and saw how the sappers were getting on with the foundations and instruments. Several of the principal men of the town came to dinner this evening, so we put on full dress and furbished up our French. The speeches were very amusing, and the way in which the Frenchmen mixed their liquors, taking sherry, hock, champagne, moselle, bitter ale, curaçoa, coffee, brandy, and then bitter ale again, was a wonderful sight. The dinner party did not break up till very late.

Monday, December 19th.—Raining, blowing, thundering, and lightning almost all day. Prospects very unfavourable for eclipse. Went up to the observatory tents and worked at the telescope and spectroscope I am to have, it having been decided in committee this morning that Huggins was to have the large telescope and equatorial. On the road home made some purchases at the shop of Moïse Ben Ichou, 22 Rue Philippe. Towards evening the rain, wind, and lightning got much more violent.

Tuesday, December 20th —A very fine morning after the rain. We went up to the observatory soon after breakfast and got the instruments adjusted. I had considerable trouble with mine, as the spectroscope was very roughly put together. The instrument I am to work with is a telescope of Grubb's make, equatorially mounted, but without clockwork   The object glass is   inches in diameter, and   feet focus. The eye-piece is removed, and in its place a spectroscope is attached. In front of the slit of the spectroscope is fastened a white card, having a small longitudinal aperture over the slit. On this card the image of the sun is well projected in sharp focus. At the eye end of the spectroscope is an arrangement for rapidly bringing a pointer on to any line in the spectrum and pricking its position on a card. After getting the telescope, &c., into good position, I practised with the recording arrangement until the clouds came over.

Walked home through the Arab quarter with Captain Noble. In evening went on shore again, and looked in at the *café chantant*. Nothing interesting or amusing. Looked in again at Ben Ichou's.

Wednesday, December 21st.—Much rain fell during the night, and the wind was very violent at times. On coming in to breakfast the prospects looked very poor indeed. It was raining hard, and there was not a break in the leaden sky. Went on shore to arrange about a photographer to come to-morrow and take a view of the totality, and also take several pictures of the instruments, tents, &c. Captain Collier accompanied me. When in the town we met a cab containing Captain Salmond, who told us that the tent had been blown down that morning and the two largest instruments had been laid completely prostrate. We took a cab and hurried up to the tents, and there found what looked at first sight a complete wreck of all our hopes, as well as instruments. The great 6-inch "Cooke" equatorial lay almost on the ground, the tube battered in, the glass of the driving clock broken, and the whole spattered over with the tenacious red clay of the country. Fortunately a sapper who was in the tent at the time and saw it falling had the presence of mind to push a large packing case under it to receive it. This protected the more delicate parts from injury, and when the matter was fairly looked in the face, and we all set energetically to work, the instrument was very soon set up again and got into working order. My telescope had been completely overthrown, but on putting that up also I found that no very great damage had deen done, and within two hours of our first witnessing the wreck we were at work with them and getting as good results as we had yesterday. Mr. Hunter's telescope had also been upset, but that not being so large had sustained no injury. During our work the clouds broke, and the sun came out. At midday it was shining perfectly clearly, and free from clouds. I put my instrument in good adjustment and brought some parts of the recording instrument back with me to the ship in order to give them some final adjustments. Went in the afternoon and finally arranged with the photographer. He is to take four views of totality with a large portrait lens, full aperture, and long exposure, to try and get some detail in the outlying streamers of the corona. He is also to take six or eight negatives of the instruments, party, &c., and is to receive for the whole £6 sterling. After dinner I was occupied for some time in adjusting the recording part of the spectroscope till I thought it worked very well.

On going on deck the stars were shining brightly, but there were several dense clouds about, and the wind seemed inclined to shift round to the north-west. We did not go on shore this evening, as we wished to keep our heads cool and collected for the important operations of the morrow.

Thursday, December 22nd, 1870.—The eventful day has arrived at last. We rose at daybreak, and breakfasted at half-past 7. The sun was shining brightly, but the wind was very high, and there were many clouds about. We all started directly after breakfast to our temporary observatory and found that the high wind in the night had done no damage. Indeed, it would have required nothing less than a hurricane to have injured the instruments, for the sappers, under the energetic direction of Captain Salmond, had built a small wooden house round the two large telescopes in the principal tent and shored up the sides so that had the tent been blown away bodily the hut would have resisted. When we got there the hut was down, and the front of the tent partly removed. The back of the tent being now fully exposed to the wind was shored up with planks placed crosswise, and scantling shoring them up on each side. The tent poles were also shored up, and extra tent ropes had been secured to pegs driven into the ground outside the ordinary row of pegs. I had a table arranged close to the eye-end of my telescope, and on it placed my reading-lamp, watch, paper for notes, pencil, opera glass, dark glasses of various shades, small telescope (2-inch, lent me by Captain Noble), together with spirit lamp and various metallic salts and platinum wires.

At about $\frac{1}{4}$ to 11 o'clock a.m. I had the telescope and spectroscope perfectly adjusted. A blank card (of which I had several properly prepared) being placed on the recording part of the spectroscope, I pointed the telescope to a white cloud, of which there were far too many about just then, and got the Fraunhofer lines beautifully sharp, D being clearly· double and the brilliancy being very great. Mr. Howlett, who had charge of the direction of the telescope, then arranged with me, in conjunction with Mr. Huggins, how we were to manage during the two minutes and odd seconds of totality. We were to try and get records of three spectra from different parts of the corona. The first was to be taken a few moments before totality, and the slit of the spectroscope was to be placed radially projecting outwards from the dark edge of the moon (the part opposite to the vanishing crescent of light) in the hopes of getting indications of corona before the actual moment of totality. The telescope was then (at totality) to be moved so as to illuminate the slit with the lowest stratum of corona, tangential to the moon's edge, and the third position was to be with the slit tangential or radial, at Mr. Howlett's discretion, but some distance from the moon so as to get a spectrum of one of the streamers some distance off if there were any. A gentleman whom we took on board at Gibraltar volunteered to act as amanuensis for me, and give me time, &c., during totality, so that if I saw anything remarkable I should not waste time by writing it down, or risk losing valuable observations by trusting to memory. Each of us being at our post, I practised several times recording spectra, Mr. Howlett moving the telescope each time in a manner which would imitate the movement required during the totality. The time required to effect the three different registrations deliberately varied from 55 to 70 seconds, but by hurrying a little I could do the three in 40 seconds.

My intention was to shade my eyes a few moments before totality to render the retina more sensitive, and then to keep at the spectroscope and record whatever I could see of the three spectra, looking out especially for Fraunhofer lines, extra black lines, and the bright lines in the green, said to have been seen by some of the American observers, repeating aloud the name of the spectrum (whether 1, 2, or 3), and any special peculiarity observed in the lines as to sharpness, &c., for the amanuensis to take down. I proposed then to snatch up the telescope, already focussed by my side, and devote 10 seconds to an examination of the protuberances and corona. Then I should take up my opera glass, which was also ready focussed by my side, and go out of the tent and get a general view of the corona, the country, horizon, and notice the reappearance of light in the western horizon, and the rapid sweep of the shadow towards our station. I had practised all these operations several times and felt sure that I could manage them in the time and have several seconds to spare—provided only the clouds did not stop all observations.

At about 11 a.m. I was watching the sun through my opera glass protected by dark glasses, when I detected a distinct indentation a little above the centre, on the right. The eclipse for which we had travelled so many hundreds of miles, and spent so much time, trouble, and money had commenced. Ten seconds afterwards a cloud came over and nothing more could be seen till 11.8, when the advance was clearly visible. At 11.10 thick clouds covered the sun for several minutes. 11.25 sky quite overcast. 11.30 clouds breaking. 11.40 sun visible, and occasionally so till 12.10, when it disappeared behind light fleecy clouds. At 12.15 the sun totally disappeared, and was no more visible till about half an hour after totality. At 12.20 the whole sky was overcast. Here and there a few patches of blue sky could be seen in various parts to windward, and over the landscape patches of sunshine were seen sweeping along as the clouds moved. At 12.28 approximately, at which time totality was to commence, the sky was anxiously scanned for blue patches. One approached, but passed too much to the north, and on going out of the tent at 12.25 I saw that there was not the least chance of the next blue patch coming across our meridian for at least a quarter of an hour, whilst it seemed certain then to pass to the north of the sun. The light was now declining

159

rapidly, and although there was no sign of break in the density of the obscuring clouds, I went to the eye-piece of the instrument and looked in on the chance of seeing something. Not a trace of a spectrum could be seen, and I had to decide rapidly whether to stay there in the absolute certainty of seeing nothing, or to go outside and at all events see something of the general effect of the approaching darkness on the landscape. Had there been the faintest chance of seeing anything with the spectroscope I should have stayed at it, but as it was I decided to go outside, where most of the observers were already.

On the distant horizon and here and there in the far east gleams of bright light and patches which looked like sunshine were tantalisingly visible. The western horizon was of a dark blue-black, the sky overhead was like indigo. Suddenly a dark purple pall seemed to rise up behind Santa Cruz, the high ground on our west, and rapidly cover us in deep gloom spreading to the east almost as far as the eye could see. The sky overhead looked as if it were crushed down on to our heads, and the sight was impressively awful. The darkness was not so great as I had expected, for at no time was I unable to read small newspaper type, or see the seconds hand of my watch, but the colour of the darkness was quite different from that of the ordinary darkness of night, being of a purple colour. The high range of mountains in the extreme south (about   miles off), which were out of the line of total phase, were visible the whole of the time, whilst some light fleecy clouds in the north, where the sky was not so thoroughly overcast, showed reflected sunlight all the time. This, however, made our darkness more impressive.

The reappearance of the light was much more sudden and striking than its disappearance. A luminous veil with a comparatively sharp upper boundary shot up from behind the western hills. It passed over us and spread its illumination towards the east before we could fairly realise the fact that the long-expected total phase of the eclipse of 1870 was over without any of our observers seeing anything of it.

Ten minutes after totality Captain Noble and I went to the telegraph office and sent messages announcing the failure of our expedition to the London daily papers. He sent a short message to the *Daily News*. I sent a message of 19 words to *The Times* (cost 20 frs. 80 c.), and one of 40 words to the *Daily*

Telegraph (cost, 41 frs. 60 c.). Owing to the rupture of the cable between Gibraltar and Lisbon, the messages had to go through Algiers, Malta, Gibraltar, Madrid, Lisbon, and Falmouth.

The instruments and observers were as follows :

The large 6-inch equatorial by Cooke, with clockwork movement, to which a recording spectroscope was attached, was in the hands of Captain Noble and Dr. Huggins. The former was to bring the different parts of the corona on to the slit of the spectroscope, whilst Dr. Huggins recorded the lines observed in the spectra.

The $4\frac{1}{4}$-inch equatorial by Grubb, with spectroscope attached, was entrusted to Mr. Howlett and myself.

A 6-inch equatorial by Slater had a polarimeter attached to it, Captain Collins being the observer.

A 2-inch altazimuth by Ladd, with polariscope, was employed by Captain Salmond.

Admiral Ommaney was to observe the phenomena generally through a 2-inch altazimuth.

Mr. Wharton, ditto, through a 3-inch equatorial.

Mr. Carpenter, ditto, through a 3-inch Dollond belonging to Mr. Howlett.

Dr. Tyndall was to take general observations by means of a 5-inch modified altazimuth by Dallmeyer.

Mr. Hunter was to sketch the corona through a $3\frac{1}{2}$-inch Grubb equatorial.

Each of these observers, with instrument in perfect adjustment, was at his post in good time, and would certainly have done excellent work had the weather been favourable.

As soon as the telegrams had been sent we returned to the tents, and after standing for one or two photographs commenced to take the instruments down and pack them up, for after our failure but one idea seemed to possess all, and that was to get away from Oran and on our homeward voyage as quickly as possible. By about 3.30 I had got my telescope packed in its case, and everything else being packed up, except the 6-inch equatorial, and that being in good progress, Dr. Tyndall and I started for a climb up to the top of the high hill at the back of Oran, intending to get to the top of Santa Cruz. We passed across the end of the Arab quarter and descended a somewhat precipitous ravine into some gardens. After crossing these we commenced the ascent, and after a very stiff bit of mountaineering for about an hour we got to the top of the high hill looking over Santa Cruz and its fort, and considerably higher than it. The time was too late to allow us to cross the gorge separating the two summits, so we descended. The mountain side was covered with cacti, wild geranium, thyme, and lavender. I gathered a handful of the latter to take to Nelly as a memento of North Africa.

Friday, December 23rd.—All the instruments, with the tents and other materials, were on board by noon this day, but Captain Henderson thought it was not advisable to start till the gale which was blowing right from the quarter whence we had to go moderated. I went with a small party to view a curious cave on the road from Oran to Masa el Kebir. It seems one mass of shells cemented together. I got two or three specimens. I then went to the photographer's to settle with him for his day's work yesterday. In case of failure he was to receive half the sum agreed upon. I therefore paid him 75 francs, and 5 francs additional for the plate box and brought away the negatives. He gave me a few pictures. On my wanderings about Oran before this I picked up what I think is some very cheap bits of coral.

In the morning I climbed over the hill on the top of which Santa Cruz is situated, trying to get from the lower fort to the upper one. I was, however, directed wrong, and after toiling up the steep mountain side for some time through heavy wind and several rain storms, I came to an almost perpendicular part of the rock which was impassable, and was obliged to give it up. I was, however, rewarded for my attempts by seeing the most beautiful rainbow I ever saw in my life. It was a complete semicircle stretching over the sea in front of me, the ends resting apparently on the shore. The principal bow was excessively brilliant in colours, the indigo and violet being very fine. Outside was a secondary bow somewhat fainter. The space between the two bows was quite transparent to objects beyond, but the part within the principal bow was filled with a uniform white veil, which partially obscured objects behind it.

During the evening the gale increased in strength, and although our ship was in the harbour and protected by the mole, she felt the force of the wind somewhat, and a hawser fastened just outside my cabin window creaked and groaned so much all night that I could get no sleep.

Saturday, December 24th.—The wind went down towards the morning and shifted a little round. Steam was accordingly got up, and at 10 a.m. we went out of the harbour of Oran, on our voyage home. On getting out into the open sea the wind proved to be very strong, and the ship tossed and pitched very much. There were many pale faces in the afternoon and gaps at the dinner table. This is Xmas Eve. The temperature is very mild, for I have been on deck this evening for some time in my ordinary indoor dress and found it quite warm enough ; in fact, I could have done with a thinner coat. The wind is just enough in our favour to allow of some sails being set, and the good ship is going along at about 8 knots an hour under sail and full steam. We shall not get to Gibraltar before Monday morning it is certain, and I shall have to eat my Christmas dinner at sea. At 11 p.m. I am writing this in my cabin under great difficulties, owing to the exceeding liveliness of the ship. I can hear the rain pattering down on the deck above me.

Good night, Nelly, God bless you. A happy Christmas to you and the dear children at home. I can hardly hope that it will be a very merry one in my absence. Mine will be anything but merry, for not having heard anything since I left the dear ones at Mornington Road on Monday, the 5th inst., I am exceeding anxious about them. I pray that I may have good news from home waiting for me at Gibraltar.

Sunday, December 25th, 1870.—I shall remember this Christmas Day as long as I live. Soon after going to bed last night the wind got stronger and blew in exceedingly strong gusts occasionally. It got right in our teeth, and although not before it had blown two of our sails to ribbons—and the waves rising with the increasing force of the gale, the pitching fore and aft became very unpleasant. The hammering of the screw on the water as the stern alternately rose out of and dipped under the waves was so great as to entirely prevent me from sleeping, and at about 7 in the morning I got up feeling worse than I had done ever since leaving England. I had no sensation of sickness, but a giddy feeling in my head and an aching pain in the back of it which was caused by the knocking and shaking it had undergone all night. On inquiring on deck I found that the ship was only making about 2½ knots an hour, and was going half-speed, the captain not wishing to force the ship against

163

the wind and waves full speed, in order to save the passengers from the tremendous knocking about that would entail, and also being desirous of not straining the ship. The wind and waves got worse as the day wore on, and by noon we were only making about 1 knot an hour. The violence of the storm still increasing, it soon became apparent that the available steam power of the ship was only about equal to hold our own position, whilst the tossing was such as to render all but experienced sailors very uncomfortable. Under these circumstances the captain decided not to battle with the elements any more, but to turn round and take shelter in Almeria Bay, where there was good anchorage, and lie there till the wind went down. I was in the saloon when this decision was arrived at, and my first intimation was hearing the officer of the watch call out to his servant to go to his cabin and secure everything movable in it by ropes and cord. He advised me to do the same, as the rolling and pitching consequent upon our altering course and crossing the direction of the waves would probably exceed anything we had yet gone through. I took the hint, and went to my cabin and made all square for any amount of rolling, short of turning bottom upwards.

On returning to deck we were making straight for land, the high mountains of the Sierra Nevada range being about 15 miles off. The wind was blowing violently on our port bow, and the great waves coming in in the same direction the ship was heeled over and caused to roll to an extent which was fearful to witness. An angle of 30 and 35 degrees was occasionally made, and once or more it seemed as if she never could have recovered, but that as the wind acted on her side now high out of the water she must inevitably have continued the heel-over and shared the fate of the *Captain*. She recovered herself wonderfully. None of the rolls were in so great an angle as one or two I measured on the night of the 9th (40°), but being now in full sunlight and the whole of the ship from one end to the other being visible at once, the effect was much more impressive.

We gradually got nearer the coast, and saw the bay open out and then the tower of Almeria became visible. Next a little village called Roquettas, and close to this a low sandy spit of land, and several vessels riding at anchor so calm and quiet as to form a most enviable contrast to our own condition. As we got nearer the waves diminished in size, and at last when the anchor was finally let go the ship was comparatively still, although the wind was rather increasing in violence.

I took the opportunity of going to my cabin and having a sleep before dinner, which was put off till six to give us all full opportunity of recovering the knocking about we had experienced before assembling for our Xmas dinner.

The dinner was in the usually excellent style served up by our cook. We had Roast and Boiled turkey, Roast beef, Plum pudding, Mince pies, and lots of other delicacies. I drank the health of the dear ones at home, silently, in a glass of champagne. I cannot say I enjoyed the dinner. This was the first I had passed away from Nelly since our marriage, and not having any news of her or the children for three weeks all sorts of horrible fancies kept coming into my mind. I trust that when we get to Gibraltar there will be good news waiting for me. Just before going to bed I heard that the wind had gone down, and the night was very fine.

Monday, December 26th, 1870.—Before rising this morning the noise on deck told me that a move was taking place, and on getting on deck we were steaming out of our haven, the sun shining on us from a cloudless sky, the wind almost nil, and the sea as calm as the most inexperienced sailor could desire. All the day we skirted the rocky coast of Spain and had a wonderfully beautiful view of the bold Sierra Nevada range of mountains, with their snowy peaks. It is expected that if we have no more head winds we shall get into Gibraltar by about midnight to-night, but the wind is shifting about to such an extent, having gone right round the compass in a few hours, that there is no calculating on a continuance of favourable circumstances.

An illustration of the very changeable character of the wind in this sea has just occurred. Since writing the above—whilst writing it in fact—I noticed the ship was gradually listing over to the starboard ; this getting stronger and stronger, the pitching and rolling nearly ceasing at the same time, I went on deck to see the reason. The wind was blowing half a gale, and at right angles to our course. All the fore and aft sails were set, and we were going on at a fine pace, being helped along considerably, and blown over to one side at the same time by the force of the wind. Ceuta light on the African coast and Europa point on Gibraltar were brightly visible ; and the Great Rock in the form of a Lion couchant, its head towards Spain, was rising up before us.   It gradually got larger and larger, until at about 11 p.m. we rounded the point and at 11.30 dropped anchor safely in Gibraltar bay.   Before going to bed wrote a letter to Nelly announcing our safe arrival here.

Tuesday, December 27th, 1870.—At anchor in Gibraltar bay. Taking in coals all the day.   It is expected that our complement of 280 tons will be on board by dinner-time to-morrow.

In the morning received a letter from Nelly, dated December 10th, but one which she was to have sent me on December 17th has not arrived, although other passengers have letters from England of that date.   In the afternoon another letter from Nelly arrived, dated December 8th, *via* France.   Gillman sent a letter dated December 10th, and also a *C.N.* dated December 9th.

Went to Post Office to inquire about letters.   Nothing for me, and I hear that the next mail from England " *via Southampton* " does not come in till Friday—a day after it is proposed that we leave.

Some of our party went to Tangiers in the *Red Pole* Gunboat, but after duly weighing the pros and cons most of us, myself included, decided that it was not worth the trouble of going about.   The boat does not start till 1 p.m.   It is a four-hours' passage.   The landing is very difficult, first in small boats, and then on men's shoulders through the surf.   The Gates of Tangiers shut at sunset, and it is doubtful whether they will be opened for the party if they do not get in before.   There is no sleeping accommodation whatever on board the *Red Pole,* and she will come back at about 11 a.m. to-morrow.

Went about Gibraltar with Captain Noble.   Tyndall made an attempt to get away by the P. & O. steamer which arrived here at 10 a.m. and left at 1 p.m., but just missed it.

Captain Collins, Mr. Smith, and Mr. Wharton started to try and get to Seville and other places, and be picked up by our ship at Cadiz.   It seems a rather wild freak.   In the evening Mr. Smith came back again, wet through, and said he had had enough and turned back—but the others had pushed on.

Towards night the rain came down in torrents and the wind rose very high.

The Eclipse was seen very partially here.   We hear rumours

of the wreck of the Sicilian party, and loss of all their instruments. This news in London will be sure to make Nelly and others in London uneasy about me, especially as they will have only just got my letter from Gibraltar dated December 14th, in which I give an account of the storm we passed through. I therefore determined to telegraph to-morrow morning to announce my safe arrival here, and saying that I have heard nothing from home since 10th inst., and that we shall be 2 days at Cadiz. This will give them an opportunity of telegraphing to me at Cadiz if necessary.

Wednesday, December 28th, 1870.—Started after breakfast with Tyndall to see St. Michael's caves. After getting the keys at the Town Major's, we went accompanied by a serjeant as guide up the west face of the rock to the entrance, which is about 700 feet above the sea. Before entering at the principal entrance we went a little higher and went down a smaller cave till we came to a deep chasm, down which we threw stones and listened to them echoing down in the far distance. A bit of magnesium wire revealed the awful precipice on the edge of which we were standing, and threw up the stalactites in the various recesses in brilliant relief. We then entered the large entrance hall of the cave, and lighting our candles explored the various recesses of St. Michael's and of Leonora's caves. The latter is entered by a narrow passage with steps cut in it. After descending 20 or 30 feet the cave widens out into a sort of hall, with beautiful nooks and corners of stalactites in every direction, a little hole in one corner being pointed out to us as the one down which we were to creep in order to penetrate further. We got on our hands and knees, and candle in hand, followed our guide. In a very short time the passage widened, and another stalactite cave containing a pool of water was visible. We were now told that the most beautiful of all the caves was still some distance lower, but the way to it was somewhat difficult. This not being considered by us as an obstacle, we decided to explore it. The passage was steep, tortuous, and narrow, running along for some distance so low that we were on our faces wriggling along like worms and accommodating our bodies to the various sinuosities of the hole, which ran sometimes up, sometimes down, till we had gone some 30 yards. It ultimately led us into a most exquisitely beautiful grotto, not so large as some of the others we had seen, but surpassing all in the fairylike delicacy of the stalactitic forms.

One part was a perfect vista of columns and pillars, delicate needles hanging from the roof, and more massive pillars rising to meet them. The junction of stalactite and stalagmite was not effected in many cases, but in others they had united, and had thickened out into a pillar as thick as one's arm. A curious appearance of the stalagmites was observed. They did not appear of uniform thickness all the way up, nor tapering gradually, as might have been supposed had the action which formed them being uniform, but they showed a rythmical action, being alternately thick and thin, the distance from maximum to minimum being sometimes 2 inches, sometimes much more. The stalactites, on the contrary, were very regular and delicate. I broke off one or two of the more accessible stalactites, and also a portion of the stone curtain which forms in a very beautiful manner by somewhat similar means. Reluctance to injure the beauty of nature's handiwork prevented us from breaking off any of the more beautiful pieces. Magnesium wire was burnt here and revealed a scene which surpasses my powers of description.

The air felt very close and oppressive, and we were glad to get on our way out   Here and there as we scrambled along, our guide warned us of going too near certain parts where there were unexplored holes and deep chasms. He said the caves were gradually being explored by convicts, who were set to this work from time to time. The deepest part yet explored extended about 700 feet down, or to about sea level, but we did not go above 200 feet. The getting back was, if anything, more difficult than entering, as it was uphill mostly, and I had my stalactites as well as candle to take care of. It was, however, at last effected, and we emerged into daylight, in what state may be imagined, our hats, clothes, and hands one mass of mud and dirt.

We made our way next to the look-out station, where we had a capital lunch of bread, cheese, butter, and shandygaff. Tyndall wishing to get specimens of a deposit of sand covering a considerable portion of the east side of the rock, clambered for some distance down the face of the almost perpendicular face, but was at last obliged to give it up. From the look-out station—almost the highest point of the rock—we descended by the staircase on Charles's Wall to nearly the bottom. In one portion of this I counted 656 steps.

After getting to the ship and cleaning myself a little, I went out again to send the telegram to Nelly. Owing to the circuitous route it has to go it will be delayed somewhat. I hope, however, that she will get it in the course of to-morrow so as to allay her anxiety.

After dinner I went to a grand ball at Sir F Williams's, the Governor, at "The Convent." The entertainment commenced with private theatricals. *Caste* was acted by amateur ladies as well as gentlemen, and was most excellently performed. One lady in particular being equal to most of the professional actresses I have seen. The play was followed by dancing. After supper I returned with Captain Noble and Ommaney.

Thursday, December 29th, 1870.—At breakfast the letters which had just arrived by P. & O. steamer were delivered. I had one from Nelly dated the 17th. Had it not been for my stupidity in giving her the dates I might have been reading hers of the 23rd instead. I trust sincerely that the six additional days' news would still have been good.

At 10 a.m. we steamed out of Gibraltar Bay, and at about 6.30 p.m. we cast anchor in Cadiz Harbour. No occurrence of importance took place this day. The sea was tolerably smooth, and the wind not high, although with our usual luck it was dead in our faces.

Friday, December 30th, 1870.—A little before 10 a.m. a boat took us ashore, and for some hours Captain Noble and Mr. Watkins and myself wandered about Cadiz. It is a somewhat large town, but not interesting. The streets are very narrow, and resemble one another greatly. The houses are beautifully clean, and have light green or blue jalousies to all their windows, giving the streets as looked at from one end a very bright appearance. The stories overhang one over the other, and as the streets at the bottom are not above 20 to 30 feet wide, they narrow at the top so much as to obstruct much light and air. This must be a necessity in the heat of summer, but in the winter it is an objection, owing to the smells which abound in all the streets. The streets are very clean, and appeared to have little traffic going on in them. Spaniards muffled up in villainous-looking cloaks thrown over their faces walked about with a stealthy tread as if they were plotting mischief (due really, I believe, to the cold weather, which, although mild to us, is very cold to them), whilst dark-eyed ladies in mantillas have flitted past, looking saucily at us. The beauty of the ladies of Cadiz did not, however, strike me as anything particular, but I think dark beauty far inferior to its fair prototype, which is so plentiful in our island, and having for so many years worshipped at the shrine of fair beauty, I may not perhaps be an impartial witness.

There does not seem to be one street especially better than the others, either in width or architectural beauty. The shops are not good, and are not overstocked with commodities, and altogether I doubt whether Cadiz would ever be thought especially worthy of visiting had not some clever person—Byron or other—discovered that *Cadiz* rhymes to *Ladies*.

The cathedral struck me as especially magnificent, and when I was wandering through it I thought I had discovered the secret of the notoriety of Cadiz. On mentioning this to my fellow-travellers in the evening I soon found out that although to my insular ideas the cathedral was most magnificent—yet, compared with that at Seville and other cities in Spain, it sank into insignificance.

At 2 p.m. I had seen quite enough of Cadiz. I purchased a Faha and a fan, and went back to the ship with no desire to spend any longer time on Spanish soil. During the evening and night all the stragglers came on board, and having been reinforced by Lord Lindsay and all of his party, together with one or two others who were coming to England as passengers, our good ship was fuller to-night than she had been since I came on board : about 55 would sit down to dinner when all arrived.

Saturday, December 31st, 1870.—New Year's Eve !

Before I was up this morning the noise of the screw let me know that we were on our homeward voyage, and on going on to deck I saw the beautifully situated town of Cadiz already some distance off. I had some conversation with Lord Lindsay about Spiritualism after breakfast. All this day the sea was very calm, and the wind tolerably favourable—sufficient to allow of three fore and aft sails being set. We went on steadily at the rate of 10 or 11 knots an hour, and at this rate we could get into Portsmouth by Wednesday. This has been a lazy uneventful day.

I am now writing this in the saloon. It is getting on for midnight, at which hour we are to have punch, and the Admiral is to go and strike sixteen bells. I cannot help reverting in thought to this time last year. Nelly and I were then sitting together in communion with dear departed friends, and as 12 o'clock struck they wished us many happy New Years. I feel that they are looking on now, and as space is no obstacle to them, they are, I believe, looking over my dear Nelly at the same time. Over us both I know there is one whom we all—spirits as well as mortals—bow down to as Father and Master, and it is my humble prayer to Him—the Great Good as the Mandarin calls Him—that He will continue His merciful protection to Nelly and me and our dear little family, and bring us together in the course of the next week to our happy home, which I shall now appreciate as I never have yet done. May He also allow us to continue to receive spiritual communications from my brother who passed over the boundary when in a ship at sea more than three years ago.

Nelly, Nelly, my own darling, God bless you. If my good wishes and prayers for your safety can fly across nearly 1,000 miles of sea to you at this moment, you must become conscious of the fact that I am thinking of you. My other New Year's Eves come back to me one by one, and all of them—but this—have been hallowed by your presence and loving words. The time by Greenwich time now is just upon midnight, and I feel that this is the time for holy communion with you and the dear children. Ship's time is twenty minutes later, and then I must shut up the book, dismiss these thoughts from my mind, and join in the revelry which will doubtless take place. Nelly darling and my dear children, Alice, Henry, Joe, Jack, Bernard, Walter, and little Nelly baby, I wish you all many, many happy New Years, and when the earthly years have ended may we continue to spend still happier ones in the spirit land, glimpses of which I am occasionally getting.

January 1st, 1871.—This was a somewhat uneventful day. The ship went on steadily on her course. Towards night the wind freshened.

Monday, January 2nd, 1871.—After a tossing night, much disturbed by the noise of the screw, we got up to find the wind blowing strongly in our faces, and the ship rolling and pitching furiously. This was altogether the most unpleasant day we have had, as the motion, compounded of a roll and a pitch with an occasional kick-up behind, precluded all comfort in sitting, standing, or lying down. I tried to lie down in the stern cabin before dinner, but was jerked off the sofa twice, and nearly thrown upright on to my feet by one tremendous lurch, followed by a kick-up. During dinner the rolling and pitching seemed to reach their maximum. Soup was emptied over our waistcoats, joints were thrown into our laps, wine and beer shot out into our faces, and those who were on loose chairs were shot off at a tangent and found themselves landed in different parts of the saloon floor among the broken plates and dishes. With the exception of one or two lurches the *Urgent* made on the night of the terrible storm , we had during dinner some of the greatest angular movements I have ever witnessed, an angle of 40° being nearly reached sometimes. Eating was almost impossible, for nearly all one's attention was required to keep the meal, &c., on the plate, and ourselves on the benches. Huggins being small and not very careful, disappeared once, plate and all, under the table.

Several of us arranged to-night to have swing cots and hammocks on the main deck, amid ship. Mine was swung there, and although the pitching and tossing was very severe all night, I had one of the best night's rest I have had since leaving Oran. This day we made very little progress—not more than 4 and 5 knots an hour. We got fairly into the Bay of Biscay in the night.

Tuesday, January 3rd, 1871.—In the night the wind went down, the sea got calmer, and all this day we went along at a good speed. It was suggested at dinner that our wives and sweethearts had at last got hold of the tow-ropes and were hauling us along.

Wednesday, January 4th, 1871.—Another fine day, during which we made good progress. At 2 p.m. we were up to Ushant Island, and altered our course straight for the Isle of Wight, which we hope to have in sight before breakfast to-morrow morning.

We are now sufficiently near England to be able to reckon tolerably closely upon the time we shall get in. By noon to-morrow it is imagined we shall be at Portsmouth, and shall be off the ship in an hour or two from then.

Thursday, January 5th, 1871.—On going on deck this morning I saw we were passing the Isle of Wight. Snow was on the hills, and the temperature was very cold. At 10.30 we were alongside the pier at Portsmouth, and after some delay on account of the multiplicity of packages I had to see to, I finally left the ship and started for London by the 3 p.m. train.

Crookes had added interest to his homecoming by sundry purchases, of which he gives the following price-list :

List of Articles purchased on Eclipse Expedition, December 1870.

Frs.

1 Burnous (white), made in Oran	20
1 Burnous, tissue made in Morocco, but worked and made up in Oran	35
1 Coloured Wool Haïk, Morocco	20
1 Worked Morocco cushion	7½
1 Necklace (Mecca)	2
1 Mouchoir (Morocco) in silk and gold thread	10
2 Kerchiefs (Tunis)	6
1 Pair Slippers (Tunis)	6
1 Pair Cups in metal stands (Morocco)	10
1 Bracelet (Morocco)	4
1 Pair Ear-rings (Morocco)	4
2 Lockets (Morocco)	8
1 Brooch (Morocco)	4
Photographs	8
1 Worked Morocco cushion	10
1 Pair Slippers (Tunis)	4
1 Pair Child's Boots (Tunis)	2
	158½
Coral (sundry pieces)	30
	188½

(188½ francs = £7 11s.)

1 Hammered Brass Waiter (Moorish)	10/-
Faha, 2/- ; Fan, 2/-	4/-

CHAPTER XII

MYSTERIOUS FORCES AND APPARITIONS

(1871–4)

WE MUST NOW DEAL with the most controversial episode in Crookes's life. It extends over the years 1870–4. It is usually described as his "conversion to Spiritualism" or as his "scientific investigation of psychic force." The two attitudes assumed towards this byepath of Crookes's career may be characterised as the Spiritualist attitude and the Rationalist attitude respectively. Let us give the two rival versions :

(*a*) *Spiritualist Version.*—Crookes was an eminent man of science inclined to agnosticism. He was an insatiable investigator, ever ready to probe into new and unknown phenomena. Modern spiritualism, born in America in 1848, presented to the world an ever-increasing array of baffling physical phenomena having within them a spiritual meaning and a message to humanity. Crookes was not primarily interested in the message, but was keenly interested in the physical phenomena, and anxious to bring them under the reign of natural law. He had a very happy manner with mediums, being courteous and gentlemanly without the least relaxation of scientific vigilance. This somewhat rare combination of qualities accounts for his marvellous and unprecedented success. His experiments with D. D. Home were classical, and

174

absolutely free from flaw. They were sufficiently rigid to stand the keenest scientific scrutiny. Their rejection by the Royal Society is but another sad illustration of the blindness to new facts sometimes shown by conservative corporate bodies. In any case, the facts convinced Crookes personally of the reality of psychic phenomena. His further experiments with Florence Cook convinced him that there was a supra-mundane intelligence behind the phenomena, and so he finally became a convinced spiritualist. His great task being accomplished, he returned to his laboratory work and gave to the world the Crookes tube and the radiometer. But he remained a spiritualist for the rest of his life.

(b) *Rationalist Version.*—Crookes, like many another physicist, had a streak of mysticism in his mental constitution. The death of his brother under tragic circumstances threw him into spiritualism. Being, like most scientific men, rather guileless himself, he fell an easy victim to the impostors who were then ministering to what had become a society craze in England. D. D. Home succeeded in gaining his entire confidence and then in duping him by his trickery. Being already a convinced spiritualist, Crookes was incapable of applying scientific tests to any matter involving his rather vivid personal feelings. In any case, he was by his training absolutely unfitted to detect the clever methods of fraud which had been evolved by mediums since 1848. So he was deceived both by Home and Florence Cook. He probably in the end suspected that all was not well, and in 1874 he decided to have done with the matter for ever. Having publicly committed himself to raps, levitations, and " materialisations," he did not like to retract. But he abruptly closed a rather unfortunate chapter in his career, and made amends by an unparalleled devotion to pure

science, which soon brought forth abundant and refreshing
fruit.

There, I think, are the two versions fairly stated.
The biographer has an anxious task in keeping the balance
even. But probably I am as capable of impartiality in
this matter as anyone living. In a book [1] published
many years ago, and now out of print, I gave the full
story of Miss Cook's " Katie King " from spiritualist
records, and sketched out a theory to account for the
phenomena, supposing them to be genuine. In 1920 I
translated Dr. von Schrenck-Notzing's *Materialisations-
phænomene* into English, and personally attended some of
" Eva C.'s " séances given before a committee of the
Society for Psychical Research. In 1921 I went to
Belfast to investigate the phenomena described by Dr.
Crawford in his three books, and wrote a report [2] giving
a judgment adverse to the claims of Dr. Crawford's medium.
I can therefore claim to have seen both sides of the question. I shall piece Crookes's story together from what
records I have. There are several letters hitherto unpublished. There is some testimony from surviving
relatives and friends. But much has been lost or destroyed. This is notably the case with Crookes's numerous
letters to Florence Cook as well as to W. H. Harrison,
editor of *The Spiritualist*, and to and from a number of
other spiritualists. These, if ever recovered, would bring
us much additional light. But it is certain, at all
events, that when in July 1870 Crookes, at the request,
it is said, of a London daily paper, announced his intention
of " investigating spiritualism, so-called," he was already
much inclined towards spiritualism. What he really

---

[1] *New Light on Immortality*, by E. E. Fournier d'Albe (Longmans, Green
& Co., 1908)

[2] *The Goligheer Circle*. Experiences of E. E. Fournier d'Albe (Watkins,
London).

intended to do was to furnish, if possible, a rigid scientific proof of the objectivity and genuineness of the " physical phenomena of spiritualism," so as to convert the scientific world at large and open a new era of human advancement.

Let us try to put ourselves into Crookes's mental attitude on both hypotheses. On the spiritualist version he would have gone through a profound spiritual crisis on the death of his brother. He would feel defeated by the powers of darkness. Being a man of great power and resource, he would rebel against the powers of darkness, and would look about for means of defying and defeating them in turn. His victory over death would be assured if he could throw a bridge across the chasm. He had heard that such bridges had been thrown already. Why, then, should he not construct a better bridge, built on scientific principles, a bridge of permanent use to mankind, the greatest feat of bridge-building ever attempted ! Having set to work on this great attempt, he proceeded as he always did in his investigations, collecting and sifting raw material, following every promising track, recovering himself when temporarily lost, and following steadily the beckoning light ahead. That light would no doubt in the end grow into a blaze, and would perhaps reveal the glories of hidden worlds to come !

In the end, having had a glimpse of those glories such as has been given to few mortals, but having entirely failed to exhibit them to his colleagues, and having been overwhelmed with aspersions and ridicule from the public, and quarrels with old friends, he would close the chapter and return to his older avocations, with the pure light of another world shining for ever on his inmost soul.

Such a state of mind would be quite consistent with what we know of Crookes's character and dominant impulses.

How are we, then, to figure to ourselves the mind of the same man on the "rationalistic" hypothesis? We must assume a predisposition in favour of the supernatural, intensified by a grave personal bereavement. Meeting others who had passed through similar crises, and had found consolation in spiritualism, he would try the same path, and, in order to succeed, would throw himself into that passive and devotional attitude alleged to be favourable to the development of spirit manifestations. He would begin, perhaps, in his intimate family circle, and observe table movements and planchette writings under conditions when deliberate deception, even in a family of mischievous young children, would appear as a monstrous absurdity. Having thus obtained a *primâ facie* conviction that "there was something in it," he would go farther, and gradually come into touch with the more advanced exponents of the new cult. He would by then consider himself capable of telling the true from the false, and would go chiefly to those mediums who inspired him with confidence. On his finding such a medium, there would be no limit to his trustfulness, and he would be his convinced and active champion. He would throw himself whole-heartedly into the conflict of opinion then raging, and would try to secure fresh evidence to support his point of view. He would be, as ever, a "bonnie fighter," and be willing to give and take hard knocks. But in the end he would find out his mistake of placing any confidence in the professional medium. He would sicken of the perpetual useless struggle, and would at length realise that science and spiritualism were incompatible and incommensurable, science being of the mind and spiritualism being of the heart. He would return to his first allegiance, and keep his private longings out of sight. He would still give all mediums the benefit of the doubt, and would not retract anything he had said. But he would "pull up short," and refuse to advance farther into the swamp.

This is, I take it, the rationalistic attitude towards Crookes's spiritualism and "psychic force." It is highly desirable that these questions be decided. Was Crookes tricked ? Why did he not go on ? Did he have doubts later ? Was he always a spiritualist ?

The name of Crookes has been used for fifty years to support spiritualism. Hardly a week passes but his name is flourished in the face of a sceptical world, often in support of the grossest fraud. The amount of harm thus done is incalculable, both to the public and to the good name of Crookes. There is no protection for the dead man's memory, for nobody could keep up with the mass of misrepresentation issued every week. There is not a single fraudulent medium who does not habitually reel off half a dozen of the most eminent names in science to support his (or her) pretensions. And the list invariably includes Crookes.

Was Crookes the founder of a new science of the supernatural, or was he an eminent physicist gone wrong ? Are we to venerate his name as the greatest genius of his time, who was the first to show the way towards the promised land of Hereafter, or was he a melancholy example of what a highly trained intellect can become—temporarily at least —under stress of the sorrows of life ?

These questions, important as they are to millions of human beings, born and yet unborn, can hardly be decided yet. But the duty of the biographer is clear. He must collect and collate what authentic information there is ; he must furnish the documents and materials for future examination and judgment, and must refrain from obtruding his private opinions *pendente lite*.

The amount of fresh material I can bring forward

is not large. Much of the material belonging to this period has been destroyed. To give an idea of the amount of destruction, I can furnish the following list of letters written by Crookes between November 1873 and December 1874 which, so far as I know, are lost :

> C. Blackburn, 13 letters.
> J. Blyton, 2 letters.
> Mme. Boydanof, 1 letter.
> Florence Cook, 1 letter (from Mrs. Crookes).
> Florence Cook, 4 letters (from W. Crookes).
> Serjeant Cox, 17 letters.
> Lady Caithness, 1 letter.
> Benjamin Coleman, 5 letters.
> E. E. Corner, 8 letters.
> Mrs. Honeywood, 7 letters.
> W. H. Harrison, 7 letters.
> Miss Kislingbury, 1 letter.
> Epes, Sarjent, 4 letters.
> C. E. Williams, 2 letters.

These seventy-three letters all bore on the events of the period and on the subject we are considering. The originals were copied into a letter-book and indexed. The copies were subsequently removed, but the index was left intact, and the above numbers are extracted from the index. I have been unable to ascertain why and by whom these copies were removed. Some light is shed on the contents of some of the missing letters by extracting the entries of incoming letters from the same persons from a register covering the period 1869 to 1872, which has fortunately been preserved. The extracts come out as follows :

Blackburn, None.
J. Blyton :
> 1871, Sept. 29th, Asking for *Quarterly Journal of Science*
> Sp. articles.
> Oct. 6, Florrie Cook.

1872, May 8, Miss Florrie Cook.
    Aug. 6, Florrie Cook's cabinet.
    Aug. 23, Answer to my letter.
    Sept. 30, For support to D. A. E. Spm.
F. Cook :
    1872, May 8, Séances.
    Sept. 11, Séances for me.
    Sept. 17, Invitation to Mrs. Crookes.
    Sept. 20, Declines invitation. Further experiments.
Serjeant Cox, 79 letters.
B. Coleman, 26 letters.
W H. Harrison, 62 letters.
Mrs. Honeywood, 28 letters.
Miss Kislingbury :
    1870, July 6, Spiritualism.
    July 9, Spiritualism.
    Aug. 4, Adamanta letter.
No letters from the remainder.

None of the above 208 letters have been preserved.

We have already seen that at the time of the Eclipse Expedition Crookes was a convinced spiritualist. He made propaganda for spiritualism on board, and if the entry " Sptl. books " means books on spiritualism, he probably took them with him for the purpose of propaganda. He also, in that beautiful New Year's Eve invocation, asks for a continuance of the " spiritual communications from my brother," and confesses to having got occasional glimpses of the " spiritland."

In order to give definite data bearing on the evolution of Crookes's attitude towards spiritualism, I may cite the following early entries from the " Index to Letters " for 1869–74, being letters received by Crookes.

    1869.
April 25, Attwell, Séance.
Oct. 4, Coleman, Visit to him.
July 9, Dyte, Dialectical Society.

July   15, Edmunds, Dr., Dialectical Committee.
Oct.   5, Guthrie, F., Visit to Coleman's.
Aug.   7, Hart, Miss, Spiritualism and report of events.
July   22, Home, D. D., Medium.
Aug.   5, Hodgson, Spiritualism.
Dec.   7, Home, D. D., Medium.
Dec.   7, Home, D. D., Medium.
Dec.   7, Home, D. D., Medium.
June   28, Hunt, T., Reporter on spiritualism.
June   30, Hunt, T., Reporter on spiritualism.
Aug.   18, Hunt, T., Reporter on spiritualism.
(No date), Mrs. Marshall, Medium.
July   10, Maurice, Mrs. M.
Aug.   1, Maurice, Séance.
Aug.   17, Maurice, Spiritualism.
Aug.   18, Maurice, Séance at Cox's.
Aug.   25, Maurice, Séance at Cox's.
Sept.   19, Maurice, Séance at Cox's.
Oct.   27, Maurice, Séance.
Nov.   8, Maurice, Séance.
Nov.   18, Maurice, Séance.
Dec.   9, Maurice, Sideboard.
Dec.   2, Morse, J. J., Spirit directions.
Dec.   4, Morse, J. J., Séance.
Dec.   4, Morse, J. J., Séance, Overton.
Dec.   16, Morse, J. J., Séance.
Dec.   16, Morse, J. J., Séance.
April   9, Dr. R. A. Smith, Spiritualism.
Dec.   10, Dr. R. A. Smith, Spirit phenomenon, speculations
              respecting visit.
July   19, Webster, Mrs. Spiritualism.
Aug.   14, Weldon, W., Spiritualism.
Aug.   11, Watts, Mrs., Spiritualism.
Sept.   8, Wallace, A. R., Spiritualism.

On arranging these letters according to date, we obtain the following numbers per month :

Jan., 0.	April, 2.	July, 5.	Oct., 3.
Feb., 0.	May, 0.	Aug., 10.	Nov., 2.
March, 0.	June, 2.	Sept., 2.	Dec., 10.

The earliest recorded communication to Crookes on spiritualism is therefore the letter received by him from his friend, Dr. R. Angus Smith, F.R.S., who took an academic interest in spiritualism, and advised Crookes in a letter dated November 6, 1871, to " keep to physics." The first communication from Mr. Daniel D. Home, " medium," appears to date July 22, 1869. In the autumn of 1869, Crookes's friends and relatives appear busily investigating professional mediums, but there is as yet no evidence of any great preoccupation of Crookes himself with the subject. Let us, therefore, continue the extracts into 1870 :

1870.

May	13,	Bird, Alice, Inquiries.
May	27,	Bird, Alice, Spiritualism article, critique.
July	11,	Bird, Alice, Miss Kislingbury.
Feb.	24,	Childs, E., Séance.
April	9,	Childs, G., Spiritualism.
May	1,	Childs, G., Séances.
May	9,	Childs, G., Séances.
July	6,	Childs, G., Burns's attack.
Oct.	4,	Childs, G., Psychic pamphlet.
Feb.	1,	Cox, Serjt., Invitation.
March	31,	Cox, Serjt., Séances.
May	10,	Cox, Serjt., Séance.
May	10,	Cox, Serjt., Acknowledging pamphlet on Spiritualism.
July	28,	Cox, Serjt., Invitation to Moat Mount.
Aug.	4,	Cox, Serjt., *re* Adare's book.
Oct.	7,	Cox, Serjt., Tea and table, Mrs. Edmunds.
Oct.	13,	Cox, Serjt., To meet Mrs. Everett.
Oct.	20,	Cox, Serjt., Discussion of lights and voice (Everett).
Oct.	31,	Cox, Serjt., Invitation to " At Home."
Nov.	5,	Cox, Serjt., To meet Mrs. Guppy.
Nov.	5,	Cox, Serjt., Invitation.
Nov.	5,	Cox, Serjt., To meet him at Everett's.

Oct.	13,	Crookes, Walter, Letter from Dan.
Oct.	19,	Crookes, Walter, Sec. to séance circle, Dan.
Oct.	31,	Crookes, Walter, Miss Fowler, *Echo*.
Nov.	8,	Crookes, Walter, To meet D. D. Home.
Nov.	20,	Crookes, Walter, Séance with Herne and Williams.
Dec.	1,	Crookes, Walter, Séance with Kate Fox.
June	4,	Douglas, Miss, Séance, &c.
Oct.	21,	Douglas, Miss, Séance with Mrs. Guppy.
July	30,	Edmunds, Dr., Visit to Serjeant Cox.
Aug.	9,	Edmunds, Mrs., Séances.
June	12,	Garbutt, Cases of levitation.
July	14,	Grubb, H., Spiritualism.
Aug.	18,	Gill, Miss, Lord Adare's book.
Oct.	8,	Guppy, Mr., Address.
[Jan.	26,	1871, Guppy, Mr., Tyndall to call on him!]
Jan.	17,	Harrison, Spiritualism.
July	11,	Harrison, W., Spiritualism.
Aug.	2,	Harrison, W., Spiritualism.
July	29,	Harrison, W., Spiritualism.
July	29,	Harrison, W., Mrs. Perrin, medium.
July	29,	Hearn, F., Medium, handbill.
July	6,	Home, D. D., Dinner.
July	6,	Honeywood, Mrs., Spiritualism with account of séance.
Aug.	10,	Houghton, Miss, Introduction to Guppy's.
July	1,	Huggins, Spiritualism, queries.
July	4,	Huggins, Wants to investigate.
July	10,	Huggins, Queries.
July	19,	Huggins, Sundry, spiritualism.
Aug.	3,	Huggins, Will visit here at séance, desirable to investigate farther.
Sept.	27,	Huggins, Spiritualism.
Oct.	15,	Huggins, When can he call?
Oct.	27,	Huggins, Miss Douglas's séance.
Nov.	7,	Huggins, Invitation to dinner.
Nov.	9,	Huggins, Invitation, &c.
Nov.	10,	Huggins, Spiritualism, Roscoe, De la Rue, Tyndall, Eclipse invitation.

June   4, Smith, R. A., Spiritualism.
June   5, Smith, R. A., Spiritualism.
June   8, Smith, R. A., Spiritualism.
June   15, Smith, R. A., Spiritualism article, Highlands.
April   6, Spence, Mary S., Spiritualism.
July   5, Spence, Mary S., Spiritualism.
July   11, Stenhouse, Spiritualism.
Aug.   3, Squire, W. S., Spiritualism.
Dec.   22, Tyndall, J., Dialectical Society.
Aug.   6, Tuson, Rapping medium.
July   9, Varley, C. F., Spiritualism.
July   7, Wallace, A. R., My article on Spiritualism.
April 21, White, Wm., Shorter, and spiritualism.
July   15, White, Wm., Shorter, and spiritualism.
July   18, White, Wm., Shorter, and spiritualism.

The number of these letters totals 108, of which 29 were received before June 1, 1870. Adding these 29 letters to the 36 already recorded for 1869, we obtain 65 letters on spiritualism received before Crookes publicly announced his intention of investigating the subject. The announcement, made first in the *Athenæum*, and then more elaborately in the *Quarterly Journal of Science* for July 1870, brought Crookes an enormous increase of correspondence on the subject.

A few words must be said concerning the *Quarterly Journal of Science*, which Crookes made his organ for expounding his views concerning spiritualism. Founded, as we have seen, in January 1864, it was edited jointly by Crookes and J. Samuelson. In 1869 it was under the sole editorship of Samuelson, with Crookes's name appearing on the publication committee. The situation in May 1869 is illustrated by the following rather pungent letter:

May 12th, 1869.

Dear Mr. Samuelson,

I am in receipt of your letter of the 10th inst., in which you say that Mr. Longman and yourself have had some conversation about my name appearing in the *Quarterly Journal of Science*, and "in conformity with my wish it will be withdrawn at the end of the year." You entirely mistake my meaning. I have no wishes one way or the other on the subject. If Messrs. Longmans think my name is of no value to the *Journal*, why retain it for three more numbers ? If on the contrary they think it is of value they are perfectly welcome to make use of it beyond the close of this year. This I have already told Mr. Longman.

What I do complain of is the way in which you and your friends have played fast and loose with me. The position was none of my seeking. *You* asked *me* to join you in the editorship, and for five or six years that union has been prominently mentioned in every advertisement. During this time I have certainly exerted myself to the utmost, and can point to a long list of valuable articles written by personal friends of my own. That my services were once appreciated by yourself and friends I may conclude from the present of shares, which you evidently think of great value. Judge my astonishment, then, when I recently found out that you and your friends had passed a resolution in October last : "That the editorship and literary management of the *Journal* shall be under the sole control of Mr. James Samuelson, of Liverpool " ! A most ungracious resolution, and one which can only be regarded as a deliberate insult offered to myself. To make matters worse, if possible, in the letter in which I first receive intimation of this resolution, you tell me that my name is now upon the *Journal* "as a matter of good feeling and courtesy," and that if I wish to withdraw I may. Viewed by the light of that resolution, no course was open to me but to take the hint thereby conveyed.

Now, I want to know in what position I stand. I am not a child to be frightened by a resolution of a board of directors, or to be soothed by the present of shares (*each carrying £10 liability*) in a company which has been losing money every year from its commencement, and which nothing short of a miracle could make pay a dividend as hitherto managed ; neither is my position in Scientific Circles here such as to render my retaining the editorship of your *Journal* of any advantage to me—indeed, I am vain enough to think that I confer more than I receive. If Messrs. Longmans and yourself wish me to retire at once, say so, but if you wish me to stay I must in common justice stipulate that the obnoxious resolution be rescinded, and my name recognised by the present proprietors in such a way as will obviate the necessity of re-opening this discussion six months hence, and will render our joint positions equal in stability as they have hitherto been in responsibility.

Truly yours,

WILLIAM CROOKES.

This letter bears all the appearance of a sharp quarrel, but the matter was eventually settled quite amicably by Crookes taking entire charge of the editorship.

He was, therefore, in sole control when he published his four articles on spiritualism. The first of these, called " Spiritualism Viewed by the Light of Modern Science," was published in July 1870. He says :

Some weeks ago, the fact that I was engaged in investigating spiritualism, so-called, was announced in a contemporary, and in consequence of the many communications I have since received, I think it desirable to say a little concerning the investigation which I have commenced. Views or opinions I cannot be said to possess on a subject which I do not pretend to understand. I consider it the duty of scientific men who have learnt exact modes of working, to examine phenomena which attract the attention of the public, in order to confirm their genuineness, or to explain if possible the delusions of the honest and to expose the tricks of deceivers. But I think it a pity that any public announcement of a man's investigation should be made until he has shown himself willing to speak out.

Criticising the claims of spiritualists, Crookes proceeds :

The spiritualist tells of manifestations of power, which would be equivalent to many thousands of " foot-pounds," taking place without known agency. The man of science, believing firmly in the conservation of force and that it is never produced without a corresponding exhaustion of something to replace it, asks for some such exhibitions of power to be manifested in his laboratory, where he can weigh, measure, and submit it to proper tests.[1]

[1] In justice to my subject, I must state that, on repeating these views to some of the leading " spiritualists " and most trustworthy " mediums " in England, they

For these reasons and with these feelings I began an inquiry suggested to me by eminent men exercising great influence on the thought of the country. At first, like other men who thought little of the matter and saw little, I believed that the whole affair was a superstition, or at least an unexplained trick. Even at this moment I meet with cases which I cannot *prove* to be anything else ; and in some cases I am sure that it is a delusion of the senses.

I by no means promise to enter fully into this subject ; it seems very difficult to obtain opportunities, and numerous failures certainly may dishearten anyone. The persons in whose presence these phenomena take place are few in number, and opportunities for experimenting with previously arranged apparatus are rarer still. I should feel it to be a great satisfaction if I could bring out light in any direction, and I may safely say that I care not in what direction. With this end in view, I appeal to any of my readers who may possess a key to these strange phenomena, to further the progress of the truth by assisting me in my investigations. That the subject has to do with strange physiological conditions is clear, and these in a sense may be called " spiritual " when they produce certain results in our minds. At present the phenomena I have observed baffle explanation ; so do the phenomena of thought, which are also spiritual, and which no philosopher has yet understood. No man, however, denies them.

The explanations given to me, both orally and in most of the books I have read, are shrouded in such an affected ponderosity of style, such an attempt at disguising poverty of ideas in grandiloquent language, that I feel it impossible, after driving off the frothy diluent, to discern a crystalline residue of meaning. I confess that the reasoning of some spiritualists would almost seem to justify Faraday's severe statement that many dogs have the power of coming to much more logical conclusions. Their speculations utterly ignore all theories of force being only a form of molecular motion, and they speak of Force, Matter, and Spirit, as three distinct entities, each capable of existing without the others ; although they sometimes admit that they are mutually convertible. . . . The increased employment of scientific methods will promote exact observation

express perfect confidence in the success of the inquiry, if honestly carried out in the spirit here exemplified ; and they have offered to assist me to the utmost of their ability, by placing their peculiar powers at my disposal. As far as I have proceeded, I may as well add that the preliminary tests have been satisfactory.

and greater love of truth among inquirers, and will produce a race of observers who will drive the worthless residuum of spiritualism hence into the unknown limbo of magic and necromancy.

The real beginning of Crookes's systematic inquiry into spiritualism for publicity purposes we may place in the summer of 1869. Dr. Angus Smith mentioned the subject in a letter in April. In June, Mr. T. Hunt commenced a series of reports on spiritualism. In July 1869, D. D. Home—greatest of mediums—arrived in London from St. Petersburg, with a letter of introduction to Crookes from the professor of chemistry at St. Petersburg University. In the same month " Maurice " arranged a number of séances with a Mrs. Marshall as medium, at some of which Crookes attended. In September, Dr. A. R. Wallace —the co-discoverer, with Charles Darwin, of Evolution— wrote to Crookes about spiritualism. In December 1869 Crookes sat with the medium J. J. Morse.

The early months of 1870 saw much further activity. Serjeant Cox organised some séances in March, and G. Childs arranged for others in April. We find Miss Alice Bird, Crookes's literary and scientific ally and faithful friend, discussing spiritualism with him in May 1870.

When the *Quarterly Journal of Science* article appeared, Crookes immediately heard from Mr. W. H. Harrison, the editor of *The Spiritualist*, who promised active co-operation. Mr. Herne, the medium, sent him a handbill announcing his séances. Dr. Huggins, F.R.S., wrote numerous letters to Crookes, and interested a number of other prominent men in the subject. Incidentally, he invited Crookes to join the Eclipse Expedition (letter of November 10th). Crookes must have found the supply of mediums an *embarras de richesse*. Soon he was able to write a glowing account of a dark séance with three mediums :

A DARK SÉANCE

April 12, 1871.

My dear Mr. Huggins,

We had the most exciting and satisfactory meeting last night I have ever known, and I would have given anything for you and Dr. Ord to have been there. In the afternoon the celebrated D. D. Home came here. He stayed to dinner, and then I took him with us to Russell Square, knowing that he would be very welcome.

You know that it is universally agreed upon by spiritualists that the phenomena are better in darkness than in light, but Home always refuses to sit in the dark, as he says it is not so satisfactory to those present. On this occasion, however, we induced him to join our dark séance as the phenomena with Herne and Williams are not strong in the light. We were arranged round the table in such a way that each medium was held by a trustworthy person and the rule was *very rigidly* enforced that all hands were to be held during the darkness. This was so strictly carried out that when any of us wanted to use his handkerchief or get his chair, a light was struck. At first we had very rough manifestations, chairs knocked about, the table floated about 6 inches from the ground and then dashed down, loud and unpleasant noises bawling in our ears and altogether phenomena of a low class. After a time it was suggested that we should sing, and as the only thing known to all the company, we struck up " For he's a jolly good fellow." The chairs, table, and things on it kept up a sort of anvil accompaniment to this. After that D. D. Home gave us a solo—rather a sacred piece—and almost before a dozen words were uttered Mr. Herne was carried right up, floated across the table and dropped with a crash of pictures and ornaments at the other end of the room. My brother Walter, who was holding one hand, stuck to him as long as he could, but he says Herne was dragged out of his hand as he went across the table. Mrs. W. Crookes, who was at the other side of the corner, kept hold all the time.

This was repeated a second time, on Home's singing again. Both mediums this time being lifted up and placed on the table. Hands being held throughout.

This seemed to entirely alter the character of the manifestations. Home's singing appeared to drive away the low-class influences and institute his own good ones. After a minute or two I suggested that we should all sing again, and proposed the song first sung, " For he's a jolly good fellow." Immediately a very sweet voice, lligh over our heads quite out of reach of anyone present even had they been standing, and as clear as a bell, said, " You should rather give praise to God." After that we were in no mood for comic songs. We tried something sacred, and as we sung we heard other voices joining in over our heads.

Then the accordion was lifted up from the table (all holding hands) and it floated about the room, sometimes going far away outside the circle and then coming gently on to some of our heads and hovering within an inch or two of our faces, and all the time playing one of the most exquisite sacred pieces I ever heard, and being accompanied by a very fine male voice. The rapidity of the movement of this instrument was most astonishing. It really seemed to be in two places at once. It came and played on my hand. I had not time to utter the words " It's on my hand," when a person at the other end of the table 9 feet off called out the same thing. Frequently this happened, and as it was playing all the time we could tell how rapidly it moved by the direction of the sound.

Then voices came and addressed us. Not rough and frightful ones like those Herne brings, but very sweet ones, whispering close into our ears in such a manner as to preclude the possibility of the mediums tricking us. One especially kept hovering about some of the company away from the mediums, and it whispered close into my wife's ear and then went over her head to the other ear, still speaking. I was served the same twice. The little hand bell was then lifted up and rising about the table, about 18 inches above it. Three persons said they saw it moving, by a luminous cloud above it, and Home said he saw a hand holding it. After that lights appeared darting about with great rapidity and leaving a tail for a fraction of a second. I saw these distinctly, as did everyone else, but on many occasions when lights appeared on persons' heads, only about half the company saw them. My eyes appeared amongst the least sensitive to these lights, but what I did see was unmistakable.

Altogether we counted about seven distinct voices.

As the evening got on the power increased, and hands came amongst us. Serjt. Cox had a book taken from his pocket, and whilst it was being removed he liberated one of his hands (joining the hands of those on each side of him and clasping the two with his other hand, so as not to leave any other person's hand free) and he caught the fingers in the act of removing his book. It was *only a hand*, there being no arm or body attached to it, and it eluded his grasp and carried the book right across the table, where it was gently laid on my wife's hand. Then hands came to nearly all of us, faces were stroked and our hands patted and on some occasions the fingers lingered long enough to admit of being felt. On several occasions I made rapid darts in front, trying to catch the arm when the fingers were touching near me, but not once did I touch anything. Things were then carried about the table from one to another. Serjt. Cox's gloves were shaken in all our faces. Home's handkerchief was gently laid on our heads, shoulders, and hands, and then gently removed and carried elsewhere.

All this time we had notes on the accordion and voices talking. On two or three occasions there was more work of this sort going on than could have been accomplished by the three mediums present, *even had they been quite free* and trying to deceive us. Thus we had two voices, the accordion moving and playing, the bell ringing, the handkerchief moving, and fingers touching us—all in different parts of the table, whilst the mediums were conversing quietly in their chairs in their own natural voices.

I feel it is impossible to describe to you all the striking things that took place, or to convey the intense feeling of genuineness and reality which they caused in our minds, but I want you to come and attend at another séance which is appointed for next Tuesday week, the 25th inst., at Russell Square, when Home has promised to come, and we are going to try and get the same party and if possible the same conditions. You must, however, prepare for the chance of a failure. Home was in wonderful power last night, but he is the most uncertain of mediums, and it is quite as likely that the next time absolutely nothing will take place. Still the combination of the three mediums ought to be battery power enough.

I won't say Excuse this long letter, for I think you will not mind the trouble of reading it ; neither will I say keep it strictly private. At the same time please be careful to whom you show it, for I neither wish to be shut in a lunatic asylum, nor to be turned out of scientific society. I am writing this, to copy, for the sake of recording my own impressions when fresh on my mind ; but Serjt. Cox has undertaken to draw up a full report of the meeting which we shall all go over, correct, and sign.

Believe me,
Very truly yours,
WILLIAM CROOKES.

W. HUGGINS, Esq., F.R.S.

Crookes's anxiety about being "turned out of scientific society" was well-founded. The majority of his colleagues were quite unprepared to accept the phenomena as genuine, and, for that matter, the majority of scientific men are in the same position to-day, fifty years afterwards. In trying to forge a link connecting science and the supernatural, Crookes failed completely. Nor did he succeed in the alternative task of extending natural law to the spiritual world, any more than Drummond succeeded fifteen years later. The two worlds remained incommensurable, just as they co-existed in the inner minds of Crookes's great critic and friend, G. G. Stokes, and in Crookes's formidable opponent, W. B. Carpenter.

Just about this time a great fillip was given to spiritualism by the report of a committee of the London Dialectical Society specially appointed to investigate the subject. This Committee included many eminent men, notably Professor De Morgan, the mathematician. Its moving spirit was, however, Serjeant Cox, and it was he, probably, who drew up the report.

The Committee concluded that " taking into consideration the high character and great intelligence of many of the witnesses to the more extraordinary facts, the extent to which their testimony is supported by the reports of the sub-committees, and the absence of any proof of imposture or delusion as regards a larger portion of the phenomena ; and further, having regard to the exceptional character of the phenomena, the large number of persons in every grade of society and over the whole civilised world who are more or less influenced by a belief in their supernatural origin, and to the fact that no philosophical explanation of them has yet been arrived at, they deem it incumbent upon them to state their conviction that the subject is worthy of more serious attention and careful investigation than it has hitherto received." [1]

This famous report, when closely examined, does not amount to much. Practically it says that there are many spiritualists in the world, many of them very intelligent and some of high social standing ; that spiritualism is a religion rather than a science ; and that the alleged phenomena are worth investigating. A modern committee of the same composition would probably come to exactly the same conclusions, though some of its members might demur to the third phrase, in view of the paucity of results achieved since 1871.

That Crookes was strengthened in his resolve to pursue the subject by this report there can be little doubt. But at the time with which we are dealing, spiritualism was very much " in the air " in England. The first English spiritualist journal, the *Yorkshire Spiritual Telegraph*, had been founded at Keighley, in Yorkshire, a town with which Crookes had much to do on questions of water supply. The horrors of the Franco-German War brought, on a smaller scale, a reaction resembling that produced by the European War of 1914–18. And from America there was a steady influx of mediums, coming as propagandists and apostles of the new religion, which many of them regarded as a modern divine revelation.

In criticism of the séance so glowingly described to Huggins, it should be mentioned that Herne and Williams were some years afterwards convicted of trickery. The phenomena described as " exciting and satisfactory " are absolutely devoid of any evidential value, and only indicate the clumsy frauds which two, at least, of the mediums were subsequently found to practise. In reading the account, we do not recognise our Crookes. There seem

---

[1] *Report on Spiritualism*, London, 1871 (Longmans & Co.).

to be two Crookeses : One, the conscientious, painstaking, accurate man of science, who excels in weighings and measurings ; the other, an impulsive, excitable hunter after miracles, whose caution and common-sense are bowled over completely in the presence of some tricksters and some devotees of a fashionable craze. One feels some gratitude towards the Spillers, Carpenters, and Ray Lankesters for their trenchant opposition, without which Crookes might never have got clear of imposture. Their criticism forced him to use scientific methods in his investigations, and applied a much-needed brake to his ardour.

D. D. Home must have been a very winning personality. Those who knew him well speak of him in very affectionate terms. The Crookes family were devoted to him. The friendship of wealthy society ladies sometimes proved embarrassing, as when, in 1866, a Mrs. Lyon, impressed by Home's psychic powers, " adopted " him as her son, lavished over £30,000 on him, and then, in a fit of temper, compelled him to return it. At the time of Crookes's first study of him, he was a young widower engaged to the sister of Crookes's Russian correspondent, Professor Boutlerow, of St. Petersburg. Crookes wrote to the latter two days after the séance above described as follows :

April 13*th*, 1871.

PROFESSOR A. BOUTLEROW,
St. Petersbourg.
MY DEAR SIR,
Your letter of the 5th inst. arrived yesterday, and before answering it I thought it better to ask one or two questions of a lady—a member of our English aristocracy—at whose house I have met Mr. D. D. Home. This I did without breaking the confidence you have reposed in me.

The life's happiness of a young lady is so serious a matter, that even had we not been acquainted through our common science,

I should have felt it my duty to pay attention to your request ; but coming as it does from one whom I have known so long and honourably by reputation, I think it best to write somewhat fully on the subject ; and it is most gratifying to me to be able to speak my honest convictions about Mr. Home, without at the same time feeling that what I write may throw obstacles in the way of his future domestic happiness.

Mr. Home has already mentioned to me and a few of his intimate friends that he hoped shortly to become your brother-in-law ; and as he is on such intimate relations with your family, it is probable that you know even better than I do his pecuniary position, and the reasonable expectations which he has of soon receiving the fortune coming to him from his first wife.

As far as Mr. Home's character is concerned, I thoroughly believe in his uprightness and honour ; I consider him incapable of practising deception or meanness. I had not the pleasure of knowing him until some time after his first wife's death, but his nature appears to me to be peculiarly domesticated and affectionate.

The society in which Mr. Home moves in London is very varied. From the fascination of his manner, as well as the wonders of his mediumship, his company is eagerly sought by all classes. He is too good-natured to refuse to give séances when he thinks that there is a chance of his advancing the cause of spiritualism, and he seldom leaves a house without having converted those, in whose company he has been, into firm friends. I should therefore say that no one has a larger circle of genuine friends than Mr. Home. It is true that they are principally amongst those who have had their attention turned somewhat to spiritualism, but these constitute a large class in England, embracing many of the nobility and gentry.

As regards Mr. Home's private life and conduct, I have seen nothing which would tend to shake the high opinion I have formed of him. He lives in comfortable apartments at the west end of London, and I have called upon him frequently at all hours of the day. In the intimate conversation of young men associating together as bachelors a considerable latitude of speech is frequently indulged in, but although I have had ample opportunities of observing him with other young men, I have never heard him utter a word which could not be repeated to a lady. On the contrary the tone of high religious morality which appears to pervade him, effectually checks all approach to light conversation.

I think, and this opinion is confirmed by what I heard yesterday, that Mr. Home's social position would be improved were he to be married. His many friends are now somewhat embarrassed to know how they can make a fitting acknowledgment to him for the charms of his company, and his readiness to place his mediumship at their disposal for séances. We can do so little in the way of return, except ask him to dinner, and this only adds to the obligations we are under. Had he a wife, however, I am sure that every one of his friends would be eager to show kindness to her. I am assured that fifteen or twenty ladies—either titled or closely related to the aristocracy—would at once call upon Mrs. Home, and this fact alone would give her an assured position in English society.

For my own part, although my own social position is cast in a lower sphere than that of many of Mr. Home's sincere friends, I can only say that my wife and sister will be as pleased to welcome Mrs. Home to our family circles as we have been to receive him amongst us ; whilst the fact that the lady is the sister of a fellow-worker in a cherished science, will be, I hope, an additional tie between Mr. Home and myself.

Since I wrote my article in the *Quarterly Journal of Science*, I have not had many opportunities of pursuing any experiments. All that I have since seen have carried my opinions further towards those held by advanced spiritualists, but before I publish anything I want to confirm what I say by an appeal to experimental evidence. Now Mr. Home has returned to London I hope to have opportunities of trying many experiments and also giving other scientific men the opportunity of witnessing these things for themselves.

Mr. Home told me of some very interesting experiments which you tried in his presence with a dynamometer and a thermometer. I should feel greatly obliged if you would send me over a detailed account of these results. I am going to try the same things the first opportunity.

With kind regards,

Believe me very sincerely yours,

WILLIAM CROOKES.

PROFESSOR A. BOUTLEROW,
    Laboratoire de Chimie de l'Université de
        St. Petersbourg.

The next day a letter appeared in the *Standard* saying that Home, during his visit to St. Petersburg, had been tried by a committee of scientific men, with purely negative results. Crookes wrote at once to defend his friend, and based his defence on a passage of one of Boutlerow's letters, saying that he (Boutlerow) had been present at a number of Home's séances and considered him strictly honest.

About this time Crookes wrote to Professor (now Sir William) Barrett, of the Royal College of Science, Dublin, whom he had met in the company of Faraday and Tyndall :

MASBRO' HOUSE,
BROOK GREEN, W.
May 15*th*, 1871.

MY DEAR SIR,

I am staying here on a visit and shall not return home till the end of this week.

I must have some conversation with you respecting these obscure phenomena. If you could help me to form anything like a physical theory I should be delighted. At present all I am quite certain about is that they are *objectively* true.

I have had all my wits about me when at a séance, and the only person who appeared to be in a state of semi-consciousness is the medium himself. The other evening I saw Home handling red-hot coals as if they had been oranges.

When I return I will ask you to favour me with a visit some evening. When are you disengaged ?

Believe me,
Very truly yours,
WILLIAM CROOKES.

Meanwhile, some of Crookes's friends were getting seriously alarmed about his " infatuation " for spiritualism and his friendship with Home. His old collaborator, John Spiller, sent him a friendly warning, which, however, Crookes took very much amiss. The latter wrote :

MYSTERIOUS FORCES AND APPARITIONS

May 24th, 1871.

MY DEAR JOHN,

Thank you for the cutting from the *Daily News* which arrived yesterday morning I would have answered your letter before but I had to be at Westminister early in the morning and was detained in court all day giving evidence.

We are very sorry to hear of Ada's indisposition and hope she will soon recover.

You reminded me a few days ago of our friendship of nearly a quarter of a century's standing. You have now for the third time given a very mysterious hint that you are in possession of a fact which would make me entirely alter my opinion about Mr. Home. Now I put it to you whether it would not be more consistent with that friendship for you to tell me fairly and definitely what you do know, rather than keep me in suspense week after week. You say it is *impossible* for you to write about it. That is a word I do not understand.

If you will give me a plain statement of facts and will not insinuate dishonest conduct on the part of myself and family, I promise you that I shall not only be very grateful to you, but will give what you tell me the most serious attention.

I ask myself : Can this be my old friend who has worked with me in unfolding the mysteries of photography ; who has helped me to track out and give to the world a new element ; and who now thinks that the anonymous scribble of an ignorant penny-a-liner will induce me to *sift* my evidence before accepting phenomena as facts !

Good heavens ! one of us must have strangely altered before advice like that could pass between us. Have I ever shown haste in forming an opinion ? Have I ever admitted a new fact in science on insufficient testimony ? Have I not rather shown undue caution in requiring that every step of an investigation should be probed in the most absolute and irrefragable manner before believing it to be true ? Have I, in all the years you have known me, made one false step in science, or had to withdraw a single statement ? Have I not ground and winnowed and sifted evidence to such a degree that along with the bad I have thrown away much of the good because it was not good enough, and only held firm to the small residuum of absolute truth ? And now I am asked to *sift evidence* on the strength of a newspaper writer's crude views, and I can still sign myself

Your old friend,

WILLIAM CROOKES.

The above dramatic letter breathes an intense conviction regarding the genuineness of the phenomena witnessed, and a consequent impatience of criticism even from friendly sources. It contains a hint of broken friendship, and is written with an air of conscious pride in past achievements. It implies that the methods which sufficed to eliminate error in the isolation of thallium would not fail to reveal sources of error in the emotional atmosphere created by the fascinating personality of D. D. Home.

The investigations proceeded. The notes regarding a séance of May 31, 1871, have been preserved. They are important as forming the basis of the article on the " Experimental Investigation of a New Force," published in the *Quarterly Journal of Science* of July 1, 1871.

MEMORANDA OF A SITTING WITH MR. D. D. HOME, AT 20, MORNINGTON ROAD, ON MAY 31ST, 1871.

From 9.15 p.m. to ——

In the Dining Room, round a table weighing 140 lbs. and supported on four legs. The room was lighted by a gas burner giving a light equivalent to about 5 standard sperm candles.

On the table were placed paper, pencils, a galvanometer, delicate thermometer, an accordion weighing 2¾ lbs., a small hand bell, and an electro-magnet. Under the table was placed a large drum-shaped cage consisting of two wooden hoops, respectively 1 foot 10 inches, and 2 feet diameter, connected together by twenty narrow laths 1 foot 10 inches long, so as to form a drum-shaped cage-frame ; round this 48 yards of insulated copper wire were wound in 23 rounds, each being rather less than an inch apart. These horizontal rounds of wire were then netted together firmly with string so as to form meshes about 2 inches long by 1 inch high. This cage was open top and bottom. The accompanying picture[1] shows the general arrangement. The height was such that it would just slip under the table, but be too close to the top to allow of the hand being introduced into the interior or to admit of a foot being pushed underneath it. In another room were two Grove's cells, wires

[1] Not reproduced here.

being led from them into the dining-room for connection with the wire surrounding the cage if desirable.

The accordion was quite new, having been purchased for these experiments at Wheatstone's in Conduit Street. Mr. Home had neither seen nor handled the instrument before the commencement of the said test experiments.

At the side of the table an apparatus was fitted up consisting of a mahogany board   inches long by   inches wide and 1 inch thick. At each end a strip of mahogany was screwed on, forming feet. One end of the board was supported on the edge of the table, whilst the other end was supported by a spring balance hanging from a substantial tripod stand. The spring balance was fitted with a self-registering index in such a manner that it would record the maximum weight indicated by the pointer. The apparatus was adjusted so that the mahogany board was horizontal, its foot resting flat on the table ; the accompanying picture shows this arrangement.

Before Mr. Home entered the room the apparatus had been arranged in position and he had not even had the object of some of it explained before sitting down. It may perhaps be worth while to add for the purpose of anticipating some critical remarks which are sure to be made, that I called for Mr. Home at his apartments, and when there he suggested that as he had to change his dress, perhaps I should not object to continue a conversation we were having in his bedroom. I am therefore enabled to state positively that no machinery, apparatus, or contrivance of any sort except what usually appertains to a gentleman in evening costume was secreted about his person.

There were present in the room :

Mr. D. D. Home (medium)

Dr. Huggins, F.R.S.	Mr. W Crookes
Mr. Serjeant Cox	Mrs. W. Crookes
Mr. Crookes	Mrs. Humphrey
Mrs. Crookes	Miss Crookes

Mr. Gimingham (chemical assistant).

Mr. Home sat in the centre of the long side of the table, close in front under the table was the cage, one of his legs being on each side ; close to him, on his left, was Mr. Crookes, and on his right Mrs. W. Crookes ; the rest of the party were seated at convenient distances round the table.

AN ACCORDION EXPERIMENT

For the greater part of the evening when anything of importance was going forward, Mrs. W. Crookes and I kept our feet respectively on the adjacent foot of Mr. Home.

Almost immediately after we sat down, loud raps were heard from different parts of the table which was also slightly moved in various directions. Several present now felt a cold air over their hands, Dr. Huggins especially. The temperature of the room was 70·5°. Mr. Gimingham said he felt his foot touched and subsequently his chair moved slightly. Mr. Home now took the accordion between the thumb and middle finger of his right hand at the opposite end to the keys (see picture—to save repetition this will be subsequently called " in the usual manner "). I having previously opened the bass [?][1] key, and the cage being drawn from under the table just so as to allow the accordion to be passed in keys downwards, it was pushed back as far as possible, but without hiding Mr. Home's hand from those next to him. The table cloth in front of Mr. Home was also turned up the whole time. Very soon the accordion was seen by those on each side to be moving about in a somewhat curious manner, but no sound was heard, and its weight being fatiguing, Mr. Home put it down.

In a few minutes Mr. Home said he felt something moving in the cage and he thereupon again held the accordion as before and it very soon commenced moving, then sounds came from it and finally several notes were sounded in succession. While this was going on Mr. Gimingham got under the table, and said he saw the accordion expanding and contracting. At the same time I saw that Mr. Home's hand which held it was quite still, his other hand resting on the table in the sight of all present, the music now ceased and raps were again heard from under the table apparently on the cage.

In course of the general conversation which had been going on Mr. Serjeant Cox referred to an incident which had occurred a few evenings before, at a séance with Mr. Home, at his own house connected with a deceased daughter named Florence. Five loud raps were immediately heard asking for the alphabet and the following message was given.

" I will do it again, you dear old Chinchilla."

Serjeant Cox said he did not understand or remember the last appellation when it was spelt out : " Write and ask R. G., she

[1] Illegible.

remembers." He replied : " Do you mean Rosa Gill ? " " Yes." [1]

The question was asked whether the spirit who gave these messages could sound the accordion, immediately this answer was given : " A man's spirit has closed the bass note."

On examining the accordion I found that the bass note was closed as stated. It was open when the accordion was playing a few minutes before and could not have got closed by any accidental striking of the instrument against the cage or floor. The bass key was then opened and the accordion replaced in the cage, Mr. Home holding it in the usual manner. It immediately, seen by those on either side, commenced to move about, oscillating and going round and round the cage in a very striking manner and playing at the same time. Dr. Huggins was now looking under the table, and said that Mr. Home's hand was quite still and the accordion was moving about, emitting distinct sounds.

A remark having been made that it might be possible to produce sounds by pressing the accordion against the floor, the remark was spelt out :

" You can see by the position of Dan's arm that he does not touch the floor."

We then saw that he could not have pressed the instrument against the floor without stooping very much and thus being instantly detected.

Mr. Home, still holding the accordion in the usual manner in the cage, his feet being held by those next him and his left hand resting on the table, we heard distinct and separate notes sounded in succession and then a simple air was played. This was considered by those present to be a crucial experiment, as such a result could only have been produced by the various keys of the instrument being acted upon in harmonious succession ; but the sequel was still more striking, for Mr. Home actually left go of the accordion, brought his hand quite out of the cage, and took hold of Mrs. W. Crookes's hand, the accordion continuing to play whilst no one was touching it. In a few seconds Mr. Home replaced his hand in the cage and again took hold of the accordion, which he said bobbed up against his hand.

Dr. Huggins, who at the early part of the sitting had complained of great cold, so much so that he had buttoned up his coat, now

[1] See extract from Serjeant Cox's letter.

complained of the heat being very oppressive and left the room for a few minutes. We took this opportunity to open the window and connect the insulated wire surrounding the cage with the terminal wires from the two Grove's cells. On Mr. Huggins's return Mr. Home held the accordion inside the cage in the same manner as before, when it immediately sounded and moved about vigorously. I here remarked that the electric current passing round the cage certainly appeared to assist, when the following message was spelt out :

"We can see it, but it does not aid us."

The accordion was now taken by an unseen power from Mr. Home's hand, which he brought quite away—Mr. and Mrs. W. Crookes as well as myself not only seeing his released hand but also the accordion floating about with no visible support inside the cage, this was repeated a second time. Mr. Home presently reinserted his hand in the cage and took hold of the accordion whilst I read aloud a few extracts from Lord Dunraven's Introduction. The accordion commenced to play, at first chords and runs, and afterwards went through the air of " Home, sweet Home." Whilst this tune was being played, I took hold of Mr. Home's right arm below the elbow and gently slid my hand down until it touched his hand and the top of the accordion ; he was not moving a muscle. The tune ceased after I touched the instrument, his left hand was on the table, visible to all, and his feet were under the feet of those next him.

Someone here remarked how beautifully the tune had been played. Immediately five raps were heard and the following given :

"Minus one note, broken." "Accident."

In a few minutes the message was given :

"We are unable to do more."

I, however, urged that we should at all events try an experiment with the spring balance apparatus which I had fitted up (see picture). Mr. Home accordingly placed the tips of his fingers lightly on the extreme end of the mahogany board which was resting on the table whilst Dr. Huggins and I sat one on each side of it watching for any effect which might be produced. Very soon the pointer of the balance descended and rose again, the end of the board oscillating slowly several times.

Mr. Home now of his own accord took a small hand bell and a little card box and placed one under each hand, to satisfy us, as he said, that he was not producing the downward pressure. The very slow oscillation of the spring balance became more marked, and Dr. Huggins, on watching the index, said that he saw it descend to 6¼ lbs., the normal weight of the board as so suspended being 3 lbs., the additional downward pull being therefore 3¼ lbs.

On looking immediately afterwards at the automatic register, we saw that the index had at one time descended as low as 9 lbs., showing a maximum pull of 6 lbs.

In order to see whether it was possible to produce much effect on the spring balance by pressure at the place where Mr. Home's fingers had been, I mounted the table and stood on one foot at that end. Dr. Huggins, who was observing the index of the balance, said that the whole weight of my body (140 lbs.) so applied, only sunk the index to 1½ lbs. or to 2 lbs. when I jerked up and down. Mr. Home was sitting in a low easy chair, and could not therefore, had he tried his utmost, have exerted any material influence on these results. I need scarcely add that his feet as well as his hands were visible to all in the room.

This experiment appears to me to be, if possible, more striking than the one with the accordion, as will be seen on referring to the accompanying picture. The board was arranged perfectly horizontally, and it was particularly noticed that Mr. Home's fingers were not at any time advanced more than 12 inches from the extreme end, as shown by a pencil mark, which with Dr. Huggins's acquiescence, I made at the time. Now the foot being also 1½ inches wide and resting flat on the table, it is evident that no amount of pressure exerted within this space of 1½ inches, could produce any action on the spring balance. Again it is also evident that when the end furthest from Mr. Home sank, the board would turn on the further edge of this foot and would slightly raise his fingers, were he therefore to have exerted a downward pressure, it would have been in opposition to the force which was pulling the other end of the board down. The arrangement was therefore that of a see-saw (36 inches in length), the fulcrum being 1½ inches from one end.

The slight downward pressure shown on the balance when I stood on the board was owing to my foot extending beyond this fulcrum.

When the *Quarterly Journal of Science* article was in

type, Crookes received the following letters, which he appended to the article. Dr. Huggins wrote :

UPPER TULSE HILL, S.W.
June 9, 1871.

DEAR MR. CROOKES,

Your proof appears to me to contain a correct statement of what took place in my presence at your house My position at the table did not permit me to be a witness to the withdrawal of Mr. Home's hand from the accordion, but such was stated to be the case at the time by yourself and by the person sitting on the other side of Mr. Home.

The experiments appear to me to show the importance of further investigation, but I wish it to be understood that I express no opinion as to the cause of the phenomena which took place.

Yours very truly,
WILLIAM HUGGINS.

WM. CROOKES, Esq., F.R.S.

Serjeant Cox wrote :

36, RUSSELL SQUARE,
June 8, 1871.

MY DEAR SIR,

Having been present, for the purpose of scrutiny, at the trial of the experiments reported in this paper, I readily bear my testimony to the perfect accuracy of your description of them, and to the care and caution with which the various crucial tests were applied.

The results appear to me conclusively to establish the important fact, that there is a force proceeding from the nerve-system capable of imparting motion and weight to solid bodies within the sphere of its influence.

I noticed that the force was exhibited in tremulous pulsations, and not in the form of steady continuous pressure, the indicator rising and falling incessantly throughout the experiment. This fact seems to me of great significance, as tending to confirm the opinion that assigns its source to the nerve organisation, and it goes far to establish Dr. Richardson's important discovery of a nerve atmosphere of various intensity enveloping the human structure.

207

MYSTERIOUS FORCES AND APPARITIONS

Your experiments completely confirm the conclusion at which the Investigation Committee of the Dialectical Society arrived, after more than forty meetings for trial and test.

Allow me to add that I can find no evidence even tending to prove that this force is other than a force proceeding from, or directly dependent upon, the human organisation, and therefore, like all other forces of nature, wholly within the province of that strictly scientific investigation to which you have been the first to subject it.

Psychology is a branch of science as yet almost entirely unexplored, and to the neglect of it is probably to be attributed the seemingly strange fact that the existence of this nerve-force should have remained so long untested, unexamined, and almost unrecognised.

Now that it is proved by mechanical tests to be a fact in nature (and if a fact, it is impossible to exaggerate its importance to physiology and the light it must throw upon the obscure laws of life, of mind, and the science of medicine) it cannot fail to command the immediate and most earnest examination and discussion by physiologists and by all who take an interest in that knowledge of " man," which has been truly termed " the noblest study of mankind." To avoid the appearance of any foregone conclusion, I would recommend the adoption for it of some appropriate name, and I venture to suggest that the force be termed the *Psychic Force* ; the persons in whom it is manifested in extraordinary power *Psychics* ; and, the science relating to it *Psychism*, as being a branch of *Psychology*.

Permit me, also, to propose the early formation of a *Psychological Society*, purposely for the promotion, by means of experiment, papers, and discussion, of the study of that hitherto neglected Science.

I am, &c.,

EDWD. WM. COX.

To W. CROOKES, Esq., F.R.S.

Huggins's letter was, as will be seen, carefully worded and did not commit him to much. He wrote again after a few days to suggest the formation of a committee of scientific men to make further investigations, but Crookes demurred to this, writing the following interesting letter :

208

June 16*th,* 1871.

DEAR MR. HUGGINS,

I regret that it is now too late to make the alteration you speak of. Printing the *Quarterly Journal of Science* myself, I have to go to press very early in order to avoid great expense and delay at the end of each quarter. The early sheets are now all worked off and the men are completing the latter portions of the *Journal.* It has to be in the hands of the public in about ten days' time and there is about a fortnight's fair work still left to be done, so we shall have a good deal of night work as it is. But really I think you are too sensitive on the subject. I hear as much as anyone of the gossip going on in the scientific world, and I can assure you that there is no odium attaching to anyone who fairly and fearlessly states his opinions on these matters. I know we have both got considerable credit for our boldness, and those who at first were inclined to ridicule have been very willing to listen when I told them of actual facts which occurred before me.

Now as to a committee. I will gladly help in its formation, but the difficulties are enormous, and the good very doubtful. I am afraid of asking Tyndall, for I don't think he would be fair. I do not mean he would he dishonest intentionally, but his mind is such that it would be impossible for him to be at all passive. He would be all the time playing tricks, pushing or pulling the table, rapping with his feet, jumping up and down, imposing conditions such as " If the spirits can rap on the table let them rap in that corner of the room " (*vide* his *Science and Spirits*) and generally interfering with the progress of the phenomena. In this way he would stop the development of force at the commencement and we should be kept at the elementary phenomena—which we could all imitate and account for by cheating—and we should never get to the absolutely convincing phenomena. The result would be that Tyndall would set his very fluent pen to work and write many magazine articles proving that we were all fools and he was the only wise man among us. Since reading his paper *Science and Spirits* I have felt very strongly that the best thing for the truth in this matter would be for Tyndall to keep out of it. I give you an extract from a letter Tyndall sent me in 1869. " More than a year ago Mr. Cromwell Varley, who is I believe one of the greatest of modern spiritualists, did me the honour to pay me a visit, and he then employed a comparison which, tho' flattering to my spiritual strength, seems

P

to mark me out as unfit for spiritual investigation. He said that my presence at a *séance* resembled that of a great magnet among a number of small ones. I threw all into confusion."

Now that expresses exactly what I mean. Tyndall would completely stop Home's power. Mrs. St. Claire would be the medium for Tyndall, but she is difficult to get at.

Again, although I say I think a committee desirable and will help, why should I take all the trouble in the matter ? By so doing I fear I should really incur odium, for I should be regarded at once as the champion of Home and of his peculiar views and should not be treated as an independent member of the committee. I know the phenomena are true. I also think that I have a very good opportunity of investigating and possibly of making discoveries in the matter. So many men are now inquiring into it now and the subject is spreading so rapidly that it cannot be kept in the gutter (i.e. in Southampton Row) very much longer. This being the case, it would be a confession of weakness and doubt on my part for me to take the initiative in calling together a scientific committee to help me. You do not ask for a committee to examine spectra. I did not ask for one to convince me that thallium was true. I will give advice as to the constitution of such a committee and suggest rules, &c., but I must sit on it as an independent member and not as a partizan of the spiritualists. Who spoke about the committee ? I have no objection to talk the matter over with anyone. I do not think a better committee could be found than our present one : you, I, and Cox, we have met and have seen a good deal, and would certainly see more. Why not add one or two good men to us, and continue our "committee meetings." I have no objection to this.

Believe me,

Very truly yours,

WILLIAM CROOKES.

On the same day Crookes wrote to Home a letter, a copy of which, by some fortunate accident, has been preserved :

June 16th, 1871.

MY DEAR DAN,

In the first place I must offer you our hearty congratulations on your brilliant success last evening. You surpassed yourself in some of the readings ! and the room must have been full enough to have satisfied all your expectations.

I want you now to do me a great favour. My paper on our experiments was sent into the Royal Society on Wednesday. Last night I heard that it was considered too important for the " papers committee " to decide upon immediately, so it was postponed till that day fortnight, when the committee will consider it carefully whether to accept or reject it. Now to reject a paper is a thing which is seldom done, and if a paper of mine were rejected it would attach a stigma to me which would be very unpleasant. I am therefore anxious to do all I can to avoid such an injustice. I want you therefore to help me by giving me three evenings between now and the 27th inst. at which I can repeat the increase of weight experiment in the presence of one or two *other* good witnesses and then send in on the 28th an overwhelming mass of evidence, which the committee can't reject. The matter is really important for me, and I hope you will therefore be able to spare me the time without much inconvenience.

Believe me,

Very sincerely your friend,

WILLIAM CROOKES.

The paper sent to the Royal Society was substantially the same as Crookes's article in the *Quarterly Journal of Science*. It was, however, not accepted by the Royal Society, in spite of Home's willing assent to a series of supplementary experiments in the presence of men of science, and Crookes felt the indignity very keenly. He then published the article also in the *Chemical News*, " the readers of which include most of the scientific men of England."

In October 1871 another blow fell. The *Quarterly Review* published an anonymous article under the title " Some Recent Converts to Spiritualism," which severely criticised Dr. Hare, Huggins, Crookes, Cox, and others. The article insisted upon special qualifications being necessary for the investigation of obscure phenomena involving anomalous activities of the brain. It went on to say :

> Any "scientific man" is popularly supposed to be a competent authority upon obscure questions, for the elucidation of which are required the nice discrimination and the acute discernment of the sources of fallacy which can only be gained by a long course of experience based on special knowledge.

After giving credit to Dr. Huggins as a distinguished amateur astronomer, the reviewer adds :

> We believe that his devotion to a branch of research which takes the keenest powers of *observation* has prevented him from training himself in the strict methods of *experimental* inquiry. . . . To him, "seeing is believing," but to those who have qualified themselves for the study of "Psychic Force" by a previous course of investigation into the class of "occult" phenomena of which this is the latent manifestation, "seeing" is anything but "believing."

Serjeant Cox fared rather worse :

> Of Mr. Serjeant Cox it will be enough for us to say that, whatever may be his professional ability, he is known to those conversant with the history of Mesmerism as one of the most gullible of the gullible, as to whatever appeals to his organ of wonder.

But for Crookes the reviewer had "no tenderness," and he attacked him very bitterly. He explained the accordion experiment as probably a case of clever ventriloquism, and the action on the board as a conjuring trick in which Home adroitly directed the attention of Crookes and Huggins on the pointer of the recording instrument instead of his own hands. He said Crookes commenced the investigation with a foregone conclusion in favour of spiritualism, and that his predisposition in that direction was due to a deficiency of early scientific training ; also, that he had no knowledge of previous work done on the subject by eminent scientific men, and that, though he was a Fellow of the Royal Society, " this distinction was conferred upon him with considerable hesitation." The reviewer claimed to " speak advisedly " in saying this.

Crookes was able easily to refute the last statement, and could point to his career at the Royal College of Chemistry in answer to the criticism of his early scientific training. But he was not satisfied with that. He carried the war into the enemy's country. He identified the reviewer as Professor W. B. Carpenter, Registrar of London University, a very distinguished biologist, and, in a special pamphlet, Crookes subjected him to a castigation such as few men in Carpenter's position have ever experienced. It shows Crookes in all his war-paint, and some of it may be quoted here :

It was my good or evil fortune, as the case may be, to have an hour's conversation, if it may be so termed when the talking was all on one side, with the *Quarterly* Reviewer in question, when I had an opportunity of observing the curiously dogmatic tone of his mind and of estimating his incapacity to deal with any subject conflicting with his prejudices and prepossessions. At the last meeting of the British Association at Edinburgh [1] we were introduced. He as a physiologist who had inquired into the matter fifteen or twenty years ago ; I as a scientific investigator of a certain department of the subject ; here is a sketch of our interview, accurate in substance if not identical in language.

" Ah ! Mr. Crookes," said he, " I am glad I have an opportunity of speaking to you about this Spiritualism you have been writing about. You are only wasting your time. I devoted a great deal of time many years ago to Mesmerism, Clairvoyance, Electro-biology, Table-turning, Spirit-rapping, and all the rest of it, and I found there was nothing in it. I explained it all in my article I wrote in the *Quarterly Review.* I think it a pity you have written anything on this subject before you made yourself intimately acquainted with my writings and my views on the subject. I have exhausted it."

[1] Crookes endeavoured to interest the British Association in his psychic work, but fared no better than at the Royal Society.

" But, sir," interposed I, " you will allow me to say you are mistaken, if—— "

" No, no ! " interrupted he, " I am not mistaken. I know what you would say. But it is quite evident from what you have just remarked, that you allowed yourself to be taken in by these people when you knew nothing whatever of the perseverance with which I and other competent men, eminently qualified to deal with the most difficult problems, had investigated these phenomena. You ought to have known that I explain everything you have seen by ' unconscious cerebration ' and ' unconscious muscular action ' ; and if you had only a clear idea in your mind of the exact meaning of these two phrases, you would see that they are sufficient to account for everything."

" But, sir—— "

" Yes, yes ; my explanations would clear away all the difficulties you have met with. I saw a great many Mesmerists and Clair-voyants, and it was all done by ' unconscious cerebration.' Whilst as to Table-turning, everyone knows how Faraday put down that. It is a pity you were unacquainted with Faraday's beautiful indicator ; but, of course, a person who knew nothing of my writings would not have known how he showed that unconscious muscular action was sufficient to explain all these movements."

" Pardon me," I interrupted, " but Faraday himself showed—— " But it was in vain, and on rolled the stream of unconscious egotism.

" Yes, of course ; that is what I said. If you had known of Faraday's indicator and used it with Mr. Home, he would not have been able to go through his performance."

" But how," I contrived to ask, " could the indicator have served, seeing that neither Mr. Home nor anyone else touched the—— "

" That's just it. You evidently know nothing of the indicator. You have not read my articles and explanations of all you saw, and you know nothing whatever of the previous history of the subject. Don't you think you have compromised the Royal Society ? It is a great pity that you should be allowed there to revive subjects I put down ten years ago in my articles, and you ought not to be permitted to send papers in. However, we can deal with them." Here I was fain to keep silence. Meanwhile, my infallible inter-locutor continued :

" Well, Mr. Crookes, I am very pleased I have had this opportunity of hearing these explanations from yourself. One learns so much in a conversation like this, and what you say has confirmed me on several points I was doubtful about before. Now, after I have had the benefit of hearing all about it from your own lips, I am more satisfied than ever that I have been always right, and that there is nothing in it but unconscious cerebration and muscular action."

At this juncture some Good Samaritan turned the torrent of words on to himself ; I thankfully escaped with a sigh of relief, and my memory recalled my first interview with Faraday, when we discussed table-turning and his contrivance to detect the part played by involuntary muscular effort in the production of that phenomena. How different his courteous, kindly, candid demeanour towards me in similar circumstances compared with that of the *Quarterly* Reviewer !

Crookes then narrates how he saw Faraday's test apparatus at the shop of Newman, philosophical instrument maker, Regent Street, in 1853, and was introduced to Faraday at the house of the Rev. J. Barlow, the secretary of the Royal Institution.

Crookes also replies to the charge that he commenced the investigation as a convinced spiritualist. " Now, let me ask," he exclaims, " what authority has the reviewer for designating me a recent convert to spiritualism ? Nothing that I have ever written can justify such an unfounded assumption."

This passage, we may say in passing, is open to misinterpretation. At the time it was written, Crookes *was* a spiritualist at heart, and was known to be such by a number of his friends, but he had not published the fact, and evidently did not intend to do so until spiritualism was officially recognised by the scientific authorities in power. He had hopes of himself bringing about this recognition, but knew that he must proceed gradually and carefully.

MYSTERIOUS FORCES AND APPARITIONS

In his pamphlet, Crookes quotes a passage from his article in the *Quarterly Journal of Science* of July 1870, as follows :

I confess that the reasoning of some spiritualists would almost seem to justify Faraday's severe statement that many dogs have the power to come to much more logical conclusions.

To Carpenter's charge that Crookes was " a specialist of specialists," and therefore unfit to report upon psychical phenomena, Crookes made the following trenchant reply, which is of additional interest as being autobiographical :

But my greatest crime seems to be that I am a "specialist of specialists." I a specialist of specialists ! This is indeed news to me, that I have confined my attention only to one special subject. Will my reviewer kindly say what that subject is ? Is it General Chemistry, whose chronicler I have been since the commencement of the *Chemical News* in 1859 ? Is it Thallium, about which the public have probably heard as much as they care for ? Is it Chemical Analysis, in which my recently published *Select Methods* is the result of twelve years' work ? Is it Disinfection and the Prevention and Cure of Cattle Plague, my published report on which may be said to have popularised Carbolic Acid ? Is it Photography, on the theory and practice of which my papers have been very numerous ? Is it the Metallurgy of Gold and Silver, in which my discovery of the value of Sodium in the amalgamation process is now largely used in Australia, California and South America ? Is it in Physical Optics, in which department I have space only to refer to papers on some Phenomena of Polarised Light, published before I was twenty-one ; to my detailed description of the Spectroscope and labours with this instrument, when it was almost unknown in England ; to my papers on the Solar and Terrestrial Spectra ; to my examination of the Optical Phenomena of Opals, and construction of the Spectrum Microscope ; to my papers on the Measurement of the Luminous Intensity of Light ; and my description of my Polarisation Photometer ? Or is my speciality Astronomy and Meteorology, inasmuch as I was for twelve months at the Radcliffe Observatory, Oxford, where, in addition to my principal employment of arranging the meteorological department, I divided my leisure time between

Homer and mathematics at Magdalen Hall, planet-hunting and transit taking with Mr. Pogson, now Principal of the Madras Observatory, and celestial photography with the magnificent helio-meter attached to the Observatory ? My photographs of the Moon, taken in 1855, at Mr. Hartnup's Observatory, Liverpool, were for years the best extant, and I was honoured by a money grant from the Royal Society to carry out further work in connexion with them. These facts, together with my trip to Oran last year, as one of the Government Eclipse Expedition, and the invitation recently received to visit Ceylon for the same purpose, would almost seem to show that Astronomy was my speciality. In truth, few scientific men are less open to the charge of being "a specialist of specialists."

These self-revelations must have staggered Professor Carpenter and his friends, and left them little to say. But Crookes finally proceeded to the offensive, in the shape of the following neat reply, which maintains the fiction of the impersonality of the "reviewer" with amusing seriousness :

The theories of the profound psychologists of Germany, to say nothing of those of our own countrymen, are made quite subsidiary to the hypotheses of Dr. William Carpenter. An unquestioning and infatuated belief in what Dr. Carpenter says concerning our mental operations has led the reviewer wholly to ignore the fact that these speculations are not accepted by the best minds devoted to psychological inquiries. I mean no disrespect to Dr. Carpenter, who, in certain departments, has done some excellent scientific work, not always perhaps in a simple and undogmatic spirit, when I "speak advisedly " that his mind lacks that acute, generalising, philosophic quality which would fit him to unravel the intricate problems which lie hid in the structure of the human brain.

One can imagine the greybeards of the Royal Society shaking their heads over the audacity of their *enfant terrible*.

The date of the appearance of the *Quarterly Review* attack is the same as that on which Crookes's third con-tribution to the *Quarterly Journal of Science* on the same subject appeared. It was entitled "Some Further Experiments on Psychic Force," and was an endeavour to meet the objections raised at the Royal Sociey, the British Association, and elsewhere. The correspondence with these bodies was published in the article, which bears all the marks of careful preparation, and is in reality the strongest literary contribution to the subject made by Crookes. He refutes Stokes's argument that the hand bell placed by Home on the mahogany board could have been used to produce the force observed. He also reveals the fact that in June 1871 he was "fitting up apparatus in which contact is made through water only, in such a way that transmission of mechanical movement to the board is impossible." As this point was eagerly seized upon by Crookes's learned critics, something more must be said about it. If a bowl of water is placed on a balance and the hand is plunged into the water, the weight of the bowl is apparently increased by 12 or 15 ounces. This is due to the hydrostatic pressure of the water displaced by the hand. Crookes's actual experiment was not so simple as this. He mounted a copper vessel with a hole in the bottom on an independent stand and half immersed this vessel in the water. The medium was supposed to immerse his hand in the water contained in the copper vessel. Crookes wrote about this contrivance in his first paper sent to the Royal Society : "As the mechanical transmission of power is by this means entirely cut off between the copper vessel and the board, the power of muscular control is thereby completely eliminated." This statement, as it stands, is not true, as a force of as much as 12 or 15 ounces can be transmitted to the board supporting the bowl by simply immersing the hand for some time in the water contained in the copper vessel. Professor Wheatstone wrote to Crookes :

In both these sentences you explain why you employed the interposition of water, and you state nothing from which I can infer that you had any other reason for doing so.

It is further evident that in the experiments first communicated to Professor Stokes, the vessel of water was not placed directly over the fulcrum of the lever ; for you say (page 28) " In my first experiments with this apparatus, referred to in Professor Stokes's letter and my answer, the glass vessel was not over the fulcrum, but nearer B." That under such circumstances a mechanical pressure is exerted on the lever when the hand is dipped in the water is an undoubted fact ; whether it produces the effect in question or not depends on the sensibility of the apparatus and the placing of the vessel. A displacement of 3 cubic inches of water would exert a pressure which, if directly applied to your machine, would be equal to 6,816 grains ; the extreme pressure of your imaginary psychic force being, according to your own statement, 5,000 grains. The fluctuation of the pressure in your experiment would naturally follow from the varying quantity of water displaced owing to the unsteadiness of the hand in the liquid.

From the above it appears to me that your experiment with the water vessel does not offer an iota of proof in favour of your doctrine of psychic force, or any disproof of the effect not being mechanical ; though it might easily lead persons unacquainted with hydrostatic laws to infer that no mechanical pressure could be communicated under such circumstances.

I cannot see what part you intended the water to play when you subsequently placed the vessel over the dead point, and it appears to me contrary to all analogy that a force acting according to physical laws should produce the motion of a lever by acting on its fulcrum.

To this, Crookes replied, *inter alia* :

It is much to be regretted that you should have selected from my pamphlet two passages occurring on page 28, and should have omitted to read the few lines which connect these passages ; otherwise it must have been apparent to you that your self-evident exposition of a well-known hydrostatic law had no bearing on the case in point.

Let me supply the deficiency. The following paragraph, from page 28 of my pamphlet, fills up the gap between the two passages you quote :

"*On the board, exactly over the fulcrum,* is placed a large glass vessel filled with water, L is a massive iron stand furnished with an arm and a ring, M N, in which rests a hemispherical copper vessel, perforated with several holes in the bottom. The iron stand is 2 inches from the board, A B, and the arm and copper vessel, M N, are so adjusted that the latter dips into the water $1\frac{1}{2}$ inches, being $5\frac{1}{2}$ inches from the bottom of I, and 2 inches from its circumference. Shaking or striking the arm M or the vessel N produces no appreciable mechanical effect on the board A B capable of affecting the balance. *Dipping the hand to the fullest extent into the water in* N *does not produce the least appreciable action on the balance.* As the mechanical transmission of power is by this means entirely cut off between the copper vessel and the board A B, the power of muscular control is thereby eliminated."

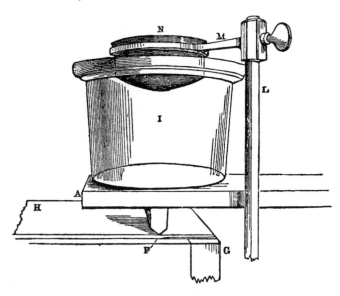

To the last paragraph of Wheatstone's letter, Crookes replied :

In this I entirely agree. I too cannot see the part the water played ; nor can I trace the analogy between the psychic force and a force acting according to known physical laws. Yet the facts recorded in my papers are true for all that.

The episode of the water experiment marks the most important epoch in the relations between official science and spiritualism. Crookes took a *faux pas*, and from that day official science turned its back upon spiritualism and all its works, nor has it felt compelled to change its attitude since then. One cannot help wondering what the world would be like to-day had Crookes carried official science with him !

Crookes's picturesque method of dealing with another type of criticism is amusingly shown in the following paragraphs appended to his article (*Quarterly Journal of Science*, October 1, 1871) :

Just before going to press I have received from my friend Professor Morton an advance sheet of the *Journal of the Franklin Institute*, containing some remarks on my last paper by Mr. Coleman Sellers, a leading scientific engineer of the United States. The essence of his criticism is contained in the following quotation :

" On page 341 (of the *Quarterly Journal of Science*) we have given a mahogany board ' 36 inches long by 9½ inches wide, and 1 inch thick,' with ' at each end a strip of mahogany 1½ inches wide screwed on, forming feet.' This board was so placed as to rest with one end on the table, the other suspended by a spring balance, and, so suspended, it recorded a weight of 3 pounds ; i.e. a *mahogany board* of the above dimensions is shown to weigh 6 pounds—3 pounds on the balance and 3 pounds on the table. A mechanic used to handling wood wonders how this may be. He looks through his limited library and finds that scientific men tell him that such a board should weigh about 13½ pounds. Did Mr. Crookes make this board himself? or did Mr. Home furnish it as one of his pieces of apparatus ? . . . It would have been more satisfactory if Mr. Crookes had stated, in regard to this board, who made it. . . . Let it be discovered that the 6-pound mahogany board was furnished by Mr. Home and the experiments will not be so convincing."

My experiments must indeed be convincing if so accomplished a mechanician as Mr. Coleman Sellers can find no worse fault with them than is expressed in the comments I have quoted. He writes in so matter-of-fact a manner, and deals so plausibly with dimensions and weights, that most persons would take it for granted that I really *had* committed the egregious blunder he points out.

Will it be believed, therefore, that my mahogany board does *weigh only 6 pounds ?* Four separate balances in my own house tell me so, and my greengrocer confirms the fact.

It is easy to perceive into what errors a " mechanic " may fall when he relies for practical knowledge on his "limited library" instead of appealing to actual experiment.

I am sorry I cannot inform Mr. Sellers who made my mahogany board. It has been in my possession about sixteen years ; it was originally cut off a length in a wood-yard ; it became the stand of a spectrum camera, and as such is described with a cut in the *Journal of the Photographic Society* for January 21, 1856 (vol. ii, p. 293). It has since done temporary duty in the arrangement of various pieces of apparatus in my physical laboratory, and was selected for these particular experiments owing to its shape being more convenient than that of other available pieces of wood.

But is it seriously expected that I should answer such a question as " Did Mr. Home furnish the board ? " Will not my critics give me credit for the possession of some amount of common sense ? And can they not imagine that obvious precautions, which occur to them as soon as they sit down to pick holes in my experiments, are not unlikely to have also occurred to me in the course of prolonged and patient investigation ?

The answer to this as to all other like objections is, Prove it to be an error by showing where the error lies, or, if a trick, by showing *how* the trick is performed. Try the experiment fully and fairly. If then fraud be found, expose it ; if it be a truth proclaim it. This is the only scientific procedure, and this it is that I purpose steadily to pursue.

It is interesting to note that in this case Crookes was correct, and so was his critic ! For mahogany varies widely in density. The usual limits are given as o·56 and o·85, compared with water. Crookes's board was unusually light, and Sellers's board unusually heavy, but both were quite within the bounds of possibility.

Failing to convince his scientific colleagues, Crookes

seems to have pursued his inquiries for his own satisfaction, while carrying out certain scientific researches which, as we shall see later, had a most profound and lasting influence upon scientific thought.

One of the mediums with whom he experimented was Kate Fox, one of the three Fox sisters who " founded spiritualism " in America in 1848. She was then thirty-eight, and was apparently of a very impulsive and tempestuous disposition. Her numerous letters to Crookes have been lost, but some of their contents may be guessed from the brief list contained in Crookes's " Index to Letters " already quoted. It should be explained that Miss Fox had for several years been befriended by Mr. Livermore, a wealthy New York widower, who " communicated " with his deceased wife through her agency. But let us quote from Crookes's " Index " :

1871.
Nov.   24, Fox, K., " Dear Wm   Guard Katie." Bella
             Livermore."
Nov.   24, Fox, K., Sick, will I call ?
Nov.   24, Fox, K., Promise to Varley after "sickness."
1872.
Jan.   Fox, K., Letter to Mrs. Gregory.   " Geary."
Jan.   Fox, K., Draft of letter for me to send to Mr. L.
             (not sent).
Feb.   8, Fox, K., Summons for cab (cost me 7/6).
Feb.   Fox, K., Penitent letter after I brought her
             home.
Feb.   Fox, K., " Can't come." Engagement.
Feb.   Fox, K., " Can't come." Not well (German).
Feb.   Fox, K., Wants Harrison's MS. of her early days.
March 6, Fox, K., With Varley's letter.   Will accept.
April   1, Fox, K., " First of April ! "
April   4, Fox, K., Account at Bowler's (statement).
April 13, Fox, K., About my threatened letter to Mr.
             Livermore.

1872.

April	16,	Fox, K., Deferring visit. Theatre.
April	16,	Fox, K., Deferring visit. Jack ill.
April		Fox, K., Jack's illness.
April	22,	Fox, K., Will reserve nights. Let us be friends.
April		Fox, K., Ill. Sick. Glasgow. Mr. Livermore.
April		Fox, K., Jencken. "Most outrageous." Ill.
May		Fox, K., From Mr. Hopps. Glasgow.
May	9,	Fox, K., News of Mr. Livermore's engagement.
May	11,	Fox, K., Mr. Townsend's letter genuine.
May	15,	Fox, K., Leaving Glasgow.
June		Fox, K., Leaving Beaumont Str. Money.
June	10,	Fox, K., Telegram from 434 Strand.
June	13,	Fox, K., Lies about telegram. Paris.
June	25,	Fox, K., Visit to Margate. Future prospects.
June		Fox, K., Illness. Paris. Abt. telegram of 10th.
Aug.	1,	Fox, K., To call on her. Ogden's character. Tempted.
Aug.	2,	Fox, K., Copy of letter from Mrs. Kane to.
Aug.		Fox, K., Sister Margaret. Letter from me to.
Aug.	28,	Fox, K., "Spirit Message from F. C. S."
Aug.		Fox, K., Letter to Jencken about report.
Sept.	1,	Fox, K., At Jones's. Jencken proposed marriage.
Sept.		Fox, K., Friendly. Séance every day, six months.
Sept.	13,	Fox, K., Returned from Jones's. Call to-night.
Sept.	15,	Fox, K., Won't stay with Ogden. Well.
Sept.	16,	Fox, K., Suddenly ill. Disappointed.
Sept.	21,	Fox, K., Ill (drugged).
Sept.	22,	Fox, K., Ill. "Will tell all soon." Drugged.
Oct.	1,	Fox, K., Answer to my letter. "Naked foot," &c.
Oct.	4,	Fox, K., Answer to my letter. "Naked foot," &c.
Oct.		Fox, K., Will come with her boxes to-night.
Oct.		Fox, K., Flower manifestations at Mrs. Parkes's.

1873.

May	1,	Fox, K., Going to Glasgow. Will visit us.
May	25,	Fox, K., Abusing Livermore. Our invitation.

1873.

May   Fox, K., Nellie's answer to the above.

May   28, Fox, K., Preparations for coming. Ogden. Penitent.

June   13, Fox, K., Lies about telegram of June 10

June   22, Fox, K., Wants money (to Mrs. H.)

July   Fox, K., Can't come. Boil. Note to Jencken.

July   24, Fox, K., Boil. My receiving Livermore. Threats.

July   25, Fox, K., Friendly. Wants to see me. News.

Oct.   9, Fox, K., About anonymous letter to Jones, with Nelly's answer to Jones.

Dec. 10, Fox, K., D. Owen's letter.

These entries are not all at present intelligible, but may possibly be unravelled by some future historian of spiritualism. From details supplied by a member of the Crookes family, I learn that Mr. Jencken, who married Kate Fox, was a former employee of Crookes's and a very honest man, who hoped to rescue Kate from her tendency towards alcoholism. It is well known that most good conjurers are almost of necessity total abstainers, so that the above statement tends to support the spiritualist version of the Fox episode. It is known that Kate Fox, like her sister Margaret (who married Dr. Kane, the Arctic explorer), at one time " confessed " that all their manifestations were trickery, though at least one of them subsequently withdrew that allegation.[1]

In November 1871 Crookes entered the lists against John Spiller, who had publicly stated that he had found out how Home worked the accordion under the table. It is of no particular interest to revive the whole controversy, but a short letter of Crookes's to the *English Mechanic* may be quoted here :

[1] Personally, I do not attach much importance to such " confessions." If we believe a medium's " confession," why not believe another medium's assertion of genuineness ?—E. E. F.

MYSTERIOUS FORCES AND APPARITIONS

Novr. 4, 1871.

To THE EDITOR OF THE *English Mechanic.*

SIR,

In your issue of the 3rd inst. a courteous correspondent, " A Fellow of the Royal Astronomical Society," draws attention to a rumour which he says is floating about in Scientific circles to the effect that Mr. Spiller was my assistant who got under the table during the progress of the accordion experiment described in the *Quarterly Journal of Science* for July last. Allow me to give an unqualified denial to the rumour. It is utterly false.

Mr. Spiller is not my assistant, neither was he present during any of the accordion or other experiments referred to in my published papers. He came to my house on one evening only, and then uninvited, when Mr. Home and some friends had been dining with me ; but on that occasion no experiments were tried and nothing worth recording took place ; in fact the experimental apparatus was not devised until some weeks after. The only other occasion on which Mr. Spiller met Mr. Home was a few evenings after at Mr. Serjeant Cox's. Having myself been one of the party on that occasion I can fully endorse what the learned Serjeant says in the *Echo* of Nov. 3. I enclose a copy of his letter.

The name of my assistant on the occasion quoted by your correspondent, is entirely unimportant. He is a youth under 18, who has been working in my laboratory for a year and a half ; during this time I have found him as intelligent and truthful a youth as ever handled a test-tube. He was present to assist me in arranging the apparatus, and is quite willing, if necessary, to give a certificate that what I printed was correct ; but, I submit, it is not customary to require a pupil to verify the truth of a statement made by his master.

It is much to be regretted that the author of these reports does not come forward, and state honestly what he has to say, in preference to setting rumours afloat behind my back, and prompting his friends, under fictitious signatures, to make false accusations against my honour and veracity. It is a cowardly, un-English trick, and one which I can scarcely believe Mr. Spiller would be guilty of.

226

If my calumniator will have the courage to give his own statements, authenticated by his own signature, I promise that my refutation shall be complete.

I remain, Sir,

Your obedient servant,

WILLIAM CROOKES.

Crookes also wrote to Huggins :

20, MORNINGTON RD., N.W.,

Novr. 6, 1871.

DEAR DR. HUGGINS,

It is very annoying that Spiller should spread these reports. I am in hopes now that I can make him put his name to them, when he shall be settled pretty quickly. I have *his own statements, at the time,* in his letters. He evidently has not kept copies of what he wrote. I don't see the least necessity of your being dragged into the controversy, and you may trust to me, but of course I can't say what the other side may do. Don't you like Serjt. Cox's style ? I thought the way he put Spiller down was particularly neat. He wrote his letter without previous consultation with me. I was surprised to see it beneath mine.

Ladd's assistant is Spiller's brother. Do you know who " B " is who fired the first shot in the *Echo* ? I consider he is Baden Pritchard, Spiller's Brother-in-Law.

I think I shall have something very important to show you soon. I have got an indicator so delicate that it will work without a medium. There is a new force, or a new form of a known force.

Believe me,

Very sincerely yours,

WILLIAM CROOKES.

A summary of Crookes's " inquiry into the phenomena called spiritual " during the years 1870–3 was printed in the *Quarterly Journal of Science* for January 1874. It was intended to be the precursor of a substantial volume, which, had it been published, would have probably become the Holy Book of spiritualism. Even as it stands, the article in question has formed the point of departure

227

for numberless later investigators. Had it never been written, it is almost safe to say that there would have been no further eminent investigators for a generation. He says that the inquiry had occupied more time than he could well afford :

Thus a few months have grown into a few years, and were my time at my own disposal it would probably extend still longer. But other matters of scientific and practical interest demand my present attention ; and, inasmuch as I cannot afford the time requisite to follow the inquiry as it deserves, and as I am fully confident it will be studied by scientific men a few years hence, and as my opportunities are not as good as they were some time ago, when Mr. D. D. Home was in good health, and Miss Kate Fox (now Mrs. Jencken) was free from domestic and maternal occupations, I feel compelled to suspend further investigation for the present.

To obtain free access to some persons abundantly endowed with the power I am experimenting upon, now involves more favour than a scientific investigator should be expected to make of it. Spiritualism amongst its more devout followers is a religion. The mediums, in many cases young members of the family, are guarded with a seclusion and jealousy which an outsider can penetrate with difficulty. Being earnest and conscientious believers in the truth of certain doctrines which they hold to be substantiated by what appear to them to be miraculous occurrences, they seem to hold the presence of scientific investigation as a profanation of the shrine. As a personal favour I have more than once been allowed to be present at meetings that presented rather the form of a religious ceremony than of a spiritualistic séance. But to be admitted by favour once or twice, as a stranger might be allowed to witness the Eleusinian mysteries, or a Gentile to peep within the Holy of Holies, is not the way to ascertain facts and discover laws. To gratify curiosity is one thing ; to carry on systematic research is another. I am seeking the truth continually. On a few occasions, indeed, I have been allowed to apply tests and impose conditions ; but only once or twice have I been permitted to carry off the priestess from her shrine, and in my own house, surrounded by my own friends, to enjoy opportunities of testing the phenomena I had witnessed elsewhere under less conclusive conditions.

My observations on these cases will find their due place in the work I am about to publish.

Crookes then proceeds to classify the phenomena as follows :

    I. Movement of heavy bodies with contact, but without mechanical exertion.

    II. Percussive and other allied sounds.

    III. Alteration of weight of bodies.

    IV. Movement of heavy substances when at a distance from the medium.

    V. Rising of tables and chairs off the ground, without contact with any person.

    VI. Levitation of human beings.

    VII. Movement of various small articles without contact with any person.

    VIII. Luminous appearances.

    IX. Appearance of hands, either self-luminous or visible by ordinary light.

    X. Direct writing.

    XI. Phantom forms and faces.

    XII. Special instances which seem to point to the agency of an exterior intelligence.

    XIII. Miscellaneous occurrences of a complex character.

The most remarkable of these classified descriptions must be quoted in full :

Class VIII.

Luminous Appearances.

These, being rather faint, generally require the room to be darkened. I need scarcely remind my readers again that, under these circumstances, I have taken proper precautions to avoid being imposed upon by phosphorised oil or other means. Moreover, many of these lights are such as I have tried to imitate artificially, but cannot.

Under the strictest test conditions, I have seen a solid self-luminous body, the size and nearly the shape of a turkey's egg, float noiselessly about the room, at one time higher than anyone present could reach standing on tiptoe, and then gently descend to the floor. It was visible for more than ten minutes, and before it faded away it struck the table three times with a sound like that of a hard, solid body. During this time the medium was lying back, apparently insensible, in an easy chair.

I have seen luminous points of light darting about and settling on the heads of different persons ; I have had questions answered by the flashing of a bright light a desired number of times in front of my face. I have seen sparks of light rising from the table to the ceiling, and again falling upon the table, striking it with an audible sound. I have had an alphabetic communication given by luminous flashes occurring before me in the air, whilst my hand was moving about amongst them. I have seen a luminous cloud floating upwards to a picture. Under the strictest test conditions, I have more than once had a solid, self-luminous, crystalline body placed in my hand by a hand which did not belong to any person in the room. *In the light*, I have seen a luminous cloud hover over a heliotrope on a side table, break a sprig off, and carry the sprig to a lady ; and on some occasions I have seen a similar luminous cloud visibly condense to the form of a hand and carry small objects about. These, however, more properly belong to the next class of phenomena.

Crookes finally discusses the various theories advanced to account for the phenomena. He maintains that some, but not all the phenomena could be explained by cheating. As regards hallucination or unconscious cerebral action, he dismisses these " very briefly " by remarking that these two theories " are evidently incapable of embracing more than a small portion of the phenomena, and they are improbable explanations of even these. He finally leaves two theories—that of spirit agency and of psychic force respectively, and concludes :

The difference between the advocates of Psychic Force and the Spiritualists consists in this—that we contend that there is as yet insufficient proof of any other directing agent than the Intelligence of the Medium, and no proof whatever of the agency of Spirits of the Dead ; while the Spiritualists hold it as a faith, not demanding further proof, that Spirits of the Dead are the sole agents in the production of all the phenomena. Thus the controversy resolves itself into a pure question of *fact*, only to be determined by a laborious and long-continued series of experiments and an extensive collection of psychological *facts*, which should be the first duty of the Psychological Society, the formation of which is now in progress.

The year 1873 did not see the end of Crookes's psychical researches. In October 1871 he had been introduced to Miss Florence Cook, of Hackney, who had developed mediumship for " materialisations " earlier in the same year. He had over forty séances with her, many of them in his own house, and became absolutely convinced of the genuineness of her phenomena. According to Florence Cook's own version, her first acquaintance with the physical phenomena of spiritualism was made during some table-tilting experiments with a schoolfellow at Hackney, in which the table rose a clear four feet from the ground. At another sitting she was carried about on her chair by some abnormal means. Continuing the séances at home, she was instructed by raps to proceed to the house of Thomas Blyton, at Dalston, the secretary of a small group of spiritualists, through whom she was introduced to a number of people interested in the phenomena. Acting on their advice, she had regular sittings in her own family, the kitchen being curtained off to form a dark " cabinet " for her, while the family sat outside on the stairs. In these circumstances she was " controlled " by an entity who called herself " Katie King," or " Annie Owen Morgan," and who endeavoured to peep out through the curtain while Florence was lying in a trance inside. The séances went on for some time, gradually developing, until the form of Katie King acquired sufficient " power " to emerge completely from the cabinet.

Several attempts were made by spiritualists to photograph " Katie King," and this was eventually accomplished on May 7, 1873,[1] by the light of burning magnesium powder.

Crookes does not seem to have taken up Miss Cook seriously until after she had been " exposed " by a Mr. Volckmann, who seized " Katie King " and found himself holding the medium dressed up. This happened on December 9, 1873, after Crookes had completed his *Quarterly Journal of Science* article. At Miss Cook's earnest request, he devoted five months to an elaborate investigation of her " materialisation." At some of these sittings, Mr. C. F. Varley, the cable electrician, and an earnest spiritualist, assisted. The latter contributed to *The Spiritualist* of March 20, 1874, an account of a seance at the house of Mr. J. C. Luxmore, J.P., 16, Gloucester Square, W., which was attended by Mr. Luxmore, Mr. and Mrs. Crookes, Mrs. Cook, Mr. Tapp, Mr. W. H. Harrison, and himself. The medium was inserted in an electric circuit, in which was also a galvanometer which could be watched outside the cabinet, so as to make sure that during the " materialisation " the medium did not leave the cabinet. The electrical arrangements were such that the medium could not have broken the circuit without the fact being indicated by the galvanometer moving over 200 divisions down the scale. The only other possibilities to be guarded against were (1) that the medium should simply join the two sovereigns serving as electrodes on freeing herself. This would have produced an upward deflection of eighty divisions, whereas the greatest fluctuation observed was thirty-six divisions downward ; (2) that the medium should substitute another semi-conductor for

[1] A full account of this episode is given in my book *New Light on Immortality.*—E. E. F.

her own body and free herself without detection. The galvanometer readings were taken every minute by Mr. Varley and recorded and timed by Mr. Harrison.

Beginning at 7.10 p.m., the galvanometer reading fell gradually and steadily by thirty divisions until 7.25. This fall was due to the drying of the blotting-paper which made contact with the medium's skin. At 7.25 a fall of thirty-six divisions occurred. At 7.27 " Katie " appeared, lifting the curtain to show herself. At 7.36 " Katie " showed her hand and arm, and the galvanometer fell another seventeen divisions. Half a minute afterwards the galvanometer rose twenty-one divisions. After that the galvanometer fell off quite gradually and steadily until 7.48 while " Katie " came and put her hand on Crookes's head and wrote with a pencil on paper. On testing the circuit with electrodes in contact at 7.36, it was found that the electromotive force of the battery (two Daniell cells) had not fallen off as much as 1 per cent.

In discussing this séance, Mr. Varley attaches great importance to the fact that the galvanometer did not vary while " Katie " was moving her arms and writing, which, had Miss Cook emerged to personate the " spirit " must have been the case. It is not stated whether " Katie's " wrists were carefully examined to see if " Katie " was the medium in disguise, with the electrodes still on her wrists. The main object of the experiment, however, seems to have been to eliminate the possibility of the medium freeing herself without detection. This would have been comparatively easy had she known what " resistance " to substitute for that of her own body, as the sovereigns which served as electrodes were only attached to her wrists with elastic rings. The value of the experiment, therefore, hinges altogether on whether Florence Cook had the presence of mind (and sufficient cleverness) to

233

substitute, say, a moist handkerchief, for her own person in the circuit. The observations would in that case be fully accounted for.

Crookes's own account of his experiments with Florence Cook are embodied in three letters published in Mr. W. H. Harrison's journal, *The Spiritualist*, of February, March, and April 1874. He describes a seance on March 12th in which Katie King walked about the room (it was Crookes's own house) and then retired into the library used as a "cabinet" where the medium was lying. He proceeds :

In a minute she came to the curtain and called me to her, saying : "Come into the room and lift my medium's head up, she has slipped down." Katie was then standing before me clothed in her usual white robes and turban head-dress. I immediately walked into the library up to Miss Cook, Katie stepping aside to allow me to pass. I found that Miss Cook had slipped partially off the sofa, and her head was hanging in a very awkward position. I lifted her on to the sofa, and in so doing had satisfactory evidence, in spite of the darkness, that Miss Cook was not attired in the "Katie" costume, but had on her ordinary black velvet dress, and was in a deep trance. Not more than three seconds elapsed between my seeing the white-robed Katie standing before me and my raising Miss Cook on to the sofa from the position into which she had fallen. . . .

I pass on to a seance held last night [1] at Hackney. Katie never appeared to greater perfection, and for nearly two hours she walked about the room, conversing familiarly with those present. On several occasions she took my arm when walking, and the impression conveyed to my mind that it was a living woman by my side, instead of a visitor from the other world, was so strong that the temptation to repeat a recent celebrated experiment became almost irresistible. Feeling, however, that if I had not a spirit, I had at all events a *lady* close to me, I asked her permission to clasp her in my arms, so as to be able to verify the interesting observations which a bold experimentalist has recently somewhat verbosely recorded. Permission was graciously given, and I accordingly did—well, as any

[1] March 29, 1874.—E. E. F.

gentleman would do under the circumstances. Mr. Volckmann will be pleased to know that I can corroborate his statement that the " ghost " (not " struggling," however) was as material a being as Miss Cook herself. . . . I went continuously into the cabinet, it being dark, and felt about for Miss Cook. I found her crouching on the floor. Kneeling down, I let air enter the [phosphorus] lamp, and by its light I saw the young lady dressed in black velvet, as she had been in the early part of the evening, and to all appearance perfectly senseless ; she did not move when I took her hand and held the light quite close to her face, but continued quietly breathing. Raising the lamp, I looked around and saw Katie standing close behind Miss Cook. She was robed in flowing white drapery as we had seen her previously during the seance. Holding one of Miss Cook's hands in mine, and still kneeling, I passed the lamp up and down, so as to illuminate Katie's whole figure and satisfy myself thoroughly that I was looking at the veritable Katie whom I had clasped in my arms a few minutes before, and not at the phantasm of a disordered brain. She did not speak, but moved her head and smiled in recognition. Three separate times did I carefully examine Miss Cook crouching before me, to be sure that the hand I held was that of a living woman, and three separate times did I turn the lamp to Katie and examine her with steadfast scrutiny until I had no doubt whatever of her objective reality. At last Miss Cook moved slightly, and Katie instantly motioned me to go away. I went to another part of the cabinet and then ceased to see Katie, but did not leave the room until Miss Cook woke up, and two of the visitors came in with a light.

In criticism of these remarkable observations it has been urged that in the former séance, at Crookes's own house, the medium was either a dummy or " Katie " herself, who had quickly " dematerialised " in the dark " cabinet " Also that in the latter séance, where there were undoubtedly two persons, one of them was a confederate of the Cook family, in whose house the " materialisation " took place. Crookes, however, speaks with unshakable conviction. He took forty-four negatives of " Katie." " But photography," he says, " is as inadequate to depict the perfect beauty of Katie's face as words are powerless to describe her charms of manner." He indignantly rejects the hypothesis of fraud. "To imagine," he says, "that an innocent schoolgirl of fifteen should be able to conceive and then to carry out for three years so gigantic an imposture as this, and in that time should submit to any test which might be imposed upon her, should bear the strictest scrutiny, should be willing to be searched at any time, either before or after a séance, and should meet with even better success in my own house than at that of her parents, knowing that she visited me with the express object of submitting to strict scientific tests—to imagine, I say, the Katie King of the last three years to be the result of imposture does more violence to one's reason and common sense than to believe her to be what she herself affirms."

Florence Cook was, at this time, secretly married to a Mr. Corner. She announced her marriage shortly after the séance above described, at which "Katie King" bade farewell to her earthly friends. She gave a number of séances in after years, but not very successfully, and it is alleged that she was eventually "exposed." But the fact is that Crookes befriended her until her death in 1902. Indeed, Crookes kept up almost all his spiritualist friendships until they were severed by death. His "crucial experiments" were publicly derided as "Crookesial experiments in sly-kick force." An American journal published one of the Katie King photographs under the title "Sir William Crookes arm-in-arm with an Angel," and the successful "duping" of a great English scientist by a girl of fifteen became a cause of merriment to the cynic and of sorrow to Crookes's well-wishers.

But Crookes never recanted, never wavered, never withdrew. He recognised in time the hopelessness of his attempt to carry the scientific world with him. Science is the knowledge of things which always happen under certain conditions. No " certain conditions " can be arranged if they depend upon the whim of an entity which does not even inhabit this world ! You cannot bring it to book, nor make it sign a document, nor sue it for damages or neglect. Even a scientific fact is not generally accepted until it becomes a habit. Crookes says of this :

The following remarks are so appropriate that I cannot forbear quoting them. They occur in a private letter from an old friend, to whom I had sent an account of some of these occurrences. The high position which he holds in the scientific world renders doubly valuable any opinion he expresses on the mental tendencies of scientific men. " Any *intellectual* reply to your facts I cannot see. Yet it is a curious fact that even I, with all my tendency and desire to believe spiritualistically, and with all my faith in your power of observing and your thorough truthfulness, feel as if I wanted to see for myself ; and it is quite painful to me to think how much more proof I want. Painful, I say, because I see that it is not reason which convinces a man, unless a fact is repeated so frequently that the impression becomes like a habit of mind, an old acquaintance, a thing known so long that it cannot be doubted. This is a curious phase of man's mind, and it is remarkably strong in scientific men—stronger than in others, I think. For this reason we must not always call a man dishonest because he does not yield to evidence for a long time. The old wall of belief must be broken down by much battering."

But it is time we left this chapter of Crookes's life behind. Volumes could—and may yet—be written about it. Nor has the time come to pronounce a final verdict upon it. The jury would inevitably disagree. It may be that future ages will regard Crookes's incursion into spiritualism as a temporary aberration, illustrative of nothing but the fallibility of human judgment. It may be, on the other hand, that history will look upon Crookes's statement of his " supernormal " observations as one of the major steps in human evolution. Half a century has elapsed. A religion counting its adherents by the million has been founded upon Crookes's " researches in spiritualism." That circumstance may, to many, be Crookes's greatest condemnation, since it may be plausibly argued that the majority of religions are built upon fallacies. But the fact remains that the " physical phenomena " of spiritualism are ultimately based upon Crookes. Later eminent workers in the same field sought to reproduce his phenomena and to corroborate them. There have been slight variations, but nothing essentially new. " For authentic materialisations," a writer recently remarked in *Light*, " we have to go back to ' Katie King.' " Crookes covered the whole ground. He marked out the boundaries of the physical basis of spiritualism, and no subsequent investigator has been able to extend them. Indeed, no subsequent investigator has been able quite to cover the same ground. Twenty-five years later, when President of the British Association, Crookes said : " I have nothing to retract. I adhere to my already published statements. I only regret a certain crudity in those early expositions which, no doubt justly, militated against their acceptance by the scientific world." That world has since become more tolerant and more elastic. Its outlook is wider, its inquiry more profound. It admits much in psychology and psycho-physiology that used to be doubtful and obscure, but it does not admit spiritualism : it does not even admit its " physical phenomenon." Spiritualism as a religion may legitimately be studied in a section of anthropology, but spiritualism as a science does not exist. To be a spiritualist, the scientist must surrender his wishes, his methods, his views into the hands of his " spirit friends " on the " other side." If he does that he may achieve a certain peace of mind, but his scientific work will be at an end. His surrender may soothe him on his death-bed, but so may almost any religion when embraced with a fervent faith. And we may expect that the world's work will best be done by those who follow the light of reason to the utmost visible horizon, content in the belief that the divine spark within us is but the promise of a greater glory as yet unrevealed.

CHAPTER XIII

THE RADIOMETER

(1874-7)

AFTER THE *Sturm und Drang* of the early seventies, Crookes steered his barque once more into the calmer waters of Exact Science. His quarrel with W. B. Carpenter, in which Crookes gave harder blows than he took, and showed all the bitterness of which he was capable, rankled for many years. It had been a battle of giants, and both antagonists had suffered. Crookes showed himself to be the more resourceful fighter of the two, and succeeded in driving Carpenter out of his attitude of rather arrogant superiority. But in one point Carpenter remained the victor. He prevented Crookes from capturing the Royal Society for the propagation of spiritualism. He kept the Royal Society out of the dismal swamp of " mediumistics," and forced Crookes and his followers to relegate the investigation of " borderland " phenomena to special societies such as Cox's " Psychological Society " and afterwards the Society for Psychical Research.

Not that Crookes's scientific work had ever been altogether interrupted. There was the *Chemical News* to keep him in constant touch with new developments in chemistry. There was the *Quarterly Journal of Science*, with its new departments of engineering, mining, and metallurgy, now exclusively under Crookes's control. Crookes's book on *Beet-root Sugar* (Longmans) appeared in 1870. His *Select Methods in Chemical Analysis* was published in 1871, also by Longmans, Green & Co., and gave much-needed information on the methods of isolating the rarer elements. His short article " On the Spectral Phenomena of Opals," [1] described a very beautiful application of his binocular spectrum microscope to the study of the light transmitted and reflected by that mysterious jewel. But Crookes's best work, carried through many years and right through the " psychic force " episode, is his monumental research on the atomic weight of thallium. It is almost as if Crookes had clung to that very unexciting and arduous piece of spade-work in order to preserve his clearness of mind amid the upheaval of transcendental physics and philosophies. It reminds the present biographer of an incident in his youth during a nightmare in which he seemed to face some kind of demon. " Well, anyhow," he said to his adversary, "*you* cannot deny that $(a + b)^2 = a^2 + 2ab + b^2$! " Whereupon the demon incontinently fled.

The method preferably used by Crookes to find the atomic weight of thallium was to take a known quantity of metallic thallium, dissolve it in nitric acid, and weigh the nitrate of thallium produced. Nitrate of thallium contains only three elements : oxygen, nitrogen, and thallium. The atomic weights of oxygen and nitrogen had already been determined with great care by Stas, with reference to the hydrogen atom as unity. In the formula for thallium nitrate Tl_2NO_3, the acid radicle NO_3 was known. Its molecular weight was taken by Crookes to be 61·889, and a comparison of the weight of thallium taken with the total weight of the nitrate gave the weight of the acid radicle ; and by using the proportion

Weight of NO_3 : weight of thallium : : atomic weight of NO_3 : at. wt. of thallium.

[1] *Quarterly Journal of Science*, October 1869.

the required atomic weight of thallium could be calculated. One such set of figures is given by Crookes as follows :

$$NO_2 \qquad Tl$$
$$55 \cdot 855834 : 183 \cdot 790232 : : 61 \cdot 889 : x$$

whence $\qquad x = 203 \cdot 642.$

In his Royal Society paper, read on June 20, 1872, he claimed that " the most extraordinary pains have been taken to secure the absolute purity both of the thallium employed and of the agents used to act upon it. The glass and other apparatus have been specially constructed for these researches, and the balances and weights have been of an accuracy never before surpassed in any research."

Weighings were made both in air and in a vacuum. The vacuum balance was made by Oertling, with a 14-inch beam and agate knife-edges and planes. It was enclosed in a cast-iron case which could be exhausted by means of an air pump. But cast iron being porous, Crookes painted it with several coats of white lead mixed with copal varnish until no more air was sucked in through the pores. The difficulty of illuminating the scale and pointer without heating it was overcome by means of a small vacuum tube mounted inside the case—a significant precursor of Crookes's subsequent activities in electric lighting. Other difficulties were met in a similar spirit :

With a rider there is some difficulty in estimating the exact point at which it rests, and it is necessary to note the oscillations, placing the rider as exactly as possible on one of the divisions of the beam. The best weighings, perhaps, will be taken when the arc is not very small.

Temperature has an effect upon the air-balance, rendering it less sensitive when increasing. This is perhaps due to the varying expansibility of the arms and the knife-edges upon which the pans are hung, or the superior and inferior parts of the beam may expand unequally. The two arms of the balance at times expand unequally ;

and in finding the true value of the weights employed in the determination, this cause of error is eliminated by following Gauss's method of interchanges—the constant friction of the forceps against the weights in transferring them from one pan to another being obviated by employing hooks of thin wire attached to the agate plane, upon which the suspension-wires of the pans could be hung. This required that the pans should not differ from each other by a quantity greater than one-thousandth of a grain.

The weights employed were of pure platinum, and were specially made by Messrs. Johnson and Matthey, of Hatton Garden—the firm which had been making Crookes's sodium amalgam.

The thallium employed was obtained from the flue-dust of pyrites burners. It was obtained in a spongy state and solidified by fusion. It was then dissolved in sulphuric acid, and so converted into " commercially pure " thallium sulphate. The preparation of " chemically pure " thallium is thus described by Crookes :

Commercial sulphate of thallium is dissolved in water, and the cold solution deluged with sulphuretted hydrogen. It is then filtered, heated to ebullition, and poured into boiling dilute hydrochloric acid. The solution is filtered whilst hot, and then allowed to cool. The chloride of thallium which crystallises out on cooling is washed by decantation until the washings are free from sulphuric acid, and further purified by re-crystallising twice from water. The chloride of thallium thus obtained is dried, mixed with pure carbonate of sodium, and projected by small portions at a time into pure cyanide of potassium kept in a state of fusion in a white unglazed crucible. The chloride is rapidly reduced to the metallic state ; the crucible is then allowed to cool, and the contents exhausted with water. The resulting ingot of metal is well boiled in water, dried and fused over a spirit lamp in an unglazed porcelain crucible with free access of air, stirred with a porcelain rod to facilitate oxidation, and finally cast in a porcelain mould. It may be preserved under water which has been boiled to expel the air. This metal was used in the determinations A and B.

Another method was the following :

Ordinary metallic thallium is fused in contact with the air, in an iron crucible made nearly red-hot, and then poured into water. The granulated metal is then exposed to a warm atmosphere to facilitate oxidation, the oxide being frequently removed by boiling out with water. When a considerable quantity of oxide (mixed with carbonate) has been obtained, the solution is heated to ebullition, and a rapid current of carbonic acid gas passed through until the liquid is quite cold, and the excess of carbonate of thallium has crystallised out. The resulting salt is re-crystallised and projected into pure cyanide of potassium kept in a state of fusion in a porcelain crucible at a dull red heat ; carbonic acid escapes with effervescence, and the metal is reduced to the metallic state. The whole is then allowed to cool, the soluble salts boiled out with water, and the lump of thallium fused over an alcohol lamp in a lime crucible, and cast in a lime mould described further on. With this ingot of thallium, the determination C was effected.

A very important result of Crookes's careful research was that the atomic weight finally ascribed to thallium was not an integral number, and its fractional portion could not possibly be due to experimental error. Crookes remarks on this point :

Professor Stas has shown the hypothesis of Prout—that the atomic weights of the elements are severally multiples of the atomic weight of hydrogen—to be without the corroboration of experimental result. This view of the hypothesis is further borne out in the present investigation ; for the number 203·642 cannot, within the limit of what has been shown to be the probable error, by any liberty be made to follow the hypothesis. Without doubt, when the atomic weights of all the metals are re-determined according to the standard of recent scientific method, it will be found that there are more exceptions to the hypothesis than are commonly considered. Marignac gives, in his confimatory discussion of Stas's experiments, and in his own results with calcium (40·21), lanthanum (94·13), strontium (87·25), analogous opposed evidence, as in the case of the weight found for thallium.

This point acquired increasing importance as time went on.

But from the careful and systemic weighings *in vacuo* there emerged a residue of facts unaccounted for by the science of the time. Crookes observed certain irregularities in the weighings when the body weighed was either hotter or colder than the enclosure. This, at ordinary pressures, could be accounted for by air-currents. But assuming these to be the sole cause, they ought to have progressively disappeared on exhausting the enclosure. This was found not to be the case. On the contrary, certain effects appeared strongest on reaching a certain degree of exhaustion. Crookes was much interested in these phenomena, and thought for some time that he had discovered a new force, which he called the X force. He hoped to shed some light on the nature of gravitation by following up this clue.

In 1873, Crookes communicated a paper to the Royal Society " On the Action of Heat on Gravitating Masses." The official abstract of this paper concludes as follows :

Speaking of Cavendish's celebrated experiment, the author says that he has experimented for some months on an apparatus of this kind, and gives the following outline of one of the results he has obtained :

" A heavy metallic mass, when brought near a delicately suspended light ball, attracts or repels it under the following circumstances :

" I. *When the ball is in air of ordinary density.*
   (*a*) If the mass is *colder* than the ball, it *repels* the ball.
   (*b*) If the mass is *hotter* than the ball, it *attracts* the ball.

" II. *When the ball is in a vacuum.*
   (*a*) If the mass is *colder* than the ball, it *attracts* the ball.
   (*b*) If the mass is *hotter* than the ball, it *repels* the ball."

The author continues : " The density of the medium surrounding the ball, the material of which the ball is made, and a very slight difference between the temperatures of the mass and the ball, exert so strong an influence over the attractive and repulsive force, and it has been so difficult for me to eliminate all interfering actions of temperature, electricity, &c., that I have not yet been able to get distinct evidence of an independent force (not being of the nature of heat) urging the ball and the mass together.

"Experiment has, however, showed me that, whilst the action is in one direction in dense air, and in the opposite direction in a vacuum, there is an intermediate pressure at which differences of temperature appear to exert little or no interfering action. By experimenting at this critical pressure, it would seem that such an action as was obtained by Cavendish, Reich, and Baily should be rendered evident."

After discussing the explanations which may be given of these actions, and showing that they cannot be due to air-currents, the author refers to evidences of this repulsive action of heat, and attractive action of cold, in nature. In that portion of the sun's radiation which is called heat, we have the radial repulsive force, possessing successive propagation, required to explain the phenomena of comets and the shape and changes of the nebulæ. To compare small things with great—to argue from pieces of straw up to heavenly bodies—it is not improbable that the attraction, now shown to exist between a cold and a warm body, will equally prevail when, for the temperature of melting ice is substituted the cold of space, for a pith ball a celestial sphere, and for an artificial vacuum a stellar void. In the radiant molecular energy of cosmical masses may at last be found that "agent acting constantly according to certain laws," which Newton held to be the cause of gravity.

Some of the experiments were highly ingenious. A balanced straw with pith balls at the ends was mounted on a pivot in its middle, and one of the pith balls was exposed successively to different coloured rays derived from sunlight. The maximum repulsion was found near the extreme visible red, which, as we now know, is near the maximum energy of radiation.

There was much discussion between Crookes and G. G. Stokes as to the title under which Crookes's complete paper should be published by the Royal Society. Among the titles suggested by Crookes were :

(1) Radiation of Direct Force.
(2) Mechanical Relations of Bodies in Space.
(3) Radiant Energy.
(4) Radiant Force.
(5) Attraction and Repulsion as the Result of Heat and Light.

The final title adopted was " Attraction and Repulsion resulting from Radiation." This was read at the Royal Society on December 11, 1873, and the apparatus was exhibited at a soirée on April 22, 1874.

One of its immediate effects was to bring a new critic into the field against Crookes, in the person of Professor Osborne Reynolds, who, on June 18th, read a paper in which he attributed the effects observed by Crookes to evaporation. He said :

When the radiated heat from the lamp falls on the pith, its temperature will rise, and any moisture on it will begin to evaporate, and to drive the pith from the lamp. The evaporation will be greatest on that ball which is nearest to the lamp, therefore this ball will be driven away until the force on the other becomes equal, after which the balls will come to rest unless momentum carries them further. On the other hand, when a piece of ice is brought near, the temperature of the pith will be reduced, and it will condense the vapour and be drawn towards the ice.

In reply, Crookes pointed out that he obtained the effect with all sorts of vacua—air, carbonic acid, water, iodine, hydrogen, etc.—and found no difference between them. Also, if the effect was due to evaporation and condensation, it should diminish as exhaustion proceeded. But as a matter of fact, the effect was obtained up to the highest chemical vacuum obtainable. Crookes concluded :

THE RADIOMETER

For my own part, I wish to avoid having a theory on the subject. As far as the facts have led me, I think that the repulsion accompanying radiation is directly due to the impact of the waves upon the surface of the moving mass, and not secondarily through the intervention of air-currents, electricity, or evaporation and condensation. Whether the etherial waves actually strike the substance moved, or whether at the boundary surface separating solid from gaseous matter, there are intermediary layers of condensed gas which, taking up the blow, pass it on to the layer beneath, are problems the solution of which must be left to further research.

A second paper on the same subject was read by Crookes at the Royal Society on April 22, 1875. He described several experiments, finally disproving Reynolds's theory, but was still more guarded concerning a theory of his own. " The facts will tell their own tale. The conditions under which they invariably occur will give the laws, and the theory will follow without much difficulty." He then added, as a supplement, the following announcement :

Since the experiments mentioned in the foregoing abstract were concluded, the author has examined more fully the action of radiation on black and white surfaces. At the highest exhaustion heat appears to act almost equally on white and on lamp-blacked pith, repelling them in about the same degree.

The action of the luminous rays, however, is different. These repel the black surface more energetically than they do the white surface. Taking advantage of this fact, the author has constructed an instrument which he calls a *radiometer*. This consists of four arms, suspended on a steel point resting on a cup, so that it is capable of revolving horizontally. To the extremity of each arm is fastened a thin disc of pith, lamp-blacked on one side, the black surfaces facing the same way. The whole is enclosed in a glass globe, which is then exhausted to the highest attainable point and hermetically sealed.

The author finds that this instrument revolves under the influence of radiation, the rapidity of revolution being in proportion to the intensity of the incident rays.

For one of his first radiometers, Crookes gives the following particulars :

TIME REQUIRED FOR ONE REVOLUTION.

Source of Radiation.				Time in Seconds.
1 candle, 20 inches off		....		182
„ 10 „		....		45
„ 5 „		....		11
2 candles, 5 „		....		5
4 „ 5 „		....		3
8 „ 5 „		....		1·6
1 candle, 5 „	behind	green	glass	40
„ 5 „	„	blue	„	38
„ 5 „	„	purple	„	28
„ 5 „	„	orange	„	26
„ 5 „	„	yellow	„	21
„ 5 „	„	light red	„	20
Diffused daylight, dull		....		2·3
„ „ bright		....		1·7
Full sunshine, 10 a.m.		....		0·3
„ „ 2 p.m.		....		0·25

The invention of the radiometer caused an immense sensation throughout the scientific and semi-scientific world. A Belgian journal, *L'Office de Publicité*, announced it in the following exuberant terms (May 16, 1875) :

Nous avons, toute chronique scientifique à part, un sujet palpitant d'intérêt, un sujet énorme, immense, prodigieux, une découverte qui changera peut-être la face du monde.

C'est ce que j'annonçais tout à l'heure en ces mots profonds :
" La science vient de faire un trou dans l'infini."

Vous voyez que ce n'est pas de la petite bière.

Le journal anglais qui raconte cette découverte, intitule son article : *La lumière, force motrice.*

Résumons-le :

On connaît, cela est élémentaire aujourd'hui, la propriété

249

qu'a la chaleur de se transformer en mouvement, mais on ne se doutait pas de l'action mécanique de la lumière. *Un savant anglais, M. William Crookes, vient de le démontrer devant la Société royale de Londres.* Or, M. Crookes n'est pas le premier savant venu ; la Société royale de Londres compte dans son sein les premiers savants du monde et elle jouit d'une certain notoriété.

Voici où en est M. W. Crookes de ses expériences. Il a pris un tube renflé à l'un des bouts. Il y a placé, sur un pivot d'une extreme ténuité, son moulin de disques de sureau, il s'est assuré d'un parfait équilibre, il a fait le vide, puis, produisant tantôt la lumière, tantôt l'obscurité, il a vu son moulin demeurer immobile tant que cette obscurité dure, et tourner dès que la lumière paraît. Il va sans dire que l'appareil est toujours soustrait par la plaque d'alun à l'influence de la chaleur. Au soleil, le petit moulin tourne si vite qu'il finit par se renverser de son pivot. Un nuage qui passe ralentit le mouvement sans l'arrêter.

Le journal anglais dit :

" La science n'a pas d'explications à donner sur cette étrange découverte qui ajoute la lumière au catalogue des forces mécaniques. On savait qu'elle exerçait une action chimique, prouvée complètement par la photographie, mais nous voyons ici le plus subtil des fluides impondérables, si fluide il y a, devenir une force physique déterminant matériellement le mouvement. Peut-on comparer l'éther, dans lequel les corps planétaires exécutent leurs révolutions et leur rotation, au vide opéré dans la boule du tube, et la lumière, que le soleil nous envoie, agirait-elle sur les planètes dans l'espace, comme elle agit sur les petits disques ? La lumière est-elle en rapport avec la force centrifuge qui balance, pour ainsi dire, celle de la gravitation ? "

An indignant Belgian, M. Bodart, of Liége, wrote to Crookes complaining that his so honourable name was being exploited for propagating a " mystification," viz. that light can produce a motive power, and asking him to contradict this " absurd " piece of news. Crookes replied with modest pride :

A HOLE IN SPACE

May 22nd,
1875.

M. ALPH. BODART, PÈRE,
Huy, près Liége.

SIR,

I beg to acknowledge the receipt of your letter of the 18th inst. inclosing a cutting from a Belgian paper, *L'Office de Publicité*, in which a discovery of mine is mentioned in very flattering terms, by the Editor, M. Bertram.

I beg to state that the facts mentioned by M. Bertram are strictly true ; I have had the good fortune to make such a discovery and M. Bertram has evidently compiled his article from one of the English papers in which the discovery has been noticed.

I remain,

Truly yours,

WILLIAM CROOKES.

A letter of more practical concern was received from Mr. John Browning, optical instrument maker, asking for instructions how to make radiometers and offering a royalty on all radiometers made. Since that day the radiometer has never been absent from the windows of prominent instrument makers.

Crookes also received invitations from all sorts of learned and semi-learned societies throughout the country asking him to give a lecture and to name his own fee.

In spite of a vast variety of experiments, no progress was made towards an explanation of the true cause of the radiometer action until the next year (1876), when Professor Arthur Schuster discovered it in the action of the residual gas contained in the bulb after it had been exhausted to the utmost vacuum attainable. This was Schuster's explanation : gaseous molecules impinge (according to the kinetic theory of gases) upon both the white and the black surfaces of the vanes. Ordinarily, they are thrown back with equal average velocities. But

251

when a source of radiant is present, the black surface gets hotter, repels the molecules with greater speed, and flies back with its own recoil. This recoil produces the rotation observed.

To prove his theory, Schuster floated a radiometer on water, and found that while the vanes rotated in one direction, the glass case rotated (though much more slowly) in the opposite direction, thus proving that the recoiling molecules communicated their force to the walls of the case.

Crookes immediately accepted this theory, and after varying the experiment, carried it out towards a complete explanation. He said (January 17, 1877) :

In a preliminary notice submitted to the Royal Society, Nov. 16, 1876, and published in No. 175 of the *Proceedings*, I gave the explanation of the movements of repulsion under the influence of radiation according to the dynamical theory of gases, first, I believe, used in this connexion by Mr. Johnstone Stoney. In this preliminary notice I brought forward experimental proof that the presence of residual gas is the cause of the movement of the radiometer, and generally of the repulsion resulting from radiation, the maximum effect being at a pressure of about 50-millionths of an atmosphere. According to the dynamical theory of gases, the repulsion is due to the internal movements of the molecules of the residual gas. When the mean length of path between successive collisions of the molecules is small compared with the dimensions of the vessel, the molecules rebounding from the heated surface, and therefore moving with an extra velocity, help to keep back the more slowly moving molecules which are advancing towards the heated surface ; it thus happens that though the individual kicks against the heated surface are increased in strength in consequence of the heating, yet the number of molecules struck is diminished in the same proportion, so that there is equilibrium on the two sides of the disc, even though the temperatures of the faces are unequal. But when the exhaustion is carried to so high a point that the molecules are sufficiently few and the mean length of path between their successive collisions is comparable with the dimensions of the vessel, the swiftly moving, rebounding molecules spend their force, in part or in whole, on the sides of the vessel, and the onward crowding, more slowly moving molecules are not kept back as before, so that the number which strike the warmer face approaches to, and in the limit equals, the number which strike the back, cooler face ; and as the individual impacts are stronger on the warmer than on the cooler face, pressure is produced, causing the warmer face to retreat.

This theory explains very clearly how it was that I obtained such strong actions in my earlier experiments when using white pith as the material to be repelled, and employing the finger as a source of heat, and how it happened that I did not discover for some time that dark heat and the luminous rays were essentially different in their actions on black and white surfaces. The explanation of this is as follows : Rays of high intensity (light) pass through the wall of the glass vessel without warming it ; they then, falling on the white surface, are simply reflected off again ; but falling on the black surface they are absorbed, and, raising its temperature, produce the molecular disturbance which causes motion. Rays of low intensity (dark heat) do not, however, pass through the glass to any great extent, but are absorbed and raise its temperature. This warmed spot of glass now becomes the repelling body, through the intervention of the molecules rebounding from it with a greater velocity than that at which they struck it. The molecular pressure, therefore, in this case streams from the inner surface of the warm spot of glass on which the heat-rays have fallen, and repels whatever happens to be in front of it, quite irrespective of the colour of its surface.

In April of the same year, Crookes varied the invention by making the vanes much lighter and making them revolve under the action of a separate fixed vane. This modification he called the Otheoscope, but the name has not survived to the present day.

In April 1877 the *Nineteenth Century* review was founded, and its first number contained an article by W. B. Carpenter called " A Lesson from the Radiometer," in which he drew a parallel between " psychic force " and the " mechanical action of light," to which Crookes had wrongly attributed the action of the radiometer. After drawing attention to his own repeated disclaimers of any desire to form a premature theory, Crookes finishes up his antagonist in the following style :

As I have said, Dr. Carpenter can draw but one lesson from the analysis of my scientific researches, and he insists that it is criminal to be " possessed " of any ideas, or class of ideas, that the common sense of educated mankind pronounces to be irrational. But the " common sense of educated mankind " at one time denied the circulation of the blood, and pronounced the earth to be the immovable centre of the universe. At the present day it upholds errors and absurdities innumerable, and " common sense " has been well characterised as the name under which men deify their own ignorance. Are scientific men never to step over a rigid line, to refrain from investigation because it would clash with common-sense ideas ? How far should we have advanced in knowledge if scientific men had never made known new discoveries, never published the results of their researches for fear of outraging this " common sense of educated mankind " ? Take the very subject which suggests the text for Dr. Carpenter's article. Can the wildest dreams of the spiritualist ask credence to anything more repugnant to " common sense " than the hypotheses imagined by science, and now held to account for the movements of the radiometer ? In the glass bulb which has been exhausted to such a degree that " common sense " would pronounce it to be quite empty, we must conceive there are innumerable smooth elastic spheres, the molecules of the residual gas, dashing about in apparent confusion, with sixty times the velocity of an express train, and hitting each other millions of times in a second. Will the " common sense of educated mankind " consider this rational doctrine ? Again, both inside this empty space and outside it, between the reader and the paper before him, between the earth and the sun, occupying all the interplanetary space further than the eye can reach or indeed the mind can conceive, there is assumed to be a *something* indefinitely more elastic and immeasurably more solid than tempered steel, a medium in which suns and worlds move without resistance. Is not such a doctrine utterly incredible to the " common sense of educated mankind " ? Yet the kinetic theory of gases and the undulatory theory of light are accepted as true by nine-tenths of the scientific men of the present day ; and doubtless in the processes of scientific evolution in the coming times many a discovery will be brought to light to give a sharp shock to " the common sense of educated mankind."

All these experiments and tests and counter-tests naturally involved a prodigious amount of constructive laboratory work. It would have been impossible for Crookes, amid his multifarious occupations, to carry these out single-handed. Much of the routine work was done by an assistant, who kept rough notebooks, some of which are still extant. A rather pathetic entry under date March 7, 1876, reads as follow :

Making radiometer for exhibition. German-silver arms answer well. Discs of rice pith.

All went on well till putting in bulb when all came to grief. Put it in two bulbs. In the first it stuck to the side. Had to be taken out, unsoldered, put in another, when the cup took a piece out of the disc. The radiometer part is still good.

This sort of thing makes me wretched. The more I tried the worse it got. I have tried hard to finish it, but find it impossible to-day.

C. H. G.

When Crookes saw this note he wrote under it, " *Cheer up !* " and sketched out a better sort of tube to tackle next day.

CHAPTER XIV

THE FOURTH STATE OF MATTER

(1877–9)

IN ORDER TO APPRECIATE the transcendent importance of Crookes's experimental discoveries in high vacua, we must bear in mind that his earlier scientific training was acquired at a time when the leaders of science were convinced that all really great discoveries had been made, and that nothing remained but to carry them " to the next decimal place." The doctrine of the conservation of energy had been thoroughly established, and had lent a sort of cast-iron consistency to a philosophic view of the universe. The elastic-solid theory of light had been accepted as the true interpretation of all known optical facts, and the vortex-theory of atoms promised to reduce the latter to mere modes of motion of an incompressible and all-pervading fluid. When Arthur Schuster in 1872 told the great Kirchhoff of the discovery made in Ireland that light falling on selenium altered its electrical conductivity, the latter merely remarked : " I am surprised that so curious a phenomenon should have remained undiscovered until now."

In chemistry the atomic theory had become so firmly established that chemists spent years of work on " exact " determinations of atomic weights—a pursuit which we now know to be futile.

In 1874 the Cavendish Laboratory was founded at Cambridge by the generosity of the Duke of Devonshire, and the new Chair of Experimental Physics was filled by Clerk Maxwell, who in the short remaining span of his life revolutionised the science of electricity and definitely turned the ideal of Cambridge University away from the conception of Todhunter, who declined to see an experiment showing conical refraction (which he had taught all his life) because " it would upset all his ideas." Thenceforth Cambridge became one of the world-foci of experimental discovery, and its greatest discoveries were made possible by the researches of Crookes in his London laboratory.

But though the seventies saw the new light which shone from Crookes's vacuum tubes, there were many other claims upon his time and energy. We have already seen how his spiritualistic investigations encroached upon his laboratory work down to 1874. From 1871 till 1880 he was also a director of the Native Guano Company, which undertook to convert sewage into saleable manure by treating it with alum, blood, charcoal, and clay, a process designated by "A B C" from the initials of the substances used. The process was the invention of Mr. W. C. Sillar, of St. Swithin's Lane. The company had asked Crookes for a professional report on the process, and Crookes gave it as his opinion that the sewage water, after this treatment, though not exactly drinkable, was sufficiently pure to throw into a river. Mr. C. A. Ionides, one of the directors, invited Crookes to join the board as director as well as chemical adviser, with a salary of £200 a year. The board also acquired Crookes's patent carbolic sulphite deodoriser for £2,200. Crookes wrote an article on sewage disposal for his *Quarterly Journal of Science*, and pointed out that the waste of manurial wealth in London was as if three million quartern loaves were daily left to float down the Thames towards the sea.

The company's works at Crossness, near the Southern Outfall on the London sewage system, dealt with about 1 per cent. daily of the total outfall, or 500,000 gallons. The works contained six settling tanks, 50 × 20 × 8 feet. The water from the last tank was made to flow out in a waterfall, with a passage under it so that visitors could admire the crystal clearness and sweetness of the water which had borne the sewage of London to Crossness and passed through the company's A B C tanks.

Comparing the A B C process with the older process of precipitation with lime, Crookes pointed out that the A B C dry " guano " sold at £3 10s. per ton, whereas the lime manure only fetched 1s. to 2s. 6d. a ton.

In 1873 Crookes invented a new process of dealing with animal refuse, chiefly fish, by treating one portion with alkali and another with acid, subsequently mixing the two portions to form a neutral salt and leaving the nitrogenous matter in a state suitable for use as fertiliser. " The moisture," says the official description of the process, " is extracted in a hydro-extractor."

Sulpho-carbolic acid or sulpho-carbolate of alumina or other acid or alkaline solutions or any desired disinfectant, deodoriser, or anti-septic may be passed through the mass in the hydro-extractor, and when the substance has been deprived of the greater part of its moisture it may be spread out to dry in the air, or it may be dried by artificial means, or it may be dried in bags by hydraulic pressure. When sufficiently dry the substance may be pulverised and sifted to remove the hard bony particles. By this treatment of the fish material the mass is divided into three classes, from which three classes of manure may be produced. The liquid portion contains as the dominant manurial element potash, chiefly required for leguminous plants ; another portion as the dominant manurial element phosphate, chiefly required for root crops ; and a third, which as the dominant manurial element nitrogen, chiefly required for cereals.

Feeling that any activity of his with regard to the

new invention might be considered to prejudice the Native Guano Company, Crookes placed his resignation in the hands of the directors, who, however, refused to accept it. Crookes thereupon founded a small company, with Messrs. Ionides and Bristowe as partners, to work the fish invention. It started on an experimental scale under the name "Crookes & Co.," but was abandoned after a few years. Crookes's connection with the Native Guano Company lasted till 1880, and he spent much time on propaganda visits to various cities, including Glasgow, Leeds, Manchester, and Paris.

The general state of Crookes's domestic affairs about this time may be judged from some letters addressed to his eldest son, Henry, born in 1859, who went on a sea-trip to Australia for the benefit of his health. Crookes provided him with many introductions, one of which was addressed to Dr. F. B. Miller, of the Royal Mint, Melbourne, a brother of Crookes's friend, Dr. W. A. Miller.

The letters to Henry reveal Crookes to us in a new light. He appears as the affectionate and indulgent father in a growing young family, surrounded by innumerable friends and relatives.

April 12*th*, 1877.

My dear Henry,

I wrote on March 30th via Brisbane giving you news up to that date. I hope you will have received it before you get this. According to the *P. & O.* handbook, the mail which takes this letter will arrive at Melbourne on the 30th of May, when you will have left us 83 days. So you and this letter should about meet. There is little news of immediate interest to tell you. We are all of us very well, which I suppose is the chief thing you care to hear. I am sorry, however, to say that my Uncle George Crookes, of Balham, my father's only surviving Brother, died on the 6th. He had been very ill and not expected to recover for some time past, so his death did not take us by surprise ; still it is always a shock to lose anyone whom we have long looked on as a permanent member of the family—one of the social landmarks in fact, and it makes serious thoughts arise, as to who is to be the next called away. And then amongst the visions which these thoughts call up there flit before one those whom we used to know and love, but who are now no more ; and then I seem to see my dear brother Phil, who was so like you in appearance and manners, and who went away for a long voyage, and now lies at the bottom of the ocean ; and then we wonder how you have been getting on for the last six weeks in which we have heard nothing of you, and we pray that you may not be taken away, but will come back to us with renewed health and vigour.

But enough of this dismal writing, I suppose you want to know how we are all getting on. Your Mamma is pretty well, rather tired just now, having been up late for the last few nights helping me to get through my letters. She and Alice are anticipating the pleasure of a trip to Brussels where I have to go on Saturday, and where I have promised to take them if they are good ! We start on Saturday morning by the day mail, via Dover and Calais, and hope to stay about ten days. I have already ordered rooms at my old Hotel, the Hotel de l'Europe, and if we have fine weather I have no doubt the ladies will enjoy the trip much. Alice is hard at work helping to pack. I am going over on Native Guano business, and am to take over a box of different A B C mixtures, to work a patent there. I am in some doubts whether the Custom House officers will let the stuff pass, for I must confess it looks suspiciously like gunpowder, dynamite, or some other infernal mixture, and if I explain it is A B C they won't understand. Joe is now writing to you, I believe, so I need not tell you about his holiday to Southsea, further than saying that it has done him a world of good, for he looks so rosy and well. It has not, however, diminished his cheekyness. Jack is sweating away at his French exercises at the table opposite me, he is working very hard, and hopes to have some tangible result to show when prize time comes. He has been to Yoxford, and Suffolk has improved him as much as Southsea has Joe. We do not know very recently how Bernie and Walter are, but we had a letter a little time ago in the curious staccato style they seem to affect at Uxbridge, saying that their holidays would begin soon. I believe they are coming home this day week. Florrie is in rude health and as rumbustical a young lady as was ever seen.

Lewis seems to make daily progress with his tongue ; he can say "poof" and "mulk" and before my box was empty he came down every morning asking for "chockaleary." But some days ago I gave him the empty box, and it is astonishing how his filial affection in the morning has declined. When he sees me he asks if I have any chockaleary, and on my saying no, he says "Then buy some." I have now a full box waiting for him, so he will restore me once more to favour. He is getting a regular little chatterbox and has to be hushed down at lunch sometimes, or we could not hear each other speak. He misses you very much, especially on a Sunday after dinner, and often talks of you and asks when you are coming back.

Charlie is at work in the laboratory. We have got a new instrument like the radiometer, but on a different principle, which I shall call the "Elaunoscope," from the Greek *Elauno*, to repel. I am keeping them quiet now, intending to bring it out at the Soirée of the Royal Society on the 25th inst. The laboratory is full of Elaunoscopes of all sorts and shapes, for, unlike the Radiometer, these can be made of an infinite variety of constructions. I have one about ½ inch diameter mounted as a scarf pin. If you have an opportunity get hold of the new periodical called the *XIXth Century* for April. My old enemy, Carpenter, has written another attack on me in it ; but as it is a tissue of false statements, and is written in the worst possible taste, I am not sorry, as he lays himself so beautifully open to an attack in reply, which I am preparing and in which I hope to cut him up, with the calmness of a vivisector, till he has not a leg to stand on. I am afraid my reply will not be out in time for you to have one sent you before you return, but it will be ready for you when you come.

Florrie and Lewis have just come in. They are happy in the enjoyment of lots of chockaleary. Lewis has written you a letter. I asked what his name was, and he told me, I guided his hand and he wrote it as close an imitation of the sound as I could get.

Mother sent me a note this morning, hearing I was going to write to you. I enclose it, as you may like to see that you are not forgotten. She and Father are pretty well now, they have both had colds. Father comes down occasionally and always asks if we have any news of you. I went with Joe to Green's on Tuesday, but they had not heard of the *Devon* since she was sighted on the

12th of March. I send you a couple of portraits for your album. Everyone thinks they are excellent likenesses. I have a couple of large ones like yours framed and hanging up in the dining-room one on each side of you. They look very nice. The Day's called in a night or two ago. They seem very well. One of them has an increase of salary or has got a better berth at another office. We have seen or heard nothing of Grün since you left.

I suppose you will have almost too many letters of introduction, but I enclose one more for you. It is to Mr. Sydney Gibbons. He is a Fellow of the Chemical Society, and public analyst to either Melbourne or some district. You may find it useful to call on him. He frequently sends papers to the *Chemical News.*

I have also copied out for you the Code of telegraphic signals, in case anything may have happened and you may have lost the copies you had with you. It is *very important* that you send us news immediately on your arrival, so don't forget to go to Reuter's office the first thing, and arrange for the message " Llandelly." I sincerely hope the other word will contain Good News.

The Fish works at Rainham are not getting on at all well. Whieldon's machinery is not equal to our expectations, and some will not work at all, whilst the hydraulic press is of no use. We have, however, got another press, worked by a screw, which will press about a ton of fish at a time, and in an hour or two. We are expecting a large press of this sort to be fitted up soon, and then we shall be able to go ahead.

An American paper, the *Popular Science Monthly,* has given a very excellent likeness of me, and has an article about me containing my Biography. I would send it, only I have only one copy.

I hope you have had pleasant weather. Since you left, we have had nothing but a succession of storms of wind, rain, thunder, and lightning. Such a turbulent time was never known. The country is now flooded almost as bad as it was in the autumn.

The Native Guano Co. are having more " Bear " circulars sent round against them. On the other hand, they are on the point of getting the process adopted at Hounslow and Isleworth. I can't think of anything else to say to you now. Mrs. Pearman is upstairs now ; she seems very well. Her two lively girls dined with us a week ago. Joe has just brought me a letter to enclose, and Alice has promised to write a few lines to-morrow. Your Mamma sends best love and loving kisses. Your grandmamma, who is as jolly and well as ever, also sends best love, and with the same from me and all of us.

Believe me, my dear Henry,
Your very affectionate Father,
WILLIAM CROOKES.

May 11th, 1877.

MY DEAR HENRY,

Since we last wrote the house and its inhabitants have been pursuing the " even tenor of our way." To us who are here on the spot there seems scarcely any difference in condition between now and then, but I suppose to you at the antipodes every little trifle of home news will be of interest. We often wonder where you are, and if our calculations are correct you will now be about a fortnight or three weeks from your journey's end, bowling along in the Trade winds in the most delightful climate in the world. We are getting very anxious now for news of you. Not a word has arrived of the ship since she left the Channel, and we may be now expecting almost every week the long-looked-for telegram.

I think we were starting, or preparing for Brussels when I last wrote. Your Mamma and Alice enjoyed the week there amazingly, but I don't think it had much effect in improving their French, for all the waiters in the hotel spoke English as well as we did. I need not say more about this, however, as I suppose your Mamma told you all about it in her last letter.

In health we are all pretty well. The weather, which has been very windy and cold, has taken a turn the last week and it is very mild and inclined to rain. To-day Joe and Jack have a holiday on account of the Athletic Sports. It is, however, pouring with rain so they can't get out and must amuse themselves indoors. I will suggest their writing to you. Your Mamma is pretty well but has a slight touch of cold. She and I were at a musical soirée last night at Mrs. Haweis's. We met Mr. Moses [1] there and he told us that Joe was working very well and getting on capitally now. He wrote an essay a week or two ago which is the best in his class and it has to be read aloud by the author and then discussed

---

[1] Mr. W. Stainton Moses, M.A., a noted spiritualist and medium. He was a master at University College School, London.—E. E. F.

in class. Joe feels quite important ! He has got a bicycling mania, and every evening he goes practising on Pennington's bicycle, and on returning gives us thrilling accounts of his adventures, interspersed with elegant phraseology expressive of delight at the mode of getting along. I can't remember how many times he has been pitched head first over the big wheel. He has just come in, and on my suggesting he should write to you, says " Oh ! I don't know what to say. I wrote once," so I suppose you won't have a letter from him this time. Alice is very well and has taken to wear a cap ! She looks like a pretty housemaid at the theatre, and I tell her she will have young gentlemen callers trying to kiss her on the stairs mistaking her for the housemaid. Whereupon she tosses her head and says " Oh, indeed, I should like to see them try ! " I believe she has copied the cap from Alicia who appeared in a cap when Alice and Joe were in Russell Square a week ago. There is no news from Bernie or Walter since they returned to Uxbridge. Florrie is happy, in having company staying in the house. Mrs. Gruneson's little boy is on a visit here, so with him and three or four from next door, and five or six promiscuous schoolfellows up the street, Florrie about keeps the nursery tea-table pretty full. Lewis is doubly happy, first there is a little kitten a few days old, to which he pays periodical visits and apostrophises in a loving tone, and secondly, because he has discarded petticoats and has a complete sailor's suit. He looks the queerest little shrimp you ever saw with his little tight knickerbockers and his hands stuck in his pockets. Our Cook has left us and last evening another one came. I hope she will be an improvement on the last. She can't, however, well help being that, for the old one was about the worst cook we have had for a long time. Charlie has been home for a week, ill with a carbuncle on the back of his neck. He is back again now, but is far from well. We are at the same old game " Log. decs." : I want to get my paper on this subject out before the Autumn. Did I tell you that Dr. Carpenter had written an article about the Radiometer in a new periodical the *Nineteenth Century* ? It is a nasty spiteful article, and I am preparing a reply to him which will I hope teach him not to attack scientific men with unfounded accusations.

I am afraid I shall not have my reply out in time to send it to you, but I will do so if I can. At Brook Green they are as well as usual. Mother has a cough, but that is not much to be wondered at considering the weather, and Father is very deaf, but in other respects they are well. Mrs. Geeves is still there and is pretty well.

Your Cousin Elizabeth is gradually getting better, but I fear she will never get about again. She cannot yet leave her room. Mrs. George (senr.) is not well, she does not rally after her sad bereavement.

We saw Frank and Annie at Brook Green on Tuesday last. They were very well.

Crookes & Co. is, I fear, in a bad way. The machinery won't answer a bit, and I think there will be a lawsuit between the directors and Whieldon about the payment of the balance they claim. Meantime another press is ordered elsewhere, and all work is stopped till that comes. Native Guano is just the same as when you left. Aylesbury is not ready yet, but we are hoping it will be at work soon.

I suppose by the time you get this you will have been about three weeks on shore. Your movements there must be entirely guided by circumstances, as the long time required for letters to pass between us entirely preclude your asking our advice as to your doings.

I hope you will have made some friends out in Melbourne or during the voyage who will be of use to you when on shore. Some of the letters of introduction will be almost certain to have proved of value. You will remember that the first and main object of the journey is *your health*, and you must let that be the chief consideration in deciding what you will do. Get as much into the country as you can. The actual city itself is not a very healthy place. If you find the money holds out and you have pleasant introductions and invitations, there is no hurry about returning at the end of the month, but if all is not comfortable, or if you are living mostly in lodgings, why the best thing you can do will be to take the first ship of Green's back to England But whatever you do keep us well posted up in your movements, so that we shall know what ship you are returning by and when to expect you. I will keep on writing by each mail till I actually know you are returning, so that in case you remain longer than we expect, you may not be without news, but as in all probability the later letters will be returned unopened, we shall not write long ones.

And now, my dear boy, Good-bye for the present. I expect

you will receive other letters by this mail, so they will supply any gaps left in my budget of news. With all love and best wishes for your happiness and safe return.

Believe me, my dear Henry,

Your very affectionate Father,

WILLIAM CROOKES.

5 p.m.—P.S. I went out after writing the above and have come in to seal up the letter. I have brought this evening's *Globe* in with me, so I send it as the very latest English news you will get. There is nothing very special in it, but you may be interested in it.

W. C.

May 25th, 1877.

MY DEAR HENRY,

It is scarcely worth while to write a long letter, as in all probability you will be on the return voyage when this gets to Melbourne ; but in case you do not leave quite so soon, I will let you know briefly how we are all getting on. Your Mamma, Grandmamma, and Alice are very well, they are out shopping just now ; and are somewhat busy in the house, as Anne, the housemaid, has gone for a week's holiday. Joe is getting on very well at school, and seems to be really working hard and trying his best to come out well in the Exams. He has a bicycle fit on, *very bad*, he has had one lent him, and he and Pennington take long rides together, and have all sorts of adventures. Joe is very anxious to coax a bicycle out of me, but I am going to tell him it depends on his progress at school, otherwise he would neglect his lessons altogether for his new hobby. Jack is at home to-day with a slight touch of earache, but it is getting all right, and he will I hope be back at school to-morrow. Bernie and Walter are well at Uxbridge, at least Walter is, and Bernie will be soon I hope, for we have just heard that Bernie has a cold and has been in bed for a day, some of us are going there to-morrow to see how he is. Florrie and Lewis, as usual, are in the most boisterous of health and spirits. Father has just been. He is in excellent health and spirits for an old man of 85, and seems as if he had 15 good years in him yet. Mother has a cough, but that is not to be wondered at considering the cold wintry weather we are having this spring. Your Cousin Elizabeth is still alive, but I fear that is all I can say, she is getting worse daily, and cannot last long now. Mrs. Geeves is going to her new house at Richmond on Tuesday. Frank and Annie are going for a few days' pleasuring to-morrow, but I don't know where. Alfred is away on a holiday. Charlie is at work at the Blowpipe and " Log. decs. " as usual and I suppose will keep at that work for some time to come, for the investigation seems likely to be a long one. As for myself, I am happy to say I am in good health, a little extra busy perhaps, just at this time of the year, but not too much so to be a trouble. I went yesterday to Green's, but they had heard nothing of the *Devon* since she was spoken in the Channel, and I was told that they allowed 90 days for her to get to Melbourne. The 90 days will be up on the 6th of June, in 12 days time, so we shall be looking out constantly for your telegram after then. I hope the news will be good. We are getting very anxious about you, it seems so long a time without any news whatever, and in the absence of anything definite there is a natural tendency in the mind to imagine all sorts of dreadful things.

All send their very best love. Sunday next will be your birthday. You may rely upon us not to forget to drink your very good health. Good-bye and God bless you, my dear boy.

Believe me, your very affectionate Father,

WILLIAM CROOKES.

Friday, June 8,
1877.

My dear Henry,

I suppose you will be on your way home by the time this letter gets to Melbourne, but in case you stay a couple of months you may still be in time for it. I therefore just write a few lines to say how delighted we all were on Monday last to get your telegram " Platinum " [1] dated Melbourne the previous Saturday. I assure you nothing more opportune could have happened. It quite put your Mamma into good spirits again, she had been rather poorly and anxious for a little time past, but this capital news seems to have quite set her up again and she is looking quite bright and well again. She immediately sat down and sent off I don't know how many postcards to all our friends and acquaintances (not forgetting Alicia), and we have been receiving congratulations ever since. I hope you received

[1] A code-word agreed upon to mean " Moderate Passage."

my return telegram giving Browne's address. I thought as he was the only friend who would be likely to treat you better than a mere acquaintance would, it might put you to inconvenience if you could not find him, so I telegraphed " Browne's address, Montague Hotel, Park st."

I told you in my last that Bernard was not very well. He got much worse, and for a week your Mamma was going to Uxbridge every day to nurse him. He is now better again and in a fair way to complete recovery. It has been a sharp attack of pleurisy. All the rest of us are, I am glad to say, quite well. We shall look out for letters anxiously now, so as to know what sort of a voyage you have had and what are your probable movements.

With all our loves,

Believe me, my dear Boy,

Ever your affectionate Father,

WILLIAM CROOKES.

About this time the Native Guano Company heard about a French method of sewage utilisation for which great efficiency and economy were claimed. It was a process invented by M. Georges Fournier, of Rheims. At the request of the company, Crookes went over to Paris to inquire into the process. Immediately on his return he wrote to Mr. Rawson to say that the process worked efficiently and cheaply, and that he had made a draft agreement whereby M. Fournier was to receive £2,500, spread over four years, for the English patent rights. This agreement was immediately ratified by the company.

The process involved the use of aluminium chloride instead of the sulphate. The chloride was derived from aluminous shale by a process which yielded readily saleable by-products. Crookes was sure that the shale could be obtained cheaply from British mines, notably in the Campsie district near Glasgow. He visited Glasgow with M. Fournier, but found it difficult to gain access to the Campsie mines. He then visited Leeds and examined the shale to be found in that neighbourhood. No shale containing iron pyrites was found, but some iron pyrites was found in the coal, and this, mixed with the shale, would give the aluminous compound required.

It was eventually found to be more advisable to form a special company to work the Fournier process. It was called the Fournier Company, Ltd., with offices at 4, Bishopsgate Street Within, E.C. Its first chairman was William Roebuck.

It was not a very favourable time for starting new companies, as the Russo-Turkish War was at its height. There was much speculation, and Crookes, too, had a " flutter." He wrote to his friend Ionides :

Novr. 28,
1877.

DEAR MR. IONIDES,

In the present dense fog which hides the political horizon I cordially approve of your precaution in selling my Dover A's whilst a small profit can be secured.

As I was going home I called on Mr. Arnold to see if I could pick up anything. There is nothing to pick up, however. He says Egypt Rails are as good as ever. Egypt is nearer England than it was six months ago Russia is not to have free passage of the Dardanelles and the Cabinet are discussing the advisability of intervening before Turkey is quite beaten. But these are mere rumours.

The most solid fact of all is what you told me about Goschen's people buying Egypts.

Believe me,
Very sincerely yours,
WILLIAM CROOKES.

The Goschen here referred to is no doubt Viscount Goschen, at that time (or shortly afterwards) British Ambassador at Constantinople.

Crookes made frequent visits to Paris and Brussels in connection with sewage precipitation business. In

Paris he took daily samples of the sewage at Genevillers, some miles outside the city. These samples were collected in long stone jars such as are used in France for storing brandy. These jars were a puzzle to the *octroi* officers at the *barrière*. Crookes and Fournier solemnly declared that they contained *eau d'égout* and not *eau de vie*, but met with a large amount of incredulity until one day when an over-zealous and precipitate officer insisted on " sampling " the sample himself ! The effort was so profound and instantaneous that Crookes wrote in a letter home : " We could have smuggled any amount of real *eau de vie* into Paris after that with impunity."

The Native Guano Company was never really prosperous. With the exception of the Aylesbury sewage works, the various works were a constant source of expense. Crookes himself drew a salary of £1,000 a year. The London office and management cost £1,370 a year, of which £280 was office rent at Victoria Chambers. The Fournier process was expected to give a new lease of life to the company, but its transfer to a separate company seems to have prejudiced the finance of the original company.

On March 11, 1880, Crookes attended his last board meeting at the Native Guano Company, and bade the directors a friendly farewell.

Amongst the responsibilities thrown upon Crookes by the editorship of the *Chemical News* and the *Quarterly Journal of Science*, the working of gold-extraction patents, the solution of sewage disposal problems, and the bringing up of a family of six children, his scientific work went on steadily. It speaks highly of his organising ability that he was able to carry so heavy a burden without a breakdown. But he was very methodical, and managed to keep his various tasks in separate compartments of his mind. Besides, he was happy in the choice of his instruments and in the treatment of his subordinates. With very few exceptions, he secured loyal and cheerful service, and rewarded it well. Since the invention of the radiometer, the expense of his research work was defrayed by successive grants from the Royal Society, so that it did not fall upon his private resources. And who could say that £600 was too much to pay for the magnificent series of investigations which produced the radiometer, revealed matter in a fourth state of aggregation (" radiant matter "), and came within an ace of discovering the Röntgen rays ?

It was Crookes who first produced a vacuum containing only one-millionth of the original amount of air. He learned, so to speak, to disport himself in that tenuous atmosphere, and revealed amazing properties of ultra-gaseous matter which not only provided us with a new and powerful instrument of research, but which, in the hands of his successors, solved the main problem of the nature of electricity and revealed its atomic or, rather, electronic structure.

Most of the research work at this period was carried out under Crookes's direction at his laboratory in Boy Court by Mr. Charles Gimingham, the young chemist who had taken part in the Home séances. Gimingham was an expert glass-blower, and Crookes entrusted him with the execution of some very complex and difficult experiments. One of his numerous letters to him reads as follows :

Aug. 15*th,*
1877.

My dear Charlie,

I have just got home all right, but tired with a 36 hours' spell of travelling. All seems right in the Laby. I think you did quite right to leave as you did, for it would have been a pity to have missed the opening meeting of the B.A. I have tried no experiments of course with the instruments, but they seem all right, and the results in the big book are very interesting and quite agree with what were formerly obtained.

What state is the otheoscope in on the pump? Is it the one the last notes are about? Is there Hydrogen in? I rather hoped you would have put hydrogen in all, as the abnormal results are likely to be better. But perhaps I shall find more about it in the book. I hope you will enjoy yourself. Write occasionally when you have time.

Very sincerely yours,
WILLIAM CROOKES.

Some days later Crookes wrote to him :

To-day is our General Meeting of the *N.G. Co.*, and I have to make a speech which will send the shares up !

In November 1877 the scientific public witnessed another skirmish between Crookes and the redoubtable Carpenter, who in 1875, when Crookes received the Royal Medal from the Royal Society, had proposed to " bury the hatchet." The occasion for digging up that implement of warfare was the publication in the United States of a letter written by Crookes about a medium, Eva Fay. The letter testified to Crookes's conviction of her *bona fides*, and Carpenter, considering that the credit of the Royal Society was at stake, immediately took up his tomahawk and wrote to *Nature* to protest. Crookes rallied his friends to his side, and repelled the attack in letters published in *Nature* on November 1 and 15, 1877. The following letters throw some interesting side-lights on the situation, which was complicated by the fact that both Crookes and Carpenter were on the Council of the Royal Society.

Novr. 9, 1877.

MY DEAR FOSTER,

You may perhaps feel interested in reading the enclosed letter which I have just received from Mr. Johnstone Stoney. I wish he would write a letter to *Nature*, but I fear his health will not permit him to do so. I was in hopes my last letter would have rendered another from me unnecessary, but Carpenter has followed his old tactics, he has quite ignored the refutations of his numerous misstatements, and has broken fresh ground. My answer to this will be very simple : " It is false."

What a nice happy family we shall make round the council table of the R.S. !

Please return Stoney's letter, and believe me,

Very sincerely yours,
WILLIAM CROOKES.

Novr. 11, 1877.

MY DEAR WALLACE,[1]

I am glad you have sent that note to *Nature*. I have just finished a long letter to *Nature*, which I should like you to see before you send in your article to Fraser.

I have got Carpenter in a cleft stick now, and I don't see how he will wriggle out.

Believe me,
Very sincerely yours,
WILLIAM CROOKES.

Novr. 11, 1877.

MY DEAR LOCKYER,[2]

I hope you will find room for this in next *Nature*. I was in hopes I should not have had to reply again, but Carpenter's last accusations were too serious to be passed over.

I hope I shall not have occasion to write more letters, but Carpenter will go on till Doomsday if you let him.

I leave London on Tuesday night. Can I have a proof on Tuesday morning ?

Very sincerely yours,
WILLIAM CROOKES.

Crookes was at this time still engaged on the disposal of the Paris sewage, and paid frequent visits to the French capital. The fish refuse disposal concern known as Crookes & Co., and financed by Mr. C. A. Ionides, was

[1] Dr. Alfred Russell Wallace.
[2] Professor (Sir) Norman Lockyer, Editor of *Nature*.

T

being earnestly pushed forward, and Crookes's eldest son was employed on the task. Crookes's natural bent towards speculation was stimulated by the critical financial aspect of international politics, and it is interesting to find him taking an amateur hand in *haute finance* for once in a while. The following letters make strange reading to those who look upon Crookes as a dreamy embodiment of the ideal of abstract science.

c/o M. FOURNIER,
42, Bd. de STRASBOURG, PARIS,
Decr. 5th, 1877.

MY DEAR MR. IONIDES,

In a letter received from my wife this morning I hear that Mr. Evans went down to Rainham on Monday and gave the engineer notice to leave at the end of the week, and told Harry to be prepared for a week's notice when he went down again. Harry asked what was to be done with the machinery. Evans said he had a person who would look after the things until they were sold. Harry says this new machine works *most satisfactorily,* so he is very much astonished at being told he is to leave so soon.

From all I can hear, Evans is not the sort of man to make the thing pay. Harry says he goes down to Rainham, just looks round, speaks to no one, and goes away. About ten days before the machine was ready he sent some fish down, which got quite putrid before it was wanted, and he then ordered it to be buried.

I have had several letters from Evans telling me of little delays in the final fitting up of the large machine, and postponing the trial which was to come off in our presence. I find there was a board meeting on Monday. I regret that I was away on that occasion, as I might have induced a different decision being arrived at. I have now a large stake in the success of the Company, and I cannot imagine that just as the New Machine proves to be a great success, it can be contemplated to sell everything off and wind up the Co.

Political affairs here are very black. The opinion of all to whom I have spoken is that a revolution is imminent. The Maréchal is very firm, and is likely to dissolve the Chamber and have another election. The Chamber has decided *not* to be dissolved, as it says the opinion of the country has just been asked and answered unmistakably. My friend Fournier knows Gambetta very well, and is also acquainted with M. Dupanloup's private secretary. He is quite willing to get me any information I want. Do you think it will be worth while telegraphing anything over to you? I shall probably be here for ten days, and during that time the crisis will have come and perhaps gone. A revolution will be certain to depress French Rentes, and our Funds will be likely to suffer a temporary fall, in sympathy. The question of the Exhibition being held entirely depends on how political affairs here go on between now and Xmas. If there is a settled Government in Paris before Xmas, there will be an Exhibition, but if all is in the present unsettled state there will be no time to organise the Exhibition, and it will be postponed.

I will write to you again as soon as I have anything to communicate. I may hear something this afternoon.

Believe me,
Very sincerely yours,
WILLIAM CROOKES.

27, RUE LEPELLETIER, PARIS,
Decr. 9, 1877.

MY DEAR MR. IONIDES,

Since I wrote to you on Friday, or rather as I was writing, the scene changed, and all the favorable hopes of political calm were upset by the Maréchal's obstinacy. You are probably as well acquainted as I am with the way affairs are going on here. This morning's news confirms what was reported last night, that MacMahon has asked Batbie to form a ministry. This will be a ministry of the Right—the Minority—and it will be a ministry of Dissolution. It is the general opinion here that the Chambers will refuse to dissolve, and then will commence a war between Maréchal and the Chambers. I don't think there will be an old-fashioned revolution, for the Army is on the side of the Republicans, but the Maréchal is obstinate, and is likely to attempt a *coup d'état*.

I hoped I should have had a letter this morning from you. I am not quite certain whether or not you wish for political news. You once told me you did not much care for it, as you did not do much with French funds, but recently you said if I would send you *facts* from Paris, you would use your *judgment* about them,

275

and the result might be to the advantage of both. I am very well pleased at the result of our first spec., and I think it is not unlikely that there may be several rises and falls in securities during the next week, and early trustworthy information may enable something else good to be done. The changes here are so sudden just now that I am sure you will agree with me that it is safer to realise early to secure a small profit, than to risk all by waiting.

I will telegraph to you late to-night to the Stock Exchange, so that you will have it waiting for you when you get to town. At about half-past ten p.m. my friend and I have arranged to go to the Petite Bourse and see one or two men he knows, who can give him the latest news. This I will telegraph.

Please note alteration of address. I shall be in and out all Monday morning if you have occasion to telegraph to me.

Believe me,

Very sincerely yours,
WILLIAM CROOKES.

On December 10th Crookes dispatched the following telegram, the innocent mystery of which can hardly have deceived the vigilance of the French telegraphists :

IONIDES,
 3, Copthall Buildings,
  Throgmorton Street, London.

In alumina to-day, William will ask iron to be acidified. Alumina will refuse. William will then make sulphate with James for the purpose of attacking alumina, with the assistance of phosphate. To-morrow phosphate will be asked to cause precipitation.[1]

Crookes also wrote :

27, RUE LEPELLETIER, PARIS,
Decr. 11, 1877.

DEAR MR. IONIDES,

 At nine this morning I sent a telegram to you as follows : " During night Batbie found difficulties. Postscript to letter not confirmed so far."

[1] In the Chamber to-day, MacMahon will ask for the Budget to be voted. The Chamber will refuse and MacMahon will then get Batbie to form a Ministry for the purpose of dissolving the Chamber, with the help of the Senate. To-morrow the Senate will be asked to cause a dissolution.

My information last night was obtained direct from the secretary of M. Dupanloup, who is a friend of Fournier's. Dupanloup is editor of the *Défense*, the organ of MacMahon, so I have no reason to doubt its perfect accuracy. During the night, however, something occurred to prevent the names being published in the *Journal Officiel* this morning.

I quite agree with you that great caution must be exercised now. A thing is apparently settled one hour, and the next all is changed ; therefore the most authentic news may be erroneous by the time it reaches you. The Bourse is the best political barometer, and the funds have been sinking.

There is this, however, to be borne in mind. Whatever the Marshal or the Chambers do, the Republicans will conquer in the end. They started with a strong majority, and the recent conduct of the Marshal has alienated many of his friends. A fortnight ago the Senate was probably on his side, now it is against him. However much the funds fall therefore, they will rise before long, for the conflict for supremacy will be short.

I will keep my eyes open, and if there is anything to be learnt before 6, I will give it in a postscript.

Believe me,

Very sincerely yours,

WILLIAM CROOKES.

It is of interest to record that on the whole Crookes was successful in his *haute finance* on this occasion, and added considerably to his fortune.

In January 1878 Crookes became a Fellow of the newly founded Institute of Chemistry of Great Britain and Ireland

At this time his eldest son, Henry, was in Paris, and the following letter breathes the spirit of true paternal solicitude :

Jany. 9*th*, 1878.

MY DEAR M. FOURNIER,

I wrote to you on December 18th about my son Henry, saying that I thought it better to defer his coming to Paris till he had had a little more experience in chemistry. I think I should like to bring him with me when I come on the 27th, and thus have him with me for a short time in Paris, so as to see how he gets on before leaving him to his own resources.

I understood you to say that our friend M. Coquerel would take him into his laboratory, and put him through the routine of the chemical work as carried on there. Henry could also sleep at M. Coquerel's and take his meals there.

Will you kindly ascertain for me whether I am right in the above suppositions, and also kindly get me answers to the following questions : (*a*) What sum of money I should pay to M. Coquerel per month for Henry's Board and Lodging ? (*b*) What are the hours of work ? (*c*) Is there any supervision kept over the young men when these working hours are over ? (*d*) Henry is young and has never been on his own account yet, except a month at Melbourne. He is very steady, but there are temptations in Paris which might prove too much for him if left entirely to himself without any restraint.

I shall be greatly obliged if you will speak to M. Coquerel about these points and let me know the answer. With kind regards to Madame Accard and yourself.

Believe me,

Very sincerely yours,

WILLIAM CROOKES.

One of Crookes's great ambitions was to have some official connection with the Académie des Sciences in Paris. He knew that it would be very useful in France, and the quarrel over thallium had blown over. He wrote to Comte du Moncel, the author of a great work on the Applications of Electricity, and himself one of the immortals, asking him to present a note on the radiometer to the Academy. Du Moncel caused him to be invited to show his radiometer to the Academy, which he did in February 1878.

The reports on work done under the Royal Society's grant were sent to Professor Stokes, who assisted Crookes with much sound advice and often suggested experiments to prove certain points. Stokes usually turned out to be right on points where he differed from Crookes. The latter, in the course of radiometer observations, had found that if the vanes were charged negatively (as in his " electric radiometer ") a dark space surrounded each vane and rotated with the vane. On November 30, 1878, Crookes sent a paper to the Royal Society under the extraordinary title, " On the Illumination of Lines of Molecular Pressure and the Trajectory of Molecules." It was really the first description of the phenomena subsequently known as those of " Radiant Matter " (and in modern times as " cathode ray phenomena "). The unusual title was probably modelled on Faraday's title of his announcement of the rotatory polarisation of light in a magnetic field, which Faraday announced under the title, " The Illumination of Lines of Magnetic Force and the Magnetisation of Light." That Crookes's description of the phenomena, and his implied theory, met with considerable criticism from Stokes is evident from the following letter :

April 22, 1878.

DEAR PROFESSOR STOKES,

I have carefully gone over my paper and corrected it in Red Ink. It is now ready to take to Mr. White. The corrections are in most instances verbal, and tend to clothe my meaning in more accurate language.

In your " Remarks " you state four times that my explanation is wrong. I see now that the *wording* of the explanation was inaccurate, but my meaning was correct. I have spoken in each case of the *radiation* of heat from the lampblack causing molecular pressure ; and you very properly correct me by saying that radiation of heat will produce cold, and hence diminished pressure ; and that the real cause is the *absorption* of heat by the lampblack.

My meaning was this :—The great absorptive power of lampblack for radiation is taken for granted. Its radiating power for heat is equally great. The absorption of radiation keeps up a constant supply of thermometric heat in the lampblack, and its radiating power for heat enables the lampblack to pass on this surplus heat to the adjacent gaseous molecules. By radiating power I meant power of rapidly communicating heat to the molecules touching it. But I see I used radiating in a wrong sense, as real radiant heat from lampblack would pass through the rarefied gas with little quenching. At the bottom of page 27 my explanation is a little clearer, and shows that I had got hold of the right idea, although in that case also I speak of *streaming molecular pressure.*

In a new subject in which several investigators are working each selects his own nomenclature, and although each may mean the same thing, their language being different may make it appear as if their meanings were different. Most of the quarrels in life are caused by two people attaching a different meaning to the same word, and I suppose the same result might equally well arise from different words being used to express the same meaning. I wish we could all agree on certain terms. Stoney uses the terms " Crookes's Force " and " Crookes's pressure." Maxwell calls it " Stress." I have used the words " Molecular pressure " and " molecular wind." " Stress " seems to be a good word, but it seems to want some addition to render it less indefinite.

Instead of speaking of " Molecular pressure streams from," &c., I suppose I am right in saying " lines of molecular pressure radiate."

I have struck out a passage on page 60.

I am in doubt as to what is best to do in reference to my explanation of the rotations of the Cup Radiometer, pp. 79–82. If I make the verbal correction about the *radiation* of heat from the lampblack, I think my explanation is correct. But you have put the explanation in much clearer and more condensed language, and it is probable that some who would not understand my explanation would understand yours.

Would it do to let mine stand (provided it will pass muster) and then give yours in a footnote?

I am getting on with the self-registering chronograph you suggested, and hope to have some results soon.

Believe me,
Very sincerely yours,
William Crookes.

Before giving Crookes's own description, it may be of interest to give the modern view of the same phenomena, so that the reader may judge how far astray Crookes went in his first theories, and to what extent he eventually approximated to the true theory as evolved in the next fifty years.

The phenomena are produced in an exhausted glass tube into which two platinum wires are fused, each bearing a plate or other metallic terminal. These terminals Crookes calls " poles." They are now called " electrodes " in accordance with the elegant Greek nomenclature used by Faraday in the case of electrolysis. Crookes's " negative pole " is now called the " cathode " and his " positive pole " the " anode."

Crookes's " dark space " of " molecular disturbance " is, according to modern views, a space filled with negative corpuscles (" electrons ") projected from the cathode under the influence of electrostatic force. Their velocity increases in their flight along the lines of electric force, and finally their speed suffices to " ionise " or split up the remaining gas molecules into particles or " ions " carrying positive and negative charges respectively. This ionisation is attended with the luminosity which limits the dark space. At extreme vacua, the electrons hit the glass walls of the tube, and their impact produces a green or blue fluorescent patch which emits X-rays.

In his first Royal Society paper on this subject, Crookes's main results are summarised as follows :

(*a*) Setting up an intense molecular vibration in a disc of metal by electrical means excites a molecular disturbance which affects the surface of the disc and the surrounding gas. With a dense gas the disturbance extends a short distance only from the metal ; but as rarefaction continues the layer of molecular disturbance increases in thickness. In air at a pressure of ·078 millim. this molecular disurbance extends for at least 8 millims. from the surface of the disc, forming an oblate spheroid around it.

(*b*) The diameter of this dark space varies with the exhaustion ; with the kind of gas in which it is produced ; with the temperature of the negative pole ; and, in a slight degree, with the intensity of the spark.   For equal degrees of exhaustion it is greatest in hydrogen and least in carbonic acid, as compared with air.

(*c*) The shape and size of this dark space do not vary with the distance separating the poles ; nor, only very slightly, with alteration of battery power, or with intensity of spark.   When the power is great the brilliancy of the unoccupied parts of the tube overpowers the dark space, rendering it difficult of observation ; but, on careful scrutiny, it may still be seen unchanged in size, nor does it alter even when, with a very faint spark, it is scarcely visible.   On still further reduction of the power it fades entirely away, but without change of form.

The official abstract says further :

From these and other experiments, fully described in the paper, he ventures to advance the theory that the induction spark actually illuminates the lines of molecular pressure caused by the electrical excitement of the negative pole.   The thickness of the dark space is the measure of the length of the path between successive collisions of the molecules.   The extra velocity with which the molecules rebound from the excited negative pole keep back the more slowly moving molecules which are advancing towards that pole. The conflict occurs at the boundary of the dark space, where the luminous margin bears witness to the energy of the discharge.

Crookes finally indulged in some daring assertions :

Rays of Molecular Light.

In speaking of a ray of molecular light, the author has been guided more by a desire for conciseness of expression than by a wish to advance a novel theory.   But he believes that the comparison, under these special circumstances, is strictly correct, and that he is as well entitled to speak of a ray of molecular or emissive light when its presence is detected only by the light evolved when it falls on a suitable screen, as he is to speak of a sunbeam in a darkened room as a ray of vibratory or ordinary light when its presence is to be seen only by interposing an opaque body in its path. In each case the invisible line of force is spoken of as a ray of light, and if custom has sanctioned this as applied to the undulatory theory, it cannot be wrong to apply the expression to emissive light. The term emissive light must, however, be restricted to the rays between the negative pole and the luminous screen : the light by which the eye then sees the screen is, of course, undulatory.

The phenomena in these exhausted tubes reveal to physical science a new world—a world where matter exists in a fourth state, where the corpuscular theory of light holds good, and where light does not always move in a straight line ; but where we can never enter, and in which we must be content to observe and experiment from the outside.

The experiments, more fully described in the subsequent Bakerian Lecture, attracted a great deal of attention, and Crookes was soon requested to deliver lectures on the same subject before two other high scientific tribunals— the Royal Institution and the British Association for the Advancement of Science. It is interesting to watch the evolution of Crookes's ideas as evidenced by the respective titles of these two lectures. At the Royal Institution on April 4, 1879, he lectured on " Molecular Physics in High Vacua "—a very sober title—whereas at the Sheffield meeting of the British Association, on August 22, 1879, he boldly chose the title " Radiant Matter." After the Royal Institution lecture he received the following letter from his friend, Miss Alice Bird, the sister of Dr. George Bird :

49, WELBECK STREET,
CAVENDISH SQUARE, W.,
April 5, 1879.

MY DEAR MR. CROOKES,

Of course you were discussed at a thousand breakfasts this morning, and as you could not hear us talk of your wonderful experiments and of the lovely way in which you acquitted yourself, George says I must send you a line. He says whatever you may have suffered inwardly, your outward calm and self-possession were perfect. I hope you are not more thin to-day, although I daresay you lost a pound or two more last night. But I am so glad you are going to Brighton, and I hope on your return you will show yourself here in much better plight.

I saw Ellen looking nervous, I thought, when she entered, but she must have been well pleased with your triumph.

I feel I only understood in a glimmering way, you seemed to me like the magician of the Future before whom no secrets are hid.

Love to you both, and best congratulations.

Yours always,

LALLAH.

We may fittingly close this chapter by a few extracts from this epoch-making discourse :

If, in the beginning of this century, we had asked, What is a Gas ? the answer then would have been that it is matter, expanded and rarefied to such an extent as to be impalpable, save when set in violent motion ; invisible, incapable of assuming or of being reduced into any definite form like solids, or of forming drops like liquids ; always ready to expand where no resistance is offered, and to contract on being subjected to pressure. Sixty years ago such were the chief attributes assigned to gases. Modern research, however, has greatly enlarged and modified our views on the constitution of these elastic fluids. Gases are now considered to be composed of an almost infinite number of small particles or molecules, which are constantly moving in every direction with velocities of all conceivable magnitudes. As these molecules are exceedingly numerous it follows that no molecule can move far in any direction without coming into contact with some other molecule. But if we exhaust the air or gas contained in a closed vessel, the number of molecules becomes diminished, and the distance through which any one of them can move without coming in contact with another is increased, the length of the mean free path being inversely proportional to the number of molecules present. The further this process is carried the longer becomes the average distance a molecule can travel before entering into collision ; or, in other words, the longer its mean free path, the more the physical properties of the gas or air are modified. Thus, at a certain point, the phenomena of the radiometer become possible, and on pushing the rarefaction still further, i.e. decreasing the number of molecules in a given space and lengthening their mean free path, the experimental results are obtainable to which I am now about to call your attention. So distinct are these phenomena from anything which occurs in air or gas at the ordinary tension, that we are led to assume that we are here brought face to face with Matter in a Fourth state or condition, a condition ás far removed from the state of gas as a gas is from a liquid.

*   *   *   *   *

I have long believed that a well-known appearance observed in vacuum tubes is closely related to the phenomena of the mean free path of the molecules. When the negative pole is examined while the discharge from an induction-coil is passing through an exhausted tube, a dark space is seen to surround it. This dark space is found to increase and diminish as the vacuum is varied, in the same way that the mean free path of the molecules lengthens and contracts. As the one is perceived by the mind's eye to get greater, so the other is seen by the bodily eye to increase in size ; and if the vacuum is insufficient to permit much play of the molecules before they enter into collision, the passage of electricity shows that the "dark space" has shrunk to small dimensions. We naturally infer that the dark space is the mean free path of the molecules of the residual gas, an inference confirmed by experiment.

I will endeavour to render this "dark space" visible to all present. Here is a tube, having a pole in the centre in the form of a metal disc, and other poles at each end. The centre pole is made negative, and the two end poles connected together are made the positive terminal. The dark space will be in the centre. When the exhaustion is not very great, the dark space extends only a little on each side of the negative pole in the centre. When the exhaustion is good, as in the tube before you, and I turn on the coil, the dark space is seen to extend for about an inch on each side of the pole.

Here, then, we see the induction spark actually illuminating the lines of molecular pressure caused by the excitement of the negative pole. The thickness of this dark space is the measure of the mean free path between successive collisions of the molecules of the residual gas.

*   *   *   *   *

THE FOURTH STATE OF MATTER

Glass is highly phosphorescent when exposed to a stream of Radiant Matter. My earlier experiments were almost entirely carried on by the aid of the phosphorescence which glass takes up when it is under the influence of the radiant discharge ; but many other substances possess this phosphorescent power in a still higher degree than glass. For instance, here is some of the luminous sulphide of calcium, prepared according to M. Ed. Becquerel's prescription. When the sulphide is exposed to light—even candle light—it phosphoresces for hours with a bluish-white colour. It is, however, much more strongly phosphorescent to the molecular discharge in a good vacuum, as you will see when I pass the discharge through this tube.

Other substances besides English, German, and uranium glass, and Becquerel's luminous sulphides, are also phosphorescent. The rare mineral Phenakite (aluminate of glucinum) phosphoresces blue ; the mineral Spodumene (a silicate of aluminium and lithium) phosphoresces a rich golden yellow ; the emerald gives out a crimson light. But without exception, the diamond is the most sensitive substance I have yet met for ready and brilliant phosphorescence. Here is a very curious fluorescent diamond, green by daylight, colourless by candle light. It is mounted in the centre of an exhausted bulb, and the molecular discharge will be directed on it from below upwards. On darkening the room you see the diamond shine with as much light as a candle, phosphorescing of a bright green.

Next to the diamond, the ruby is one of the most remarkable stones for phosphorescing. In this tube is a fine collection of ruby pebbles. As soon as the induction spark is turned on you will see these rubies shining with a brilliant rich red tone, as if they were glowing hot. It scarcely matters what colour the ruby is, to begin with. In this tube of natural rubies there are stones of all colours —the deep red and also the pale pink ruby.

Radiant Matter comes from the pole in straight lines, and does not merely permeate in parts of the tube and fill it with light, as would be the case were the exhaustion less good. Where there is nothing in the way, the rays strike the screen and produce phosphorescence, and where solid matter intervenes they are obstructed by it, and a shadow is thrown on the screen. In this pear-shaped bulb the negative pole (*a*) is at the pointed end. In the middle is a cross (*b*) cut out of sheet aluminium, so that the rays from the negative pole projected along the tube will be partly intercepted by the aluminium cross, and will project an image of it on the hemispherical end of the tube, which is phosphorescent. I turn on the coil and you will all see the black shadow of the cross on the luminous end of the bulb. (*c, d*). Now, the Radiant Matter from the negative pole has been passing by the side of the aluminium cross to produce the shadow ; the glass has been hammered and bombarded till it is appreciably warm, and at the same time another effect has been produced on the glass—its sensibility has been deadened. The glass has got tired, if I may use the expression, by the enforced phosphorescence. A change has been produced by this molecular bombardment, which will prevent the glass from responding easily to additional excitement ; but the part that the shadow has fallen on is not tired—it has not been phosphorescing at all, and is perfectly fresh ; therefore, if I throw down this cross—I can easily do so by giving the apparatus a slight jerk, for it has been most ingeniously constructed with a hinge by Mr. Gimingham—and so allow the rays from the negative pole to fall uninterruptedly on to the end of the bulb, you will suddenly see the black cross (*c, d*) change to a luminous one, because the background is now only capable of faintly phosphorescing, whilst the part which had the black shadow on it retains its full phosphorescent power. The stencilled image of the luminous cross unfortunately soon dies out. After a period of rest the glass partly recovers its power of phosphorescing, but it is never so good as it was at first.

*   *   *   *   *

RADIANT MATTER IS DEFLECTED BY A MAGNET.

I now pass to another property of Radiant Matter. This long glass tube is very highly exhausted ; it has a negative pole at one end and a long phosphorescent screen down the centre of the tube. In front of the negative pole is a plate of mica with a hole in it, and the result is, when I turn on the current, a line of phosphorescent light is projected along the whole length of the tube. I now place beneath the tube a powerful horseshoe magnet ; observe how the line of light becomes curved under the magnetic influence, waving about like a flexible wand as I move the magnet to and fro.

*   *   *   *   *

THE FOURTH STATE OF MATTER

During these experiments another property of Radiant Matter has made itself evident, although I have not yet drawn attention to it. The glass gets very warm where the green phosphorescence is strongest. The molecular focus on the tube, which we saw earlier in the evening is intensely hot, and I have prepared an apparatus by which this heat at the focus can be rendered apparent to all present.

I have here a small tube with a cup-shaped negative pole. This cup projects the rays to a focus in the middle of the tube. At the side of the tube is a small electromagnet, which I can set in action by touching a key, and the focus is then drawn to the side of the glass tube. To show the first action of the heat I have coated the tube with wax. I will put the apparatus in front of the electric lantern, and throw a magnified image of the tube on the screen. The coil is now at work, and the focus of molecular rays is projected along the tube. I turn the magnetism on, and draw the focus to the side of the glass. The first thing you see is a small circular patch melted in the coating of wax. The glass soon begins to disintegrate, and cracks are shooting starwise from the centre of heat. The glass is softening. Now the atmospheric pressure forces it in, and now it melts. A hole is perforated in the middle, the air rushes in, and the experiment is at an end.

I can render this focal heat more evident if I allow it to play on a piece of metal. This bulb is furnished with a negative pole in the form of a cup. The rays will therefore be projected to a focus on a piece of iridio-platinum supported in the centre of the bulb. I first turn on the induction-coil slightly, so as not to bring out its full power. The focus is now playing on the metal, raising it to a white-heat. I bring a small magnet near, and you see I can deflect the focus of heat just as I did the luminous focus in the other tube. By shifting the magnet I can drive the focus up and down, or draw it completely away from the metal, and leave it non-luminous. I withdraw the magnet, and let the molecules have full play again ; the metal is now white-hot. I increase the intensity of the spark. The iridio-platinum glows with almost insupportable brilliancy, and at last melts.

*　　*　　*　　*　　*

It may be objected that it is hardly consistent to attach primary importance to the presence of *Matter*, when I have taken extraordinary pains to remove as much Matter as possible from these bulbs and these tubes, and have succeeded so far as to leave only about the one-millionth of an atmosphere in them. At its ordinary pressure the atmosphere is not very dense, and its recognition as a constituent of the world of Matter is quite a modern notion. It would seem that when divided by a million, so little Matter will necessarily be left that we may justifiably neglect the trifling residue and apply the term *vacuum* to space from which the air has been so nearly removed. To do so, however, would be a great error, attributable to our limited faculties being unable to grasp high numbers. It is generally taken for granted that when a number is divided by a million the quotient must necessarily be small, whereas it may happen that the original number is so large that its division by a million seems to make little impression on it. According to the best authorities, a bulb of the size of the one before you (13·5 centimetres in diameter) contains more than 1,000000,000000,000000,000000 (a quadrillion) molecules. Now, when exhausted to a millionth of an atmosphere, we still have a trillion molecules left in the bulb— a number quite sufficient to justify me in speaking of the residue as *Matter*.

To suggest some idea of this vast number I take the exhausted bulb, and perforate it by a spark from the induction-coil. The spark produces a hole of microscopical fineness, yet sufficient to allow molecules to penetrate and destroy the vacuum. The inrush of air impinges against the vanes and sets them rotating after the manner of a windmill. Let us suppose the molecules to be of such a size that at every second of time a hundred millions could enter, how long, think you, would it take for this small vessel to get full of air? An hour? A day? A year? A century? Nay, almost an eternity! A time so enormous that imagination itself cannot grasp the reality. Supposing that this exhausted glass bulb, endued with indestructibility, had been pierced at the birth of the solar system; supposing it to have been present when the earth was without form and void; supposing it to have borne witness to all the stupendous changes evolved during the full cycles of geologic time, to have seen the first living creature appear, and the last man disappear; supposing it to survive until the fulfilment of the mathematician's prediction that the sun, the source of energy, four million centuries from its formation, will ultimately become a burnt-out cinder ; supposing all this—at the rate of filling I have just described, 100 million molecules a second—this little bulb even then would scarcely have admitted its full quadrillion of molecules.

But what will you say if I tell you that all these molecules, this quadrillion of molecules, will enter through the microscopical hole before you leave this room ? The hole being unaltered in size, the number of molecules undiminished, this apparent paradox can only be explained by again supposing the size of the molecules to be diminished almost infinitely—so that instead of entering at the rate of 100 millions every second, they troop in at a rate of something like 300 trillions a second. I have done the sum, but figures when they mount so high cease to have any meaning, and such calculations are as futile as trying to count the drops in the ocean.

In studying this Fourth state of Matter we seem at length to have within our grasp and obedient to our control the little indivisible particles which with good warrant are supposed to constitute the physical basis of the universe. We have seen that in some of its properties Radiant Matter is as material as this table, whilst in other properties it almost assumes the character of Radiant Energy. We have actually touched the border land where Matter and Force seem to merge into one another, the shadowy realm between Known and Unknown, which for me has always had peculiar temptations. I venture to think that the greatest scientific problems of the future will find their solution in this Border Land, and even beyond ; here, it seems to me, lie Ultimate Realities, subtle, far-reaching, wonderful.

> Yet all these were, when no Man did them know,
>   Yet have from wisest Ages hidden beene ;
> And later Times things more unknowne shall show.
>   Why then should witlesse Man so much misweene,
>   That nothing is, but that which he hath seene.

CHAPTER XV

ELECTRIC LIGHTING

(1878-89)

THE BRITISH ASSOCIATION LECTURE on "Radiant Matter" was the climax of Crookes's career of scientific research. When all else is forgotten, those great days of his life will remain glorious and imperishable. He was then forty-seven, and the rest of his life—another forty years—was spent in consolidating what he had built. .

We must now turn to his activity in promoting and developing the electric light. The only practical method of electric illumination known in the seventies was the electric "arc" between carbon poles, such as is still used for lantern and kinematograph projection. The Elder Brethren of Trinity House had experimented with the arc for purposes of lighthouse illumination as far back as 1855, but the method was then judged to be inferior to colza oil. In October 1866 Crookes contributed a glowing article to the *Quarterly Journal of Science* on "A New Era in Illumination—Wilde's Magneto-electric Machine." Wilde's machine was the first to use an electromagnet excited by a separate magneto-electric machine driven by the same power. The two machines mutually increased each other's power, and very strong magnetic fields were thus generated. Crookes reported concerning the output :

ELECTRIC LIGHTING

With the three armatures driven at a uniform velocity of 1,500 revolutions per minute, an amount of magnetic force is developed in the large electromagnet far exceeding anything which has hitherto been produced, accompanied by the evolution of an amount of dynamic electricity from the quantity armature, so enormous as to melt pieces of cylindrical iron rod fifteen inches in length and fully one quarter of an inch in diameter, and pieces of copper wire of the same length and one-eighth of an inch in diameter. With this armature in, the physiological effects of the current can be borne without inconvenience ; immediately after fifteen inches of iron bar had been melted, the writer grasped the terminals, one in each hand, and sustained the full force of the current. The shocks were certainly severe, but not inconveniently so.

When the intensity armature was placed in the 7-inch magnet cylinder, the electricity melted 7 feet of No. 16 iron wire, and made a length of 21 feet of the same wire red-hot. The illuminating power of the current from this armature was of the most splendid description. When an electric lamp, furnished with rods of gas carbon half an inch square, was placed at the top of a lofty building, the light evolved from it was sufficient to cast the shadows of the flames of the street lamps, a quarter of a mile distant, upon the neighbouring walls. When viewed from that distance, the rays proceeding from the reflector have all the rich effulgence of sunshine. With the reflector removed from the lamp, the bare light is estimated to have an intensity equal to 4,000 wax candles. A piece of ordinary sensitised paper, such as is used for photographic printing, when exposed to the action of the light for twenty seconds, at a distance of 2 feet from the reflector, was darkened to the same degree, as a piece of the same sheet of paper was when exposed for a period of one minute to the direct rays of the sun at noon on a very clear day in the month of March. The day on which the writer saw the machine at work (towards the end of June), the midday sun was shining brightly in at the window. He took the opportunity of roughly comparing the intensity of the sun with that of the electric light armed with the reflector. From a comparison of the shadows thrown by the same object, it appeared to him that the electric light had between three and four times the power of the sunlight. That the relative intensities were somewhat in this ratio, was evident from the powerful scorching action the

292

electric light had on the face, and the ease with which paper could be set on fire with a burning-glass introduced in the path of its rays. The extraordinary calorific and illuminating powers of the 10-inch machine are all the more remarkable from the fact that they have their origin in six small permanent magnets.

But it was not till 1879 that the era of electric light began in earnest. Arc lamps were extensively used for lighting open spaces before then, but it was the " incandescent " electric lamp or glow lamp which made electric lighting practically universal, and Edison's master patent, which was brought to England in 1879, proved to be the real solution of the problem of what was then quaintly called " splitting the electric light," i.e. dividing up the intensely luminous arc into a number of lamps of moderate power suitable for home use.

It is true that the arc is still supreme for purposes requiring great " intrinsic brilliancy," i.e. candle-power concentrated in a small and intensely luminous area. Since no optical system of mirrors or lenses can ever increase intrinsic brilliancy, the arc is still first in the field in searchlights and other optical projections. The reason is that, of all electric conductors, carbon has the highest melting-point. The arc carbon, before melting, attains the highest temperature which is artificially procurable, though it is still about 1,000° cooler than the sun. It is also the most economical light, but requires a great deal of attention.

The problem of incandescent lighting was to find a " wire " which could be heated to incandescence with a moderate current, and which would continue to glow for a reasonable number of hours before burning out. It would naturally have to be enclosed in a glass bulb free from oxygen, so as to prevent actual combustion. This again would necessitate the wire being short, and *that* implied a high electric " resistance " (i.e. a low conductivity).

No metallic wire known at that time fulfilled the last condition (the tungsten and osmium of the modern " metallic
filament lamp " could not at that time be drawn into fine
wire), so a diligent search had to be made among vegetable
fibres for a suitable material for making carbon filaments.
This was eventually done by Edison in America and Swan
in England, and the modern name " Ediswan " still com-
memorates the happy combination of Anglo-Saxon genius.
But before this came about other inventors, including
Crookes, made a determined bid for supremacy. Crookes
wrote in 1878 to Professor (now Sir William) Barrett :

Oct. 19, 1878.

DEAR MR. BARRETT,
       I shall be very pleased to hear further particulars about
Edison's new discovery on the Electric light, and might perhaps
be able to induce some friends to go into the matter commercially,
if he does not do as most patentees are so fond of doing—asking
a prohibitory price.
       I don't think there is field yet for the telephone to be used
commercially. I don't think Bell's company is doing much.
       When are you coming to London ? I leave for Paris on
the 29th.

Believe me,
Very sincerely yours,
WILLIAM CROOKES.

Crookes evidently underrated the importance of
Edison's work (and also of the telephone), and was inclined
to favour certain French efforts in the same direction.
He wrote to G. Fournier, asking him to procure him a
Reynier lamp through M. du Moncel. But the lamp
proved " an utter failure," and there the matter rested
for a time. M. Fournier had been instrumental in getting
Crookes's eldest son, Henry, an appointment in the Paris
laboratory of Professor Wurtz.[1] The second son, Joseph,

---

[1] Not Henry Wurtz, but Charles Adolphe Wurtz, French chemist.

was also in Paris, in the works and under the care of a M. Coquerel.

A significant letter to Dr. A. R. Wallace written about this time shows that Crookes was anxious not to reopen the quarrel with Dr. Carpenter :

Dec. 6/'78.

Dear Mr. Wallace,

Your letter is very good, and if there were a row going on with Carpenter nothing could be more appropriate. But I think it best to let sleeping dogs lie, and would rather not take the initiative in reviving the subject.

I had a rather important paper before the Royal Society last night ; there was a good discussion, in which Carpenter joined ; although he did not propose to bury the hatchet, he was particularly civil and complimentary, so under the circumstances I should be glad if your letter were withheld for a more convenient opportunity.

Believe me,

Very sincerely yours,

William Crookes

An element in the situation which perhaps inclined Crookes to be pacific in that direction was the attack made upon his determination of the atomic weight of thallium by Professor (afterwards Sir Henry) Roscoe. He wrote to the editor of *Nature* :

Decr. 16, 1878.

My Dear Lockyer,

I wish I could have accepted your telegraphic invitation but I am like the Starling, " I can't get out." I must avoid cold and night air for a few days, having a warning which I dare not neglect, or I shall be laid up.

I have written to Roscoe about the ingot of thallium.

He must have put himself to great trouble to spread the knowledge he thought he had got of the great impurity of my atomic weight thallium, far and wide on Thursday, for I have heard of it in three separate quarters already !

I wonder whether he would be equally keen in advertising anything greatly to my credit if he happened to find it out.

There's a deal of human nature in scientific men !

The beginning of 1879 made two landmarks in Crookes's career. The *Quarterly Journal of Science*, which he still owned and edited, became a monthly, and was subsequently sold to Mr. J. W. Slater. Also, Crookes was elected a member of the Athenæum Club, on the proposal of Professor Stokes. It was an honour he greatly appreciated, as it fulfilled an ambition he had long cherished. He also resigned his seat on the board of the Alizarine and Anthracene Company, but was induced to re-occupy it on being promised a fixed fee of £50 per annum and a guinea for every meeting he attended. This episode seems to have been a test of the enhanced market value of his name and expert advice.

The first incandescent lamps were worked with primary batteries. These were not economical, and were liable to run down in the middle of an experiment. Crookes recognised the importance of magneto-electric generation of currents for electric lighting, and lost no opportunity of emphasising it. On January 17, 1879, Tyndall delivered a lecture on electric light at the Royal Institution, and exhibited a Delahaye incandescent lamp lent to him by Crookes. It was admired for the steadiness of its light, but was rather expensive to run with batteries, and Crookes made inquiries in Paris about working it with a small Gramme machine and a gas engine.

In order to strengthen his footing in Paris, Crookes made another move to secure his election as a Corresponding Member of the *Institut de France*. He wrote to M. Antoine d'Abbadie :

Novr. 15th, 1879.

MY DEAR SIR,

Thank you much for your friendly letter. It has been my great ambition for years past to be considered worthy of election among the Corresponding Members of the *Académie des Sciences*, and should you succeed in procuring my election I shall consider it one of the highest scientific honours to which I can attain.

Professor Wurtz has invited me over to show my experiments on Radiant Matter at the Sorbonne on the 10th of January, 1880. I hope to have the pleasure of seeing you on that occasion.

This ambition, however, was fated not to be realised for another twenty-seven years.

A friendship which ripened about this time was with Dr. Johnstone Stoney, Professor of Natural Philosophy in the Queen's University of Ireland. This Queen's University was one of the many ill-fated attempts to solve the question of University education in Ireland. Dublin University had long been a Protestant Episcopalian (" Church of Ireland ") preserve, and even when religious tests were abolished it retained a " Protestant atmosphere "[1] which did not suit the Catholic hierarchy, ever anxious for the " faith and morals " of its flock. The hierarchy, therefore, denounced the Queen's University as " godless " because it did not embody any religious training or observance in its programme. Dr. Stoney was a distinguished physicist, who was the first to postulate the atomic structure of electricity and to invent and use the word " electron " for the charge carried by each " ion " in a solution under electrolysis. At the time with which we are dealing Dr. Stoney was Hon. Secretary of the Royal Dublin Society, a " learned society " best known for its encouragement of horse-breeding and agriculture and " science as applied to the arts." The members of the Society were anxious

---

[1] Some Irish wit in later years suggested that the best instrument for getting rid of this " atmosphere " was a vacuum cleaner !

to see Crookes's experiments with his famous " bottles full of nothing." That they were willing to pay for the privilege is evident from the following correspondence :

57, RATHMINES ROAD,
DUBLIN,
22 *Nov.*, 1879.

DEAR MR. CROOKES,

There are many persons over here who are most anxious to know more about your interesting researches and to have an opportunity of hearing your lectures. I therefore brought forward a motion at the Royal Dublin Society, and the result is you will have a letter from our Hon. Secretary, Mr. Stoney, inviting you to come over in the spring to lecture. Knowing the press of work you have on your hands and your multitudinous engagements, I feel almost afraid you will not be able to accept the invitation, but I write now to beg you will at least give the matter careful consideration, for there are great numbers here who will be dreadfully disappointed if you could not come, and you will really do good service to us at the Royal Dublin Society if you can manage it. We are now reorganising the Society, and your coming over to lecture would be a wonderful feather in our cap.

I may add that if you require any touching up of your apparatus after you come my workshop and any of my men are at your disposal as a labour of love.

With kindest regards to Mrs. Crookes,

I am,
Yours very truly,
HOWARD GRUBB.

Dec. 2/'79.

DEAR GRUBB,

I did not answer your letter of the 22nd ult. as I waited to receive Mr. Stoney's official invitation, and having received it I took some time to think it over before answering. I have just written to him, and as you will probably see my letter I need not further refer to it except to say that I am afraid my pecuniary demands may be considered somewhat high ; this, however, is unavoidable, for a great part of the apparatus is injured each time it is exhibited so much as to require re-exhaustion before it can be used again. I have also asked for the time of the lectures to be after Easter.

You are very kind in offering me assistance in case of any untoward accident to any of the apparatus. I will not hesitate to avail myself of your kindness, although I hope that it will not be necessary to trouble you.

It occurs to me that you may be able to supply an electric light for projections on the screen and a few extra Grove's batteries for working the coil and electro-magnets. It is too cumbersome and expensive to bring such things from London.

My wife joins me in kind regards to Mrs. Grubb, hoping she is better than she was when you were in London, and

Believe me,

Very sincerely yours,

WILLIAM CROOKES.

Dec. 2/'79.

My dear Stoney,

I am in receipt of your letter of the 27th ult. requesting me to give one or, if possible, two lectures this session before the Royal Dublin Society, and asking me to name the honorarium.

In the first place I may say that it would give me great pleasure to visit Dublin and deliver two lectures before the Royal Dublin Society, and I highly appreciate the honour which the Society has done me in making such a request ; but may I ask for the date to be postponed to as late a period as possible. I have not been very well lately, and I find that travelling about in this inclement weather to fulfil the lecture engagements into which I have already entered is almost too much for me. I have to lecture in Paris in the middle of January, and I am afraid I shall not be fit for much going about for some time after my return. I should therefore prefer to fix a date after Easter. As regards the honorarium, I wish I were in such a position as to enable me to consider the honour of delivering the lectures a sufficient recompense, but unfortunately I cannot follow my own inclinations in this matter. My usual fee for a lecture is 25 guineas and my expenses, but where two lectures are given with a short interval between them I have received 35 guineas for the two. The expenses additional are likely to be high, as I find it necessary to bring an assistant, whilst the carriage of the apparatus, which is both heavy and bulky, is likewise costly.

The loss of Clerk Maxwell is indeed a heavy blow to science. I feel it particularly so, as he had reported to the Council of the Royal Society on all my papers before they were printed, and his reports were generally sent to me afterwards. I was always much struck with the rapidity and clearness with which he seized upon the essential points in the development of a discovery, and he always suggested some new and original experiment to decide a doubtful point.

Believe me, very sincerely yours,
WILLIAM CROOKES.

After his successful research of 1878 and 1879 Crookes anticipated little difficulty in obtaining a renewal of the substantial grant from the Government funds administered by the Royal Society. His formal request was made in the following letter :

Decr. 27, 1879.

PROFESSOR STOKES, Sec. R.S.

MY DEAR SIR,

Will you be kind enough to bring before the Government Fund Committee of the Royal Society my application for a sum of Three Hundred Pounds, to assist me during the next year in continuing my Researches on " Molecular Physics in High Vacua."

During the present year I have had two papers published in the *Philosophical Transactions*, and a third is now going through the Press. A paper on " Electrical Insulation in High Vacua " has also been published in the *Proceedings*.

I had hoped to send in a paper on the Viscosity of gases at various exhaustions before the close of the year. I have been engaged on this research for many years. Most of the work is now done, but the time occupied in reducing the numerous observations is greater than I had anticipated.

I have recently commenced working with a new Induction Coil giving a 20-inch spark of great thickness. This brings out strongly many phenomena which I have before only faintly seen.

The phosphorescent light obtained when the negative discharge falls on some bodies in a high vacuum sometimes gives characteristic spectra of lines and bands. An examination of these Spectra promises good results in the detection of known chemical elements and the discovery of unknown elements.

I remain,

Very sincerely yours,

WILLIAM CROOKES.

Crookes's relations with the French Academy of Sciences were greatly strengthened by his demonstration of Radiant Matter at the Sorbonne. The Academy awarded him a gold medal and a special prize of 3,000 francs. He received the official intimation from Professor J. B. A. Dumas, a former Minister of Agriculture and a noted chemist. He attended the sitting of the Academy on March 1, 1880, to receive the awards.

The speech in which Crookes acknowledged the distinction conferred upon him by the Académie des Sciences is one of his finest flights of eloquence, a speech marvellously adapted to the occasion and the atmosphere. Here it is in full :

I must beg your kind indulgence if I find a difficulty in adequately expressing my feelings on this occasion. To every student, every investigator who is seeking to obtain a deeper and clearer insight into the recesses of Nature, any recognition of his labours on the part of the Academy of Sciences is indeed a most cherished reward. That I should have been counted worthy to receive the formal approbation of this learned body, whose present is in harmony with its glorious past, and which reckons on its muster-roll so many of the brightest names in every department of science, is an honour which I cannot appreciate too highly, and which assuredly I can never forget. I am aware that your approval is not lightly given ; that your judgments are no less competent than impartial, and that whilst you have generously recognised the merit of strangers— and even in times past of the natives of hostile countries—you have never stooped to affix the stamp of your sanction upon base metal.

ELECTRIC LIGHTING

If there be any element of truth in the old proverb—" Nolo laudari nisi a laudato "—surely the highest aim of ambition must be to receive the praises of those who are themselves most praised, and most praiseworthy.

There is one circumstance which, to my mind, invests the honour you have done me with a charm which, coming from another quarter, it could scarcely have possessed in an equal degree. Without wishing to undervalue the share taken in the progress of science, either by my own countrymen or of the Germans and Italians, I cannot help saying that France may be regarded as the classical land of those studies which lie, as it were, on the boundary-line between Physics and Chemistry. I am sure there is no need for me to prove or to illustrate this assertion at length. I will merely mention the ever-honoured names of Gay-Lussac, of Thénard, of Berthollet, of Guyton-Morveau, of Lavoisier and Laplace, of Regnault, of Dulong and Petit, and of many others, both dead and living. The Gaseous State of Matter has been a peculiar subject of French men of science, from the day when Maitrel d'Element first devised the pneumatic trough, and when Jean Rey first suggested the part played by atmospheric oxygen in the phenomena of combustion, on to those recent and splendid researches by which M. Cailletet has swept away the old notion of permanently incondensible gases, and has shown the continuity of the states of Matter.

What I have done, and what I have attempted to do, lies on this same boundary-line between Physics and Chemistry. Whilst MM. Cailletet and Pictet have been engaged with the transition from gases to liquids, I have been engaged with aëriform bodies at the opposite limit, and have been seeking to investigate their apparent passage into a new state. Hence my work, such as it is, may be regarded as a supplement to the researches of French *savants*. Surely, therefore, I must feel especially gratified and encouraged at the honour which the first Scientific Society of France, the recognised embodiment of French Science, has so kindly bestowed upon me.

I am fully aware, however, that in my case you have most generously taken the will for the deed. I can say that I have striven loyally and sincerely for the truth ; but no one can be more fully conscious than I am myself how little I have effected compared with what remains unsolved and unachieved. As I have proceeded in my investigations I have seen as it were new and unexplored

302

regions opening out to the right and the left. To some of these I may return, and endeavour to bring them within the control of science. Who shall dare to say what new treasures of truth, and even of practical applications, may there be awaiting the patient inquirer ! Whether by me or by some more fortunate and worthier successor, the task will ultimately be accomplished.

Gentlemen, I am no orator, and I will intrude upon your indulgence no further than to express the hope that the most fraternal relations, the most friendly rivalry, may ever prevail between the men of science of France and of England. May we labour side by side in our great life-task—the Interpretation of the Universe.

It was in 1880 that Crookes moved from 20, Mornington Road into a new and very large house, No. 7, Kensington Park Gardens, in which he lived for the remainder of his life. He bought it from Mr. Batley, of West Hall, Byfleet. The terms may be gathered from the following letters :

21st Feb., 1880.

Messrs. E. L. F. Swain,
   Notting Hill, W.

Gentlemen,

   7, *Kensington Park Gardens.*

   I hereby offer to purchase this freehold house for the sum of Seven thousand five hundred pounds (£7,500), to include the tenant's furnishing fixtures as per inventory, subject to a proper agreement to purchase to be prepared by your client's, Mr. John Batley's, solicitors, and to complete the purchase on the Thirty-first day of March next, when possession is to be given me. I also agree to purchase at a valuation in the usual way the Venetian Blinds, Cornices and Poles (except three Portière Rods), Three Pier Glasses in Drawing-room, Bronzed Figure and Pedestal in Hall.

(*Signed*) William Crookes.

The house at 7, Kensington Park Gardens, was the first house in England to be lighted by electricity. There was a well-equipped laboratory devoted entirely to research, and Crookes and his sons worked at the development of the incandescent lamp. Crookes became a director of the Electric Light and Power Company, and took out his first patent for incandescent lamps on June 15, 1881. The specification says :

This invention has for its object improvements in the construction of electric lamps and apparatus for electric lighting.

My electric lamp is of the class in which a carbon filament is enclosed in a vacuous glass vessel, and rendered incandescent by the passage of an electric current.

In the construction of an electric lamp, I take paper, cotton, linen, or cellulose in other convenient form, and having purified it by means of hydrofluoric acid, I treat it with a solution of ammoniacal oxide of copper, which as is well known has a solvent effect upon cellulose.

The material paper by preference having been so prepared is thereby rendered more dense, and the carbon filament which it yields has also a firmer consistency. The prepared paper is punched out to the form of a horseshoe or other suitable form and then carbonised. I obtain a good electrical connection between the film and the conductors by means of which the current is to be passed through it by the electro-deposit of copper or other metal upon both the film and the conductor at the point of junction, or where the two come together.

I afterwards improve the state of the carbon film by igniting the film electrically in a vacuous vessel, also containing a hydrocarbon, of which the boiling-point is high and the vapour tension at ordinary temperatures very low. The effect is similar to that obtained when the more volatile hydrocarbons are employed and introduced in minute quantity into the exhausted receiver ; but my process is simpler and much more convenient. Electrical ignition also expels the trace of copper originally contained in the carbon film, or it may have been previously extracted by means of solvents.

The film is enclosed for use within a glass enclosure within which a high vacuum is obtained. For the purpose of turning the current on to and from the lamp I employ a plug similar in some respects to the plug of an ordinary cock, and I have a series of contact

304

springs which in one position of the plug all bear upon it ; the current is then full on to the lamp. The plug, however, is so formed that as it is turned round it passes out of contact first with one spring and then with another until it leaves all of them, and then the lamp circuit is broken.

So long as the plug remains in contact with one or more of the springs, although it is out of contact with some of them, the circuit is not broken, but the current has to pass through resistances which are interposed between the springs. Thus by turning the plug the amount of electricity passing through the lamp can be varied and the light adjusted as may be desired.

It may appear strange that the ingenious device described in the last paragraph is not in current use. But for this there are two reasons. In the first place, the " resistances " get hot, and that means a risk of fire which the insurance companies do not like to face. In the second place, an electric lamp burning low is a very uneconomical source of light.

In 1881 there was an electrical exhibition in Paris, and Crookes was very naturally placed on the jury for awarding prizes.

On August 31, 1881, Crookes took out another patent for " Electric Lamps " (No. 3799). The specification says :

This invention relates to that class of electric lamps in which the light is produced by the heating of a continuous conductor, which is usually composed of a fine filament of carbon enclosed in a glass globe, and protected from combustion by being surrounded by a vacuous space or by an atmosphere that will not support combustion.

To make the lamp I take a cylinder of glass of any convenient size and form, and after drawing down one end in the blowpipe to make the top of the lamp, I draw the other end out into a wide neck and seal off its end in the blowpipe. While the glass is still soft (or after reheating it), I press the end inwards with a two-pointed brass tool like a two-pronged fork, so as to make a hollow projection stretching a convenient distance into the neck of the tube, and terminating in two hollow points of glass. When the glass is cool I cut the tube in two across the neck at such a distance from the end as to leave the two glass points projecting from the bottom piece of the tube. These points are now opened in the blowpipe or by breaking off their ends, and the conducting wires are passed through the openings so made and sealed in the glass by means of white enamel, or what is technically known as arsenic glass. The top of the cylinder, which was at first drawn down in the blowpipe, is now finally sealed up and made into a hemispherical end before the blowpipe. The carbon filament can now be attached to the conducting wires and the two portions of the tube are then sealed together at the edges where the tube was cut open.

At some convenient stage of the above process a small glass tube is sealed into the lamp, preferably at the lower edge, where the glass is bent inwards for the purpose of exhausting the lamp and filling it with a suitable protective atmosphere. When the lamp is completed this is sealed off.

The hollow at the base of the lamp may be filled with plaster of Paris or any other like cement, for the purpose of giving strength to the lamp and protecting the conducting wires and their attachments. The pores may, if thought desirable, be filled in with a fusible cement.

If a carbon filament in a lamp breaks, the neck of the lamp can be again cut across and a fresh filament attached, and the lamp sealed up and exhausted or filled with gas as before.

This patent was applied for in Great Britain, Austria, Italy, Spain, Denmark, Norway, Sweden, Russia, and the United States. These patents he offered to the (Brush) Electric Light and Power Company, and he did his best to prove to them that his lamps were the best. He established lamp works in Battersea and placed his son Henry in charge of them as manager. He obtained the necessary glass from France through M. Fournier.

Meanwhile Edison was " getting busy " even in England, and Crookes, after another year's work, thought it best to seek the advice of the Gramme Electrical Company, of New York, to work his patents in America, more especially as he had been refused a patent for one of his English improvements in favour of Mr. Weston.

One cannot help sympathising with Crookes in his struggle to hold his own in the big battle for the capture of the enormous trade in store for a successful solution of the electric light problem. He certainly strained every nerve and worked every lever to the desired end. He showed considerable astuteness and *savoir faire*, and made the fullest use of his advantages. He also made some money, but, as in all his industrial enterprises, he fell short of complete success. Possibly he was not sufficiently unscrupulous to take an unfair advantage. Possibly he just lacked the " flair " which is said to be the natural endowment of the financial genius. The following letters speak eloquently of the fierce struggle :

April 7, 1882.

My dear Professor Tresca,

Since writing to you a few days ago I have had a further communication from Professor Barker, in which he sends another set of "conclusions" for my approval. I think they express the results better and fairer than did the "conclusions" I sent in my former letter, and I therefore enclose them in the hope that they will still be in time to add to the report of the Committee on Incandescent lamps. Will you please therefore destroy the former page of "conclusions" and substitute this for it ?

Professor Barker tells me that at one of the meetings of the Jury in Paris a resolution was passed in which complimentary reference was made to my researches on high vacua and an opinion was expressed that had it not been for my researches, incandescent electric lamps could not have been brought to their present state of excellence.

I shall esteem it a great favour if you could let me have a copy of that resolution if it exists. It was quite unknown to me at the time, but I need hardly say that any expression of that kind emanating from so distinguished a body of scientific men would be regarded by me as a very high honour, which I should be reluctant to lose.

ELECTRIC LIGHTING

MY DEAR PROFESSOR BARKER,

Your letter of the 24th ult. is to hand. I have already replied to your letter of the 14th March. You have correctly interpreted my objection to the 4th conclusion, and on comparing the " conclusions " you now send with those in the original report I have no hesitation in saying I am perfectly satisfied with them as now proposed, and have signed and sent them on to Tresca with a request that he will substitute them for those already in his hands. I think they will still be in time, for they are a slow moving body in France. My original objection was purely one of abstract fairness to the low-resistance lamps of Maxim, and was in no wise prompted by any desire to make these lamps out better than they deserve. It is true that when in Paris I was officially connected with the Electric Light & Power Generator Co., which subsequently purchased the Maxim patents, but my connection with that company has now practically ceased owing to many reasons, the chief being my great dissatisfaction with the management of the company, and the way my advice has been disregarded in some important particulars.

You may be aware that I have taken out patents for improvements in incandescent lamps myself. The E. L. & P. G. Co. were in treaty with me for the working of these patents, but we could not arrange satisfactory terms, and so the negotiations are broken off.

I have not quite made up my mind what to do with my patents. I once thought that it would have been a good thing for all parties if Edison and I were to have joined our forces. I think that might have been effected through Mr. Lowrey if I had not then been in some degree connected with the Maxim patents, but now Edison has formed a large company in England and is pushing his lamps here, so it seems to be too late for a fusion of interests. Friends of mine think that my series of patents are sufficiently important to justify the formation of a separate company to work them ; this, however, is a matter of finance, and I leave this to them, as I am not so much in my element in forming a company as I am in my laboratory making a new lamp. I know I have got hold of a good thing and can afford to wait.

Can you tell me anything about that resolution of the Jury in which my researches on the Vacuum were referred to ? I should

308

be very reluctant to lose it, for coming from so eminent a body of men I should look upon it as one of the highest honours I could have. Who would be the proper person to write to about it ? I have written to Tresca, but he does not answer letters very promptly.

P.S.—I send you a private memo. I have got up about my patents. I have taken the liberty to refer to some of your measurements. Please consider this as quite private, as I am applying for American Patents.—W.C.

Crookes was very bitter about what he called Weston's " piracy " of his lamp invention, and accused the same person of having similarly appropriated Crookes's use of thoria for absorbing the residual gases in the bulbs of incandescent lamps. In a letter to Colonel Hazard, of the Gramme Electrical Company, of New York, he gave vent to his feelings :

I hope I am not too sensitive as regards credit of discovery. I have not the least objection to anyone making what use he likes of any of my scientific discoveries, but in common honesty he ought to acknowledge the source of his information. When Mr. Edison does me the honour to make use of a discovery of mine on the vacuum he gives me credit for it, and I am quite satisfied. When Mr. Swan uses the fruits of my researches he also states to whom he is indebted, and I have no fault to find. When Mr. Brush patents one of my discoveries in connection with Radiant Matter, and applies it to Incandescent lamps, he likewise gives me full credit for the discovery, and I ask no more. When the Jury of the Electrical Exhibition at Paris, after discussing the merits of the Edison, Swan, Maxim, and Lane-Fox incandescent lamps, say in their official report : " None of them would have succeeded had it not been for these extreme vacua which Mr. Crookes has taught us to manage," I am abundantly rewarded for all I have done, as a father feels proud at the success of his children. But when a man deliberately appropriates the discovery of another, swears that it is his own, and shelters himself behind an unjust patent law, I feel that I have a right to speak out and appeal to that which is above local laws and national feelings—to that common feeling of justice and honesty which has made the whole civilised world brand a pirate " *hostis humani generis.*"

Crookes was the principal witness for the defence in the case of Edison and Swan *v.* The Brush Company, in 1888, which marked a definite stage in the march of " Ediswan " towards supremacy. On February 5, 1889, Crookes wrote to his patent agents :

Messrs. CARPMAEL & Co.

> DEAR SIRS,
>
> I write to inform you that I have sold my three patents, 2612^{81}, 3799^{81}, and 1079^{82}, to the Anglo-American Brush Electric Light Co., Belvedere Road, Lambeth, and I shall be obliged if you will in future send notices of fees falling due to them. I have sent on your letter of the 1st inst.

The sale of these patents marks the end of the long fight. Crookes retired from the field so hardly won by Edison and Swan. He retired with honour, but with little profit. He became a director of the Notting Hill Electric Light Company, which served his own district, and fought its battles in Parliament and elsewhere. Later he became chairman of the company, and under his guidance it flourished amazingly, its dividends going up from *nil* in 1891 and 1 per cent. in 1894 to 6 per cent. in 1901 and $7\frac{1}{2}$ per cent. in 1905.

CHAPTER XVI

THE ORIGIN OF THE ELEMENTS

(1890)

ON FEBRUARY 17, 1881, Crookes presented to the Royal Society a report of his experiments on the viscosity of gases at high exhaustions. These viscosities were determined by suspending a mica vane horizontally by means of a glass fibre attached to its centre. The fibre and vane were suspended in a vacuum tube and connected with the air pump by means of a flexible glass spiral, thus eliminating all greased or cemented joints. The swings of the vane about the vertical axis of the fibre were observed by means of a mirror and scale. The difference in the logarithms of successive scale deflections (the "logarithmic decrement") was taken as a measure of the relative viscosity of the residual gas in the tube at the pressure studied. The result was a confirmation of Maxwell's law that the viscosity of a gas is independent of its pressure.

The investigation was very laborious, and the result, though of high scientific value, was not such as to appeal to the popular imagination. The great advantage which Crookes derived from the investigation was a thorough schooling in advanced physical as distinct from chemical research. He owed much to Sir George Stokes's friendly guidance, and was very conscious of such indebtedness, which he acknowledged on many occasions.

The next investigation which the Royal Society authorised (and subventioned) Crookes to carry out consisted in the application of spectroscopy to the study of the phosphorescent light shown by many chemical substances under the influence of " radiant matter " (cathode rays). He was much struck by the frequent appearance of a " citron " band in the spectrum of this light. It was shown most clearly by zircon, but after experimenting for a year with 10 lbs. of zircon crystals from North Carolina, Crookes found that the mysterious spectrum band was really due to an earthy residue of a complex character containing oxides of thorium, cerium, lanthanum, didymium, yttrium, and probably some still rarer earths. Crookes finally traced the citron band to yttrium, and described the yttrium cathode-ray spectrum as " the most beautiful object in the whole range of spectroscopy." [1]

Crookes probably hoped to repeat his thallium success by discovering a new element indicated by the " citron " band or an element producing the double orange band seen in samarskite and other bodies under "radiant matter " bombardment. He found, however, after years of research, that the phosphorescent spectrum was much more complex than the spectrum of an incandescent gas. The conclusion thus reluctantly arrived at is given in his own words before the Royal Society on June 18, 1885 :

One important lesson taught by the many anomalies unearthed in these researches is, that inferences drawn from spectrum analysis *per se* are liable to grave doubt, unless at every step the spectroscopist goes hand in hand with the chemist. Spectroscopy may give valuable indications, but chemistry must after all be the court of final appeal.

Chemistry, however, by itself would have been helpless to solve the difficulties had it not been possible to appeal at every step to

[1] Bakerian Lecture, 1883.

WILLIAM CROOKES, F.R.S., 1889.

ÆT. 57.

To face p. 312.

the radiant-matter tube and to the spectroscope. The problems to be solved are so new as to be entirely outside the experience of laboratory work. A double orange-coloured band shows itself in a faint emission spectrum obtained under novel circumstances. On further examination the band, or one not far from it, is seen to occur in minerals of very divergent kinds, and apparently irrespective of their chemical constitution or locality, as well as in laboratory reagents and chemicals of assured purity. This band is sometimes accompanied by bands in other parts of the spectrum, and occasionally shifts its place to the right or to the left. Frequently the orange band disappears and a citron-coloured band takes its place. Chemical research continued for a longer time than most chemical researches require fails to throw any light on the subject. These being the conditions of the problem, the very last explanation likely to occur to the inquirer would be that these elusive shifting bands were due to the presence of two elements almost universally distributed, and that these two elements should be yttrium and samarium—yttrium, one of the rarest of known elements, and samarium almost unknown at the time its spectrum reaction was first discerned.

In 1886 a French Academician, M. Lecoq de Bois-baudran, announced the spectroscopic discovery of a new element which he called dysprosium.[1] This discovery, following upon the announcement in June 1885 by Dr. Auer von Welsbach, that he had, after many hundred fractional crystallisations, succeeded in splitting up didymium into two distinct elements which he called neodymium and praseodymium respectively, led Crookes to announce somewhat precipitately that he had discovered nine new elements by the fractionation of earths contained in the minerals gadolinite and samarskite.[2]

It was a somewhat daring " extrapolation " of the method of phosphorescent spectra. Welsbach has since been confirmed, but Crookes has turned out to have been mistaken. Gadolinium and samarium are still enumerated

[1] *Comptes Rendus,* vol. 102 (1886), p. 1003.
[2] *Roy. Soc. Proc.,* vol. 40, No. 245, June 9, 1886.

among the elements, while didymium is not. Lecoq de Boisbaudran observed the sparks between a platinum wire and an acid solution of the salt he was studying. He attached little importance to the " radiant matter " method, knowing that phosphorescence spectra are enormously influenced by very slight impurities.

We may safely say that the chief scientific problem which occupied Crookes's mind from 1880 to 1890 was the inner connection between the elements established as independent entities by the chemical science of his day. We shall see how this preoccupation eventually ripened into a new theory of the Genesis of the Elements.

But the same period is filled with preoccupations of a very different character. There was the *Chemical News*, now his only journalistic responsibility. There were books—a new edition of *Select Methods of Chemical Analysis*, of Ville's *Artificial Manures*, of *Dyeing and Calico Printing*. There were many financial and industrial enterprises, with the usual result !

On January 16, 1884, Crookes lost his father, aged ninety-two. Next day he wrote to Mr. D. D. Home, with whom he had not corresponded for many years. The letter was as follows :

Jany. 17*th,* 1884.

MY DEAR DAN,

Long silence does not make me forget my friends, and I am glad to find that your feelings for us are the same as ever they were.

I can only write you a very short letter as you want an answer by return of post. I had the great misfortune to lose my dear Father yesterday, and you can imagine the grief that has thrown us all into, and the access of occupation so sad an event throws on the shoulders of the nearest survivors. He was the best father that ever lived, and his sons have everything to thank him for. He died literally of old age at 92 years, but retained his consciousness and mind to the last.

I am very sorry to hear that Aksakoff has acted in such a manner to you. I have quite given up spiritualism now, and therefore have not followed the recent progress of events as I ought had I seen the papers devoted to the subject. My belief is the same as ever, but opportunities are wanting. I almost forget what took place in London when Aksakoff came and asked me about mediums. The Pettys were then making some stir, and I believe I told him that he should go and see them, or something to that effect—so he may perhaps be technically correct when he says I introduced them to him. I know nothing of any " M. Clayes."

In your extract you had therefore better omit what I have struck through in red ink. I return the printed cutting.

My wife desires kindest regards to you and yours, in which I join. Pray excuse this hurried letter, but I am overwhelmed with work.

Believe me,
Very sincerely yours,
WILLIAM CROOKES.

It is evident from this letter that Home was the first to write ; also that Crookes was not in the mood to embark on any new mediumistic investigations.

The next few years were, as we have already seen, devoted to a strenuous effort to keep in the front rank of the electric light development.

In 1886, Crookes commenced his work on the daily analysis of samples of the London water supply in collaboration with Dr. Tidy and Dr. Odling, at the London Hospital. This work was carried out on behalf of the Associated Water Companies of London. After Dr. Tidy's death, Crookes carried on the work in conjunction with Sir James Dewar, at No. 14, Colville Road, and this partnership, which was very remunerative, lasted till 1913.

In 1886 Crookes was elected President of the Chemical

Section of the British Association. In his opening Address he put forward an elaboration of Prout's hypothesis that all elements are originally derived from one substance. Prout supposed this substance to be hydrogen, but Crookes thought it was a more primitive and subtle substance, which he called *protyle*.[1] He sent the proof of his address to Miss Alice Bird, his literary critic and muse, the sister of Dr. George Bird. On receiving her corrections, he wrote to her :

ATHENÆUM CLUB,

PALL MALL, S.W.,

July 16th/'86.

MY DEAR LALLAH,

Thanks, many, for the proof. I have revelled in the corrections, and have sent them off post haste to the printer. May I ask for the same kind offices for No. 2 ?

I hope I may prove to be the Star you suppose me. But Lucifer was a Star, and we hear of Falling Stars, which are somewhat plentiful in September. *Absit Omen.* Wordsworth comes in *well*. The same quotation was once used appropriately by Faraday (or by Tyndall in reference to Faraday).

I see George's judicious emendations all over the proof. Thanks to him also.

With love, in which Nellie joins.

Believe me, my dear Lallah,

Very sincerely yours,

WILLIAM CROOKES.

On April 1, 1888, Crookes bought a Hammond typewriter, and immediately proceeded, at the age of fifty-six, to acquire the art of typewriting. His youthful vigour is evidenced by the fact that he soon became an adept at it, and outstripped his speed of handwriting. In a couple of days he wrote very neatly to his legal adviser :

[1] More correctly *prothyle*, see p. 378.

April 3rd, 1888.

MY DEAR MR. MORRISS,

I send herewith a letter just received from the gentleman I spoke to you about the last time I saw you. He is desirous of my becoming a Director of his company, "the Havelock" mine, and I think he would have little difficulty in getting the Articles altered to suit any stipulations you might propose, as they are not yet finally settled. Is the fee out of all proportion to the work to be done ? I think, considerating the shares are not yet allotted, and are as likely as not to be of no value, the fee is not outrageous. But in this I will be guided by what you say is safe for me to do ; in writing please word your letter so that I can show it to Mr. Fradd. If you think an interview will be better before you write, I am at your service almost any afternoon.

Believe me,

Very sincerely yours,

WILLIAM CROOKES.

Hurrah for the typewriter !

Meanwhile, Crookes's old friend, J. B. Spence, had been working out a new method of gold extraction, on which Crookes remarked that it would not work properly unless manganese dioxide were added to the ferric chloride used. After some negotiations, Crookes took up Spence's method and applied for a patent, of which he and Spence became joint proprietors. With the help of some City friends, of which Major Ricardo Seaver was the leader, a company was formed under the name of the " Gold Mines of France Association," to acquire and work the patent. In order to obviate certain legal difficulties, it was proposed that Crookes should sell his French patent to the *Caisse des Mines*, and that the association should secure an option to buy it from the latter. Mr. Morriss, however, advised Crookes that this could not be done so long as Crookes was a director of the association. Crookes decided to resign, but the association refused to accept it, and Crookes did not insist, as Major Seaver had offered to buy the rights from the Caisse des Mines with the aid of some French capitalists. Crookes placed his share at £20,000 cash. Provisional works were established at Bankside and gave excellent results, the work being in charge of a Mr. Richardson, who had acted as assistant to Spence. Matters did not go at all smoothly. Spence and Richardson seem to have made great efforts to work independently of Crookes, and matters came to such a pass that Crookes eventually offered to Seaver to invent a still better process to be worked in their joint names. In May 1889, the Spence-Crookes process was found to have been anticipated, and everything fell through.

About this time there was a revolution in Brazil, and the Emperor Dom Pedro II, who had been very friendly to Crookes, was obliged to abdicate. Crookes used to send him copies of his scientific papers, and even samples of his apparatus, as is shown by the following letter :

Decr. 20th, 1878.

MAY IT PLEASE YOUR MAJESTY,

When you honoured me with a visit in London you were good enough to express a wish to receive accounts of any further researches I might make in Molecular Physics.

A fortnight ago I presented to the Royal Society a paper describing some recent experiments on the Illumination of Lines of Molecular Pressure, and the Trajectory of Molecules. The full paper will not be printed for several months, but I enclose herewith a proof of the abstract, which I have marked in places corresponding to the three sketches of instruments also enclosed in this letter.

By this Mail Steamer I have forwarded to Your Majesty a small wooden box, containing three instruments like the sketches. They illustrate the chief points in my new discovery, and I venture to hope that you may be pleased to experiment with them, and may obtain similar results to those described in my paper.

The instruments require to be connected with an induction coil capable of giving sparks in air about 40 millimetres long, and the

negative pole should be connected with the pole marked " negative "
on the sketch.

No. 1, The Electrical Radiometer, shows the rotation very
well. The exhaustion is carried to the point at which the " dark
space " is flattened against the glass bulb.

No. 2 shows the projection of Molecular Shadows, and the
deflection of the rays by magnetism. A small permanent magnet
is powerful enough to show this, although the deflection is seen
better with an electro-magnet.

No. 3 shows the concentration of heat where the molecular
rays meet at a focus ; the strip of platinum will be raised to a red
or white heat. It is possible that the travelling and shaking may
have bent the concave negative pole so as to make the focus fall
on one side of the platinum. If this be the case the platinum will
not be heated, but it can be immediately raised to redness by bringing
a magnet near, so as to deflect the focus onto the strip of platinum.

It is not good to keep the coil at work many minutes at a time,
or the poles will become hot, occluded gas will be given off from
them, and the vacuum will be spoiled.

Trusting that the instruments will arrive safely.

I have the honour to remain,

Your Majesty's Obedient Servant,

WILLIAM CROOKES.

P.S.—Although I feel unworthy of the high distinction Your
Majesty said you had conferred upon me, in the gift of the Order
of the Rose, I should show a want of respect were I to appear careless
about any part of the distinction.

I therefore venture to inform Your Majesty that the promised
Decoration has probably miscarried, as I have not yet received it.

I hope that my appreciation of the great honour conferred upon
me may be my excuse for troubling Your Majesty in this matter.

W. CROOKES.

We must now turn to another episode in Crookes's
versatile career. It was connected with the manufacture
of aluminium, a metal at that time in very slight use, in
spite of its lightness and portability. Considering the
enormous development of the aluminium industry in

modern times, it seems strange that Crookes could say in 1888 that "a ton of aluminium would flood the market for many years," but nobody could anticipate that the main difficulties—its difficulty in soldering and its sensitiveness to the attack of soda—would be overcome so soon. Consulted by Sir Douglas Galton with regard to Graetzel's and Kleiner's processes, Crookes wrote (May 16, 1888) that no company could pay its way unless it could sell a ton of aluminium per month.

About the same time Crookes heard of a new method for winning aluminium, invented by Professor Riatti, of Milan, through a man of the name of Ostertag, who was a sort of broker employed by Riatti to negotiate for the sale of the invention. Finding the latter unsatisfactory to deal with, Crookes asked Riatti to negotiate with him direct. After some hesitation, Riatti made up his mind to do this. But a Mr. Mallory went to Milan before Crookes found himself able to go, and bought up Riatti's process. The transaction turned out to be unsatisfactory, and Mr. Mallory himself consulted Crookes as to what ought to be done. Crookes proposed to buy patent rights in the United Kingdom in his own name at any reasonable price.

These negotiations, however, again fell through. Riatti and Mr. Mallory fell out and went to law, and Crookes was unable to make any satisfactory arrangement, although he visited Milan in broiling heat in an effort to do so.

Crookes's "operations" with companies were of a very diversified nature. He was a director of the London-Mexican Prospecting and Finance Company, Ltd., with a fee of £250 a year. He had £7,400 invested in Mexican loan. He also had money in the Vogelstruis and the South African Gold Mine Investment Trust. He put £500 into the Linotype Company, but withdrew it on the first opportunity. On the recommendation of Mr. Horatio Bottomley, he joined the board of Messrs. Kegan Paul, Trench, Trübner & Co. A letter on this subject is of interest :

Nov: 15th, 1889.

HORATIO BOTTOMLEY, Esq.,
   12 & 14, Catherine Street,
      Strand, W.C.
  DEAR SIR,
      When you first spoke to me about joining the Kegan Paul, Trench, Trübner & Co. Company, I said the first condition must be that Messrs. Davidson & Morriss should look over the Prospectus and Articles of Association on my behalf, and advise me that they were such that I could safely join. You promised to see them and place the matter before them and communicate with me. At the meeting yesterday I asked if you had seen them, and you said you had done so and it was all right. On my seeing them to-day both the partners tell me they have not seen the prospectus in my interest Mr. Davidson saying that acting as he does for the underwriters he cannot advise me, or even speak to me on the subject, and Mr Morriss saying that he has not seen any of the papers, and doubts whether he could act in my behalf if he had, as his firm are acting for the other side.

As I only consented to affix my name to the prospectus on the strength of what you said to me, I am at a loss to reconcile your and their statements, and am anxious to know how I stand.

            Believe me,
                Very truly yours,
                   WILLIAM CROOKES.

In 1888, Crookes was awarded the Davy Medal of the Royal Society, and his graceful speech on receiving the medal is worth recording :

In returning thanks for the health of " The Medallists," Mr. Crookes said :

In the absence of Professor Huxley, who unfortunately is unable to be present this evening, I have been asked by our President to return thanks for the Medallists.

THE ORIGIN OF THE ELEMENTS

The Professor, in whose shoes I now stand, has lately put it on record that for twenty years he never rose to make a speech without his tongue cleaving to the roof of his mouth through sheer nervousness. I, on whom his duties have fallen, but, alas! not his mantle, crave your indulgence if I show that nervousness for which I can quote so excellent an authority, and beg you to bear with me patiently in what must be as great a disappointment to you as it is to myself.

Although I feel it embarrassing to take the place of a philosopher so renowned as Huxley, I am bound to say that even he cannot receive the Copley Medal—the ancient olive crown of the Royal Society, as it was called by Sir Humphry Davy—without regarding it as the culminating honour of his scientific career. I am also bound to say that the award of this Medal to such a man as Huxley confers at least as much honour on the Royal Society as on its renowned Professor.

As regards the other recipients of the Medals, Professor Tacchini, the Romford Medallist, Sir Ferdinand von Müller, and Professor Osborne Reynolds the Royal Medallists—co-workers with myself in very varied departments of science—judging from my own feelings at the present moment, I cannot do better than express on their behalf the sincerest gratification at the high recognition you have bestowed on their scientific work.

For my own part, as the Davy Medallist, I may perhaps be permitted to speak at somewhat greater length. Thirteen years ago in this very room, at the Anniversary Dinner of the Royal Society, I had the honour to return thanks to this Society for conferring on me the distinction of a Royal Medal. The President, Dr. Hooker, in handing to me the Medal, referred to the then novel instrument the Radiometer, and in the course of his remarks he made the following observation :

> " It is the mystery attending this phenomenon that gives it its great importance. There is evidently some action going on with which we are not at present acquainted ; and there is no saying what a thorough investigation into the cause of the phenomenon may lead to."

From that date to the present I have not ceased to work at the elucidation of the mysteries accompanying high vacua. During this interval of thirteen years the theory of the radiometer has been established on a firm basis. Since the advent of the radiometer the otheoscope has been devised and its action elucidated ; lines of molecular pressure in high vacua have been illuminated so as to render visible to the bodily eye those actions which hitherto had been seen only by the eye of faith ; molecular shadows have been projected ; luminous streams have been deflected and magnetised ; platinum has been melted by molecular impact ; the existence of matter in a fourth state—"radiant matter"—has been demonstrated ; and the strikingly brilliant phenomena of phosphorescence in high vacua have been discovered.

The results of these investigations I have embodied in ten memoirs and Bakerian Lectures which have appeared in the *Philosophical Transactions,* in eighteen papers which have been inserted in the *Proceedings of the Royal Society,* and in fifteen other papers published elsewhere.

In the spectroscopic examinations which these researches necessitated I constantly met with a bright citron-coloured line, sometimes strong, at others only a ghost of a line, but always in the same place, and presumably due to the same cause. The hunting out of this phantom line has been a work of many years. At last I found it to be due to yttria, while a corresponding red line was tracked to samaria.

Here a new field, indeed I may say a new world, of research unfolds itself to the chemist. The phosphorescent light from yttria gave in the spectroscope a complicated and beautiful series of coloured lines. But these lines varied in different specimens of the same earth from different localities, some being strong and others weak almost to extinction. Chemical tests showed that in each case the earth was free from known impurity, and for a long time the interpretation was inexplicable. I was confronted with a series of inscriptions from the molecular world, written in a strange and baffling tongue, and requiring a Rosetta stone to give the first step to decipherment. How the key of the mystery came into my hands, and what resulted therefrom are now matters of history. The meaning of the lines, their positions and variations is gradually becoming clearer and clearer, and to me this is what they seem to be saying :

They tell me that in matter which responds to every test of

323

an element which possesses an atomic weight and combines with other matter in definite proportions, there still exist minute shades of difference among which a process of selection is possible.

They tell me that the time-honoured distinction between elements and compounds, as it exists in the Lavoisierian and Daltonian chemistry, must be modified to make room for a vast array of intermediate bodies, or as I have elsewhere called them meta-elements, capable of isolation and identification.

They tell me that there exist degrees of rank in the hierarchy of the chemical elements, some being, so to speak, more elemental than the others, some still waiting to be discovered, whilst others may have disappeared and are no longer existent under the conditions in which we are now placed.

They tell me, again, that our chemical elements are products of the action of forces on a state anterior to matter, which I have designated as Protyl ; that these elements owe their present stability in that they are the outcome of a struggle for existence, a Darwinian development by chemical evolution ; that just as in the organic world we have the " survival of the fittest," so here we have the " survival of the most stable " or possibly of the " most inert."

There are other lessons as yet only indistinctly whispered, lessons which, possibly to-day but certainly in days to come, will be seized on and interpreted.

But I will no longer weary you with speculations which have not yet received full and final confirmation. It only remains for me to thank you for the high distinction of the Davy Medal—a distinction all the more welcome as it has only once previously been awarded to one of our own countrymen. I deeply appreciate the kind reception you have given me and the very gratifying remarks with which the presentation has been accompanied.

Refreshed and stimulated by your friendly words of encouragement, I shall continue to apply myself to what Darwin called " the vivid pleasures of investigation." I shall attack with renewed energy the problems of radiant matter spectroscopy, and do my utmost " to pluck out the heart of the mystery."

In 1890, Crookes was elected President of the Institution of Electrical Engineers. His Inaugural Address, delivered on Jan. 15, 1891, was described by Sir William

Thomson, afterwards Lord Kelvin, as placing Crookes
" not second to Faraday, not second to Joule, but following
both and crowning both." On this occasion he again
put forward his memorable hypothesis concerning the
Genesis of the Elements. He said, in concluding his
address :

> To show how intimately chemistry and electricity interlock,
> I may here remark that one of the latest theories in chemistry
> renders such a division of the molecule into groups of electro-positive
> and electro-negative atoms necessary for a consistent explanation
> of the genesis of the elements. This is so important that I may
> be excused for digressing a little into this development of theoretical
> chemistry.
>
> It is now generally acknowledged that there are several ranks
> in the elemental hierarchy, and that besides the well-defined groups
> of chemical elements, there are underlying sub-groups. To these
> sub-groups has been given the name of " meta-elements." The
> original genesis of atoms assumes the action of two forms of energy
> working in time and space—one operating uniformly in accordance
> with a continuous fall of temperature, and the other having periodic
> cycles of ebb and swell, and intimately connected with the energy
> of electricity. The centre of this creative force in its journey through
> space scattered seeds, or sub-atoms, that ultimately coalesced into
> the groupings known as chemical elements. At this genetic stage
> the new-born particles vibrating in all directions and with all veloci-
> ties, the faster-moving ones would still overtake the laggards, the
> slower would obstruct the quicker, and we should have groups
> formed in different parts of space. The constituents of each group
> whose form of energy governing atomic weight was not in accord
> with the mean rate of the bulk of the components of that group,
> would work to the outside and be thrown off to find other groups
> with which they were more in harmony. In time a condition of
> stability would be established, and we should have our present series
> of chemical elements, each with a definite atomic weight—definite
> on account of its being the average weight of an enormous number
> of sub-atoms, or meta-elements, each very near to the mean. The
> atomic weight of mercury, for instance, is called 200, but the atom
> of mercury, as we know it, is assumed to be made up of an enormous

number of sub-atoms, each of which may vary slightly round the mean number 200 as a centre.

We are sometimes asked why, if the elements have been evolved, we never see one of them transformed, or in process of transformation, into another. The question is as futile as the cavil that in the organic world we never see a horse metamorphosed into a cow. Before copper, e.g., can be transmuted into gold, it would have to be carried back to a simpler and more primitive state of matter, and then, so to speak, shunted on to the track which leads to gold.

This atomic scheme postulates a to-and-fro motion of a form of energy governing the electrical state of the atom. It is found that those elements generated as they approach the central position are electro-positive, and those on the retreat from this position are electro-negative. Moreover, the degree of positiveness or negativeness depends on the distance of the element from the central line ; hence, calling the atom in the mean position electrically neutral, those sub-atoms which are on one side of the mean will be charged with positive electricity, and those on the other side of the mean position will be charged with negative electricity, the whole atom being neutral.

This is not a mere hypothesis, but may take the rank of a theory. It has been experimentally verified as far as possible with so baffling an enigma. Long-continued research in the laboratory has shown that in matter which has responded to every test of an element, there are minute shades of difference which have admitted of selection and resolution into meta-elements, having exactly the properties required by theory. The earth yttria, which has been of such value in these electrical researches as a test of negatively excited atoms, is of no less interest in chemistry, having been the first body in which the existence of this sub-group of meta-elements was demonstrated.

I frankly admit I have by no means exhausted the subject which daily and nightly fills my thoughts. I have ardently sought for facts on which to base my theory. I have struggled with problems which must be conquered before we can arrive at exact conclusions —conclusions which, so far as inorganic Nature is concerned, can only be reached by the harmonious interfusion—not confusion— of our present twin sciences, electricity and chemistry. Of this interfusion I have just endeavoured to give you a foretaste. In elaborating the higher physics, the study of electrical phenomena must take a large, perhaps the largest, share.

We have invaded regions once unknown, but a formidable amount of hard work remains to be completed. As we proceed we may look to electricity not only to aid, as it already does, our sense of hearing, but to sharpen and develop other powers of perception.

Science has emerged from its childish days. It has shed many delusions and impostures. It has discarded magic, alchemy, and astrology. And certain pseudo-applications of electricity, with which the present Institution is little concerned, in their turn will pass into oblivion.

There is no occasion to be disheartened at the apparent slow pace of elemental discovery. The desponding declare that if Roger Bacon could revisit " the glimpses of the moon," he would shake his head to think we have got no further—that we are still in a haze as to the evolution of atoms. As for myself, I hold the firm conviction that unflagging research will be rewarded by an insight into natural mysteries such as now can scarcely be conceived. Difficulties, said a keen old statesman, are things to be overcome ; and to my thinking Science should disdain the notion of finality. There is no stopping half-way, and we are resistlessly driven to ceaseless inquiry by the spirit " that impels all thinking things, all objects of all thought, and rolls through all things."

In this address Crookes attained the greatest height to which science could soar in those days. To our descendants, no doubt, his outlook and his reasoning will appear strangely limited. He was, of course, still under the influence of that school of philosophy which looks for " origins," not only of organic life, but of the inorganic universe. A later school will probably take the universe for granted, as something eternally given, so that its " evolution " proceeds in waves rather than a single bound. It aids us nothing to assign a time limit to the Past of the Universe, and leave the æons preceding that Limit an aching void. The relativity of time and space forbids a limitation which would reduce all existence to a vanishing threshold separating two eternities. But for his time, and in his place, Crookes figures as a man standing on the heights, with an outlook wide as the starry heavens.

That the high literary quality of the address was largely due to the inspiration of Crookes's " muse "— Miss Alice Bird—is evident from the letters he wrote to her concerning the proofs :

Dec. 18th, 1890.

MY DEAR LALLAH,

Your corrected proofs came to hand in the very nick of time, for I was just sending a batch of alterations to the printer, and I at once transferred those from your proof and sent him such a proof as would make him tear his hair. I have not got clean proofs again, and one of them I send herewith, hoping that you may still be in the most amiable frame of mind that prompted you to write so nice a letter with the sheets. Will you look over them like the good angel that you are, and let me have them again *before* 1890 *is out* ? I sympathise with your lack of envelopes. I have known what it is, and on several occasions my correspondents have sent me packets which have arrived in a hopeless state of disorganisation and half the contents gone. Alice and Nina are hard at work doing the diagrams, of which there are more than two dozen. Just now the skating somewhat interferes, but they have worked hard and deserve a holiday. Oh ! this weather, it will be the death of us all if it keeps on much longer. Nothing but incessant fog and darkness, and now the hardest frost we have had in London for very many years. Thames frozen over and ice in the parks six inches thick. Send George back to us soon. Nellie and I are living on his old prescriptions. At present they have worked all right, but we may make some awful mistake soon, and then he will be vexed. According to all accounts, however, you are not much better off, seeing what a correspondent from Alassio (I don't in the least know where Alassio is) writes to the P.M.G. If Italy is so bad, France must be worse. We are looking forward to a very doleful Xmas. So many of our family scattered over the world. We see in the paper this morning that Joe's ship got to Adelaide yesterday. We have had one or two letters from him, and he seems in good spirits about his health. He gives some amusing descriptions about his fellow-passengers, and he has not suffered much from storms he encountered in the Bay. Henry writes more cheerfully, he is digging wells for the pioneer force to Mashonaland, along the road. Bernard is very well, but we do not hear very frequently as he is so far from a civilised place. He is surveying in a wild country that has never been visited by a white man, and has to keep a sharp look-out for tigers. He says, however, that tigers are as common there as they are in Regent Street, and he objects more to the leeches, which stick on him a dozen at a time when he is walking through the marsh and jungle. We hope Laura has got safely to you, and that you are better and will return home quite set up and strong, ready for the dissipations of a London season. When you write to or see George, give him my love, and tell him how much I appreciate his terse and vigorous English. Some of his corrections were so good. I laughed aloud when I deciphered them. One lot I could not read a bit so I sent the hieroglyphics to the printer, and he returned such a delicious sentence back. How he managed it I can't imagine. Now your writing is as legible as my own. Now I must close as it is very late, and Nellie is sitting patiently by my side, and I do not want to be too extra late. I ought to apologise for not writing with my own hand, but really I am so accustomed to the typewriter now and get on so quickly with it that I feel almost unable to write without it.

With kind love, in which Nellie joins.

Believe me,

My dear Lallah,

Very sincerely your friend,

WILLIAM CROOKES.

Dec. 22nd, 1890.

MY DEAR LALLAH,

I have found a couple of delicious quotations, from the classics, which I am anxious to drag in. I send you the paragraphs where I have brought them in, and I want your *imprimatur* for them with corrections, especially in the concluding sentences, which seem rather " with extended sweetness long drawn out." I hope you received my packet and were not horrified at the coolness of my request.

Our weather now is " Awful." No other word expresses it to my mind, although Lewis, who has just arrived and is skating every day, says it is "Jolly." A very heavy fall of snow lasting for two days disorganised the traffic, and then came a thaw. Last night it froze hard, the thermometer sinking down to unknown depths. This morning the streets are like glass, and everyone is falling about and breaking a limb or two. I am going out as soon as I finish this, so what condition I shall be in on my return I don't know. The fog which partially left us for a couple of days has now returned, and we are experiencing an arctic night. How are you, and how are Laura and George? We are better in health here than could be expected, but considering that winter has scarcely begun, the potentialities of the future are infinite.

Good-bye, take care of yourself. I have just thought of another experiment which will probably throw more light on these puzzling atoms and their habits. I must go and start it before I go out to post this.

Nellie joins me in kind love.

Believe me,

My dear Lallah,

Ever sincerely yours,

WILLIAM CROOKES.

Crookes also wrote, a few days after the event, in a more cheerful mood :

Jan. 25th, 1891.

MY DEAR LALLAH,

The first quiet hour since the eventful evening of my address. I decided to a letter to you. I have sent you a copy of the *Electrician* (by registered post as the P.O. authorities are fond of confiscating newspapers) containing a lot that is complimentary about me. How I wish it were all true. The address was a grand success. I do not think that any of my Royal Institution lectures eclipsed it in that way. After it was over Sir William Thomson proposed a vote of thanks to me in flattering terms, and then the proceedings ended, and I went home happy, but very tired. It is not usual to see ladies at the meetings, but Nellie wanted to hear me, so they made up a small party of about six or eight.

I hope you will like the portrait in the *Electrician*. It is generally thought to be a very good likeness. And now I want to hear about yourself. There is a great chance of this letter not reaching you, as you may have flitted from the cold tempestuous shores of the Mediterranean and have come north to warmer climates. The weather to-day is simply perfect. A clear sky, a bright warm sun, no mud or dirt, all the snow washed away by a day's warm rain, and a south wind which makes it a pleasure for us to be out. If it would only last ! Where are you now ? When do you propose to return, and most especially how is your health after the unprecedented experiences of an arctic winter you have gone through ? I suppose George will be back soon, but you will probably stay.

Nellie sends her kindest love to you, and with the same from me,

Believe me,

My dear Lallah,

Very sincerely yours,

WILLIAM CROOKES.

The work of these brilliant years was beset with some anxieties, mainly arising out of Crookes's gold-mining ventures.

Ever since the old gold-amalgamation days, Crookes had maintained a scientific and financial interest in Welsh gold-mining. This interest was strengthened by the taste for mining developed by his eldest son, Henry. In August 1888, Crookes received from Mr. Pritchard Morgan, of Dolgelly, five hundred fully-paid shares in the Mount Morgan West mine. Crookes immediately started active mining operations under his own management, acquiring the lease of the Cefn Coch mine from the Office of Woods for fifteen years. He appointed a young Oxford man as manager, and endeavoured to direct the operations from London for the next seven years, with occasional visits to Dolgelly during the summer. A certain amount of gold was actually produced at Cefn Coch, and in 1894 Crookes told his patent agent that he would not proceed with his patents for gold extraction as he could spend his money to greater advantage in his own gold mine. But in December of that year a gale blew off the roof of the

331

office and the powder magazine, whereupon the watchman fled and the mine was left unprotected. Crookes estimated that the gold produced in six years was worth £500, but his outlay had exceeded £2,000. He therefore decided to sell the mine. In February 1895 a Mr. Johnson acquired half the rights in the mine for £2,250, but as Crookes considered this sum as purchase money pure and simple, whereas Mr. Johnson regarded it as working capital, there was some disagreement, and Crookes refused to finance the mine any longer, more especially as Mr Johnson was much the wealthier partner. " In the game of planking down £100 notes I should get tired long before you would "—as Crookes put it to him in his graphic way. The difficulties increased when a dispute arose between Johnson and Crookes's friend, C. R. Williams, of Tyn-y-Groes, concerning a right of way, whereupon Crookes wrote the following mordant epistle to his lawyer :

July 27th, 1895.

DEAR MR. MORRISS,

WILLIAMS *v*, JOHNSON (CROOKES dragged in).

How is it possible I can join Johnson in an affidavit saying what I intend to do with the water ? My intention is to leave it severely alone. Johnson intends otherwise.

Cannot you and Whale and Johnson settle it between you ? I am laid up here suffering from a carbuncle, and it will be very injudicious for me to leave home on Monday. Complete rest is what I want, and a holiday as soon as possible. This worry is one of the predisposing causes of it. I *will not* sign an affidavit saying it is my intention to use the water for driving a wheel or any such purpose. Johnson *may* be able to make me pay half the expenses of his lawsuit, but there is no obligation for me to commit perjury to oblige him.

There will be some fun if I am put in the witness box !

In the end, Johnson paid Crookes £1,000 for his remaining share in the ill-fated mine. On his own showing, this was about all the profit Crookes made out of his Welsh gold mine after seven years' work and worry. No wonder he came to the conclusion that he could make money more easily in other ways.

The most extraordinary company case in which Crookes ever was concerned was that of the Proprietary Gold Recovery Company, which owned the MacArthur Forrest process, and to which Crookes handed over a new patent for keeping the amalgamating mercury clean by means of a solution of mercury cyanide. He charged nothing for the rights, but was made a director of the company without qualification by shares.

We may imagine his surprise when he was summoned to the Guildhall to appear against his co-directors for a contravention of the Companies Act ! He immediately wrote to his legal advisers :

Oct. 29*th*, 1894.

Messrs. DAVIDSON & MORRISS,
    40 & 42, Queen Victoria Street,
        Mansion House, E.C.

DEAR MR. MORRISS,

I have had this put into my hands this evening, just after dinner as a digester. I did not kick the clerk.

What am I to do ? I have not the faintest idea what it is all about, and was not aware that we had swerved the hair's-breadth from the path of virtue and Companies Acts.

Please observe,

1. I am a witness *on the other side*. I am therefore to give evidence against my co-directors. If they have to go to Holloway, how do I stand ?

2. I received the munificent fee of four Shillings. I told the clerk that my fee was a little larger than that, and asked him for the balance of £52 6s. od., but he said he had not quite so much by him. Can I not refuse to give evidence unless my fee is paid in advance ?

3. The thing appears to be a *criminal* matter from the wording of Note 1.

I do not know who Messrs. Gruggen and Williams, or Alexander Crookshank are. I never heard of them.

Now this may be a very serious thing, and I should like to see you about it. I will stay at home all the day to-morrow, and shall be much obliged if you will send me an early wire telling me what time I can come to your office and see you about what is to be done.

He also wrote to a co-director :

Oct. 29th, 1894.

E. J. PAPE, Esq.,
    26, Portland Place, W

    DEAR MR. PAPE,
           This evening I have had a summons to attend at the Guildhall on Wednesday next at 12 noon, with all the books, &c., of the Proprietary Gold Mining Co., about a charge of non-compliance with the Companies Acts. I am summoned as a witness for the complainant against the Company.

I have not the remotest idea what this is about. Can you tell me ? It is most annoying that almost my first introduction to the business of the Company should be a criminal prosecution. Have you had a similar summons ? I am very busy on Wednesday, and it will be most inconvenient for me to attend.

Kindly reply at your earliest convenience.

The explanation of Crookes being summoned as a witness was that he had been previously employed in a case involving the MacArthur Forrest process, in which his fellow-witnesses were Professors Dewar and W. C. Roberts-Austen. To the latter he wrote :

Nov. 15th, 1894.

MY DEAR ROBERTS-AUSTEN,
           I have had some talks with Dewar on the subject of our fees. I enclose a list of my consultations and attendances. I propose to send in a note of fees for professional assistance, 150 guineas, not giving details. Dewar says he has attended more frequently, and will make a claim for 200 guineas

Our opinion is that we, who may be looked upon as the leaders in the Profession, owe it to our fellows to charge high, and so do our best to counteract the lowering of prices which is so prevalent in the lower branches.

Crookes was not legally involved in the trouble of his co-directors as he held no qualifying shares ! In the next few months the company went into liquidation, and Crookes thought he might as well retrieve his patents from the wreck. He offered the Official Receiver £20 for the Transvaal, New South Wales, and American patents, and the offer was accepted. So Crookes was well out of a rather unpleasant business, with three patents to the good !

Another venture concerned the glazing of white bricks with a glaze which would not subsequently crack. It speaks volumes for Crookes's prestige that a company offered to pay him £1,000 in shares in advance and a fee of £500 for the discovery of a suitable glaze and another £1,000 if successful. After twelve months Crookes considered that he had completely solved the problem set to him, and applied for a patent for the glaze. It consisted of :

Red lead, 30	Potassium carbonate, 5
Silicon, 30	Potassium nitrate, 2
Kaolin, 14	Oxide of tin, 2
Sodium carbonate, 5	Arsenic, 2·7

Sodium chloride, 1·8 parts.

To be used on a brick made of London clay 30, sand 30, and chalk 40 parts.

In this business Crookes was largely associated with Mr. A. P. Sinnett, the theosophist, who had persuaded him to underwrite Horatio Bottomley's " International Newspaper Company." When this enterprise was exposed by the *Financial News*, Crookes wrote to Sinnett : " Unless

335

Bottomley has a complete reply in to-morrow's issue, I fear it will be all U P with our little spec. in underwriting, and we shall have to eat our Christmas dinner in dust and ashes."

Instead of which Bottomley relieved the anxieties of the "amateur underwriters" by one of those adroit manœuvres of which he was a master, and Crookes, in order to express his appreciation, wrote to him :

Dec. 27th, 1889.

H. BOTTOMLEY, Esq.,
    12 & 14, Catherine Street,
      Strand, W.C.

    MY DEAR SIR,
        Mr. Sinnett has told me in how handsome a manner you have offered to relieve us amateur underwriters of some of the inconvenience of our rash attempt at speculation, by placing at our disposal some of your fully-paid shares in "The International Newspaper Co.," if we take up the shares underwritten. Allow me to express my high sense of the kindly feeling which has prompted you to make this offer. For my part I will gladly accept the £400-worth of shares you so kindly offer to make my lot up to £2,000-worth, and I have accordingly sent the application forms to the bank for 80 ordinary and 80 preference shares, with a note saying I propose to pay up in full on allotment.

I hope the International Newspaper will prove to be so great a success that our only feeling of regret will be that we did not individually invest a still larger amount of capital in it.

Crookes's own work in connection with the glaze came practically to an end in 1895, just before his departure for South Africa. The enterprise remained in Sinnett's hands, who, however, had no experience in company management, and could make nothing of it.

Several attempts made about this time to draw Crookes into discussions on psychic matters were met by an attitude of good-natured refusal, as when he wrote to the editor of *Light* :

Dec. 6th, 1894.

E. DAWSON ROGERS, Esq.,
Office of *Light,*
2, Duke Street, Adelphi, W.C.

DEAR SIR,

I am sorry I cannot accept your polite invitation to write something respecting Professor Barrett's lecture. I am so overwhelmed with pressing engagements which occupy all my time, that I regret to say I have not even read the lecture with the attention it deserves, and I could not possibly spare time required for the concentration of thought and close attention which I should have to devote to the subject in order to do justice to a review of a lecture by a scientific man of Professor Barrett's eminence.

In this connection a hitherto unpublished telegram on psychic questions is of some interest. Professor van't Hoff, the eminent Dutch chemist and originator of the modern theory of electrolysis, telegraphed to Crookes asking him point-blank whether he still adhered to the view that the psychic phenomena Crookes had described in 1870–4 were genuine. Crookes replied by telegram on January 14, 1890 :

Professor Vanthoff, Amsterdam. Published opinion private test séances unchanged.—CROOKES.

It is open to doubt whether this telegram covers " Katie King's " farewell séances at Dalston on May 13th and 24th, 1874.

An important success achieved by Crookes about this time was the winning of the Cordite Case (Nobel *v.* Anderson), in which Crookes appeared as an expert witness for the Government. The case lasted from June 1893 to February 1894, and Crookes received a fee of £500.

Crookes's far-flung fame brought him some annoyance from importunate American publicists, as is shown by the following letter addressed to one of them :

z

THE ORIGIN OF THE ELEMENTS

May 19*th*, 1888.

H. Vincent Smith, Esq.,
     61, Broomfelde Road,
       Clapham.

Dear Sir,

I am sorry you have so completely misunderstood the purport of the few remarks I made when you called on me the other Sunday. I am sure no one could accuse me of want of courtesy or inhospitality to foreign men of science who come to England with introductions to me, but if you knew how perpetually I am intruded on by newspaper writers, interviewers, and suchlike nuisances of modern American civilisation, you would I am sure excuse me, even if in some cases, as with you, my caution errs in the wrong direction. But I don't think I did so very wrong after all. I asked if you had any letter of introduction. "No." Did you know any scientific men with whom I was acquainted? "No." Did you cultivate any special branch of science? "No." You came without any special preparation of the above kind and said you wanted to ask me a particular question. I answered the question as well as I could. What more could I have done under the circumstances? My remonstrances only commenced when you began to put me through a cross-examination on the subject of your inquiry.

We can easily guess what that subject of inquiry was. It no doubt concerned Crookes's psychical researches, which he had for many years thrust into the background in order to leave the ground clear for pure science, in which he was sure of a friendly and appreciative hearing.

Crookes's activity was prodigious. He never confided his voluminous correspondence to a secretary. He did all his own typewriting in copying ink, and press-copied all his letters himself. From 10 to 1 each day he sat at his desk, paying occasional brief visits to his laboratory to see how things were going. After lunch he would go out to the various societies. At 7 he would sit down to dinner. From 8 to 9 he would sit in his library, and would read or doze. At 9 he would go to the laboratory

338

to work till after midnight. Before retiring to bed he would go out to post his letters for the night collection. For forty years he hardly varied this routine. He had no recreations and no holidays, with the exception of an annual fortnight, usually spent in the Channel Islands. He indulged in no sport. He did nothing to " keep fit " except work, and if " keeping fit " means keeping in prime condition for one's chosen work and mode of living, it must be acknowledged that he succeeded very well.

CHAPTER XVII

THE END OF THE CENTURY
(1892–9)

A S EARLY AS 1892, Crookes felt the influence of the new view of electricity which in a few short years was to sweep all before it. He spoke of an atomic theory of electricity as " not yet definitely proved, although not improbable." In an article published in the February number of the *Fortnightly Review*, entitled " Some Possibilities of Electricity," he wrote :

Helmholtz considers it to be probable that electricity is as atomic as matter, and that an electrical atom is as definite a quantity as a chemical atom. This, however, must not yet be regarded as a certainty, for it is possible that all the facts at present known may be explicable in another way. If an atom of matter is endowed with the property of taking to itself one, two, three, or more units of electricity, it does not follow that electricity is atomic. Imagine the atoms of matter to act like so many bottles, capable of holding one, two, three, or more pints. Imagine electricity to be like water in the ocean, which for the purposes of this argument may be considered inexhaustible and structureless. One of the atomic " bottle " elements dipped into the ocean would certainly take to itself, one, two, three, or more pints of water, but it would by no means follow that the ocean was atomic in that it was capable of being divided up into an infinite number of little parcels, each holding a pint or its multiple.

For this and other reasons I think we must accept the hypothesis of the atomic character of electricity as not yet definitely proved, although it is not improbable.

I have spoken of the "ether"—an impalpable, invisible entity, by which all space is supposed to be filled. By means of the ether theory we can explain electrical phenomena, as well as those appertaining to the phenomena of light.

And here follows a forecast of wireless telegraphy, several years before either Lodge or Marconi took out their first patents :

Until quite recently we have been acquainted with only a very narrow range of ethereal vibrations, from the extreme red of the solar spectrum on the one side to the ultra-violet on the other—say, from three ten-millionths of a millimetre to eight ten-millionths of a millimetre. Within this comparatively limited range of ethereal vibrations and the equally narrow range of sound-vibrations all our knowledge has been hitherto confined.

Whether vibrations of the ether, longer than those which affect us as light, may not be constantly at work around us, we have, until lately, never seriously inquired. But the researches of Lodge in England, and of Hertz in Germany, give us an almost infinite range of ethereal vibrations or electrical rays, from wave-lengths of thousands of miles down to a few feet. Here is unfolded to us a new and astonishing world—one which it is hard to conceive should contain no possibilities of transmitting and receiving intelligence.

Rays of light will not pierce through a wall, nor, as we know only too well, through a London fog. But the electrical vibrations of a yard or more in wave-length of which I have spoken will easily pierce such mediums, which to them will be transparent. Here, then, is revealed the bewildering possibility of telegraphy without wires, posts, cables, or any of our present costly appliances. Granted a few reasonable postulates, the whole thing comes well within the realms of possible fulfilment. At the present time experimentalists are able to generate electrical waves of any desired wave-length from a few feet upwards, and to keep up a succession of such waves radiating into space in all directions. It is possible, too, with some of these rays, if not with all, to refract them through suitably-shaped bodies acting as lenses, and so direct a sheaf of rays in any given direction ; enormous lens-shaped masses of pitch and similar bodies have been used for this purpose. Also an experimentalist at a distance can receive some, if not all, of these rays on a properly-constituted instrument, and by concerted signals messages in the Morse code can thus pass from one operator to another. What, therefore, remains to be discovered is—firstly, simpler and more certain means of generating electrical rays of any desired wave-length, from the shortest, say, of a few feet in length, which will easily pass through buildings and fogs, to those long waves whose lengths are measured by tens, hundreds, and thousands of miles ; secondly, more delicate receivers which will respond to wave-lengths between certain defined limits, and be silent to all others ; thirdly, means of darting the sheaf of rays in any desired direction, whether by lenses or reflectors, by the help of which the sensitiveness of the receiver (apparently the most difficult of the problems to be solved) would not need to be so delicate as when the rays to be picked up are simply radiating into space in all directions, and fading away according to the law of inverse squares.

Any two friends living within the radius of sensibility of their receiving instruments, having first decided on their special wave-length, and attuned their respective instruments to mutual receptivity, could thus communicate as long and as often as they pleased by timing the impulses to produce long and short intervals on the ordinary Morse code. At first sight an objection to this plan would be its want of secrecy. Assuming that the correspondents were a mile apart, the transmitter would send out the waves in all directions, filling a sphere a mile in radius, and it would therefore be possible for anyone living within a mile of the sender to receive the communication. This could be got over in two ways. If the exact position of both sending and receiving instruments were accurately known, the rays could be concentrated with more or less exactness on the receiver. If, however, the sender and receiver were moving about, so that the lens device could not be adopted, the correspondents must attune their instruments to a definite wave-length, say, for example, fifty yards. I assume here that the progress of discovery would give instruments capable of adjustment by turning a screw or altering the length of a wire, so as to become receptive of wave-lengths of any preconcerted length. Thus, when adjusted to fifty yards, the transmitter might emit, and the receiver respond to, rays varying between forty-five and fifty-five yards, and be silent to all others. Considering that there would be the whole range of waves to choose from, varying from a few feet to several thousand miles, there would be sufficient secrecy ; for curiosity the most inveterate would surely recoil from the task of passing in review all the millions of possible wave-lengths on the remote chance of ultimately hitting on the particular wave-length employed by his friends whose correspondence he wished to tap. By "coding" the message even this remote chance of surreptitious straying could be obviated.

This is no mere dream of a visionary philosopher. All the requisites needed to bring it within the grasp of daily life are well within the possibilities of discovery, and are so reasonable and so clearly in the path of researches which are now being actively prosecuted in every capital of Europe, that we may any day expect to hear that they have emerged from the realms of speculation into those of sober fact. Even now, indeed, telegraphing without wires is possible within a restricted radius of a few hundred yards, and some years ago I assisted at experiments where messages were transmitted from one part of a house to another without an intervening wire by almost the identical means here described.

The discovery of a receiver sensitive to one set of wave-lengths and silent to others is even now partially accomplished. The human eye is an instance supplied by Nature of one which responds to the narrow range of electro-magnetic impulses between the three ten-millionths of a millimetre and the eight ten-millionths of a millimetre.

Here we have surely a most remarkable forecast— remarkable for its accuracy and detail. For the programme thus outlined was carried out to the letter within the next five years.

A great event in Crookes's life was his first visit to South Africa, which took place in 1895, just after the Jameson Raid. He combined pleasure with good business, as shown by the following letter :

Sept. 25th, 1895.

J. S. MacArthur, Esq.,
    The African Gold Recovery Co.,
        23, College Hill, E.C.

Dear Sir,
        Referring to the proposal made by you to me in your letter of the 6th inst., on behalf of the African Gold Recovery

Company, Limited, that I should go to Pretoria to give evidence on the forthcoming trial relating to your gold extraction patents (the case is, I understand, known as " Hay *v.* the African Gold Recovery Company "), I am willing to undertake the journey, which I need not point out will occupy a considerable time and take me away from my usual professional pursuits and engagements, on the following terms, which are substantially the same as those offered by you in the letter above quoted :—

1. I am to start not later than the third week in October next.

2. I am not to be called upon to be absent for more than three calendar months from the date of my starting.

3. I am to be paid a fee of one thousand guineas for my loss of time and professional assistance, whereof half is to be paid previously to my starting and the remaining half on my return to England.

4. In addition, first-class passage money, railway fares, coach hire, &c., are to be provided for myself and wife going and returning, and the best hotel accommodation for both of us during our absence abroad.

5. It is also to be understood that so long as I arrive at Pretoria in good time for the trial I am to be at liberty to choose my own date of departure and the manner in which I break my journey " en route " for Pretoria and returning home.

I shall be glad of a letter signed by some one competent to bind the Company, agreeing to the above terms.

Crookes was accompanied by his wife, and left the house in charge of his daughter, with power to sign cheques in his absence. He booked a three-berth first-class cabin on the *Hawarden Castle*. The trial at which he was to act as expert witness was postponed twice, and he had to change over to the *Tantallon Castle*, which sailed on November 29, 1895. After a very enjoyable trip, lasting four months, Crookes returned to London, where he arrived on April 5, 1896. On the way north from Madeira the temperature dropped 50 degrees F. in twenty-four hours. Crookes reached home suffering from a severe cold, which developed into bronchitis. On April 18th he wrote,

"Here I am in bed most of the day, losing all my accumulated good health."

He had visited the Kimberley diamond mines on his way to Pretoria, and delivered two lectures on these mines at the Imperial Institute on Nov. 16th and Dec. 7th, 1896. As usual, he enlisted the services of Miss Alice Bird for a literary polish. He wrote to her :

Oct. 16th, 1896.

My dear Lallah,

I have been contemplating coming over to see you some lunch time, but I have had a bad cold for some days past, and that has made me keep indoors. I want to ask a favour of you and George, and it is much easier to do so personally than by letter.

I am writing a lecture on the Diamond mines of Kimberley, to be delivered at the Imperial Institute on Nov. 16th (a month to-day) and I am not at all pleased with it as a literary production as it stands. I am having it set up in type, as I go along, and I should be greatly indebted to you if you would extend the same favour to this as you have given to former articles of mine. May I send you the proofs as they are set up ? Anticipating that you will not absolutely refuse to infuse some artistic merit into my crude production, I venture to send you slip I, which I may tell you in confidence, I don't like one little bit, but one must begin by a little generality and bring in the Imp. Inst. somehow. Nellie is wondering whether you ever received her invitation to dine here on Thursday next, the 23rd. I am so sorry George cannot come with you. We are only tolerably well here, colds are in the ascendant. Nellie is high busy about Alice's wedding, which is to take place in Feb. next.

With our united love to you and George,
Believe me, my dear Lallah,
Very sincerely yours,
William Crookes.

After receiving her reply, he wrote again :

Nov. 6th, 1896.

My dear Lallah,

Your corrections are admirable, and suggestions most valuable, but they make me despair of ever writing decent English !

345

A change has come over the spirit of my lecture. I have decided to give only one at the I.I. and to confine it to the commercial and manufacturing side only. Then I propose to give a lecture on Diamonds, their origin and artificial production, at the Royal Institution in the winter. This necessitates cutting the lecture shorter and cutting off much of the interesting tail. I have been rearranging the lecture and have told the printer to send you a complete set of corrected and rearranged sheets by to-night's post if possible. If you would go over it again with your usual critical acumen and ruthless severity " your petitioner will ever pray, &c." I hated the conclusion, and gladly adopt your finish. I will certainly come for that lunch and chat before many days are over, but I am so busy I cannot get away just at present. I am glad you like the meteoric origin.

With love to yourself and George.

Believe me,

Ever sincerely yours,

WILLIAM CROOKES.

Crookes enlivened the rather technical lectures by some stories of the native labourers, one of which is here quoted :

As a rule the better class of natives—the Zulus, Matabeles, Basutos, Bechuanas—when well treated, are very honest and loyal to their masters. An amusing instance was told me of the devotion of a Zulu. He had been superintending a gang of natives on a small claim at the river washings near Klipdam. It yielded but few stones, and the owner—my informant—sold the claim, handing over the plant and small staff, our friend the Zulu continuing to look after the business till the new man took possession. In the course of a few months the purchaser became very dissatisfied with his bargain, not a single diamond having turned up since the transfer. Soon after this the Zulu came to his old master in a mysterious manner, and, laying a handful of diamonds on the table, said : " There, Baas, are your diamonds ; I was not going to let the new man have any of them ! "

At the conclusion of the lectures, Crookes made a reference to the artificial production of diamonds :

Interesting as is the story of the diamond industry in South Africa, quite another question fixes the attention of the chemist. The diamonds come out of the mines, but how did they get there ? How were they formed ? What is their origin ? An illustrious Frenchman, M. Moissan, aided by the electric furnace, has given to the scientific world what may turn out to be the true solution of the mystery.

It is well known that the diamond is pure carbon in a crystalline form. It is also known that iron at a high temperature readily dissolves carbon ; on cooling, it deposits the carbon in the form of black opaque graphite. M. Moissan considers that if the carbon could be deposited from the iron at high temperature and under great pressure the carbon would separate out in the liquid form, and on solidifying would assume the form of diamond. He tried the experiment, cooling the iron very quickly ; on dissolving away the metal and examining the residue with a strong microscope, crystals of diamond—clear, white, and transparent—were revealed.

Applying these facts to the occurrence of diamonds at Kimberley, it is not difficult to imagine that masses of iron saturated with carbon existed formerly at a sufficient depth below the present mines, where temperature and pressure would be sufficient to produce the reactions which laboratory experiments show to be probable. It would take too long on this occasion to enter into the subsequent actions by which masses of iron were cooled, disintegrated, and raised to the surface with their precious burden of gems.

The processes at work on the earth for the production of diamonds it has recently been ascertained also take place in celestial bodies. Certain meteorites have been shown to have tiny diamonds embedded in their substance—that is to say, diamonds actually manufactured in the heavens. The most striking diamantiferous meteorite is the one which fell in Arizona, called the " Cañon Diablo " meteorite.

One of the few regrets which beset Crookes concerning his South African trip was that he was away from the country when the first news of the discovery of Röntgen rays flashed out into the world. He would, had he been in England, no doubt have thrown himself into the investigation of their properties with the zeal he displayed when argon and helium were discovered. The former was isolated simultaneously from the atmosphere by William Ramsay and Lord Rayleigh in 1894, and it was Crookes who, comparing the spectra of samples received from each discoverer, proved their identity. He did a similar service to Ramsay in 1895, when the latter found a new gas evolved from a rare mineral called cleveite, and Crookes was able to prove that this gas was identical with the " helium " first found by Norman Lockyer in the sun ! One of his letters generously congratulates Ramsay on the coveted distinction of being made a corresponding member of the Académie des Sciences, a distinction which was the object of one of Crookes's greatest ambitions. The letter runs as follows :

July 24th, 1895.

MY DEAR RAMSAY,

I must congratulate you on an announcement I have only just seen in the *Comptes Rendus*—that you have been elected a corresponding member of the Académie des Sciences. I am very glad this honour has fallen to you, as no one deserves it better.

I do not understand how you come in the place of Frankland. I thought the election was a life one, not dependent on "good behaviour." How is it that Frankland has ceased to be a correspondent ? Has he resigned ? He is not dead.

I have filled three tubes with your helium, and am now taking a photograph of one of them. The gas appears to be very pure. Certainly it is the purest you have yet given me.

Crookes prepared a full account of the helium spectrum for *Nature,* and wrote to the editor :

August 20th, 1895.

MY DEAR LOCKYER,

Your letter is just in time to catch me before I start for the Channel Islands. I can let you have a few pounds of Samarskite, to extract the gases from, if you will return me the mineral after you have done with it.

I have been working night and day to get in type a paper on the spectrum of helium before my holidays. I will send you an early proof to-morrow, and I should much like to see it in *Nature* if you can see your way to insert it. It will appear in the *Chemical News* on Friday, but my circulation is not to the same class of researchers as that of *Nature*, and having taken a great deal of trouble about it I want the results to get to the right people.

My address will be : Dixcart Hotel, Sark, near Guernsey.

Believe me,

Very sincerely yours,

WILLIAM CROOKES

In 1897 Crookes was elected President of the Society of Psychical Research. His Presidential Address was delivered on January 29, 1897, but was not printed until some weeks afterwards. Before it went to press, he took Miss Bird's advice on its composition. He wrote :

7, KENSINGTON PARK GARDENS,
LONDON, W.
Feb. 8th, 1897.

MY DEAR LALLAH,

I have altered one or two passages in the address after reading Oliver Lodge's criticisms on them. Before sending it to press I should like you to look over the parts so altered. I have sent a copy, on which I have drawn the pen down the sides of what has been altered. Kindly run over it. I hope George is recovered and able to get about now the weather seems inclined to pick up. I so much regret he could not be present at the Wednesday's function. That he is able to write I have evidence in a characteristic postcard this morning. Yes ! I have seen the *Speaker* ! and am almost inclined to forgive its politics. Have you seen the *Spectator* ? It has come to grief over a part of the Address, putting ideas and words in it which I am quite innocent of. At the same time I must not complain, for it is more appreciative than I could have expected. We have had two or three letters from Alice. They seem to be very well, but complain of the dullness of Bath in the pouring rain, and have fled to Torquay to get out of it. We expect them back on the 15th (this day week). With love to you and George.

Believe me, ever affectionately yours,

WILLIAM CROOKES.

P.S.—Can you tell me where I shall find the lines beginning : " There is more faith in honest doubt . . ." ? Is it Tennyson's " In Memoriam " ?

In the address, he put forward the view that we may assume a principle of guidance without the expenditure of energy, and therefore not bound by the law of its conservation. He concluded as follows :

> An omnipotent being could rule the course of this world in such a way that none of us should discover the hidden springs of action. He need not make the sun stand still upon Gibeon. He could do all that he wanted by the expenditure of infinitesimal diverting force upon ultra-microscopic modifications of the human germ.
>
> In this address I have not attempted to add any item to the sound knowledge which I believe our Society is gradually amassing. I shall be content if I have helped to clear away some of those scientific stumbling-blocks, if I may so call them, which tend to prevent many of our possible coadjutors from adventuring themselves on the new illimitable road.
>
> I see no good reason why any man of scientific mind should shut his eyes to our work, or deliberately stand aloof from it. Our *Proceedings* are of course not exactly parallel to the *Proceedings* of a society dealing with a long-established branch of science. In every form of research there must be a beginning. We own to much that is tentative, much that may turn out erroneous. But it is thus, and thus only, that each science in turn takes its stand. I venture to assert that both in actual careful record of new and important facts, and in suggestiveness, our Society's work and publications will form no unworthy preface to a profounder science both of Man, of Nature, and of " Worlds not realised " than this planet has yet known.

Shortly afterwards Crookes was instrumental in averting a split in the Chemical Society nearly brought about by the precipitate action of a number of the Fellows putting up Professor William Ramsay for the Presidency against the Council's nominee, Professor James Dewar. Crookes wrote a long letter to Ramsay :

March 29th, 1897.

My dear Ramsay,
    I wish to appeal to you to try and allay the excited feelings which are doing incalculable injury to the Chemical Society, and are giving rise to a feud which will require years of tact and conciliation to allay.

I can with more appropriateness make this appeal, enjoying as I do the friendship and confidence of the leaders on both sides. I will say nothing here one way or the other about Professor Dewar, as, in my judgment, the personal feelings which his nomination has evoked sink into insignificance in comparison with the enormous injury the Society will experience by a continuance of the present unfortunate differences.

It is true that the Office of President of the Chemical Society confers honour on the holder, but in this there is only a limited amount of reciprocity. A President endowed with all social and scientific virtues can only carry on the honourable traditions of the long roll of eminent men whom he succeeds ; while a President with all the faults your over-zealous friends have erroneously conferred on Professor Dewar, in private and public, can have no power to bring permanent discredit on the Office he temporarily holds ; while the fierce light which beats against a man in so important a position cannot fail to bring into relief and intensify any faults of tact, manner, or scientific procedure he may possess.

You have now achieved so eminent a position in science ; Academies, Societies, and Universities have vied with one another in conferring well-merited honours on you ; and you have so many firm friends in the Chemical Society, that you can afford not to push to the bitter end the advantage an indiscreet minority of your friends has given you.

Had no controversy arisen at the present time, and were the Council's list accepted in the manner and spirit in which it always has been, it is an absolute certainty that in a few years you would have been voted into the Presidential Chair without a dissentient voice. The course I wish to propose to you will therefore make no real difference ; while it will allay the feud now unhappily arising, will get us all out of a difficulty, and it will redound to your own honour and dignity immeasurably more than would be the case were you to pluck the fruit before it is ripe, and accept office under the strained circumstances which will surely arise if the house list is not carried on Wednesday.

What I want you to do, and what I urge with all the force I can bring to bear, as an old friend, as your senior in age and experience, as one who in past controversies has seen the advantage of leaving a road open behind an opponent, and as one who has already filled the coveted Presidential Chair—in all these capacities I entreat you to take the first opportunity on Wednesday next, before the ballot commences and before another speaker contributes fuel on one side or the other, of saying that much as you appreciate the honour some of the Fellows present wish to confer on you, you feel that such an unprecedented step as dividing the meeting on this point will do incalculable harm to the Chemical Society. The common interests we all have at heart are immeasurably greater than any opinions which may exist as to the respective merits of two men of science ; you therefore beg leave to decline the honour, you refuse to accept office if elected, and you entreat your friends, in order to heal the breach which otherwise may bring about the disruption of the Society, to vote for the list sent out by the Council.

By acting in this manner you will reap popularity, you will come out of the battle laden with honour, you will earn the gratitude of all who have the well-being of the Society at heart by getting them out of a deadlock, and you will convert many opponents and some neutrals into firm friends, who subsequently will bring you into the coveted chair unanimously and with flying colours.

The Presidential Chair of the Chemical Society is a seat beset with too many anxieties at the best of times to be a very pleasant one, and I cannot conceive a more unpleasant position than that of one who is forced into it by indiscreet friends in the face of a practically unanimous Council.

In the letter you sent to your nominators on the 19th inst., you announce your intention of only holding office until a President is found who will command the unanimous suffrages of the Fellows. Perhaps it has not occurred to you that under the very awkward circumstances of the Council's nominee for the Presidency not being elected on Wednesday, the present Council cannot propose anyone but Professor Dewar to take your place. It was with much deliberation, and after the merits and demerits of many qualified men had been scrutinised, that the choice fell on Professor Dewar, and it will be impossible, with any regard to their dignity as a body and their feelings of duty to the 2,000 and odd Fellows they represent, for them to propose any other name. The Council may be outvoted on Wednesday next in the meeting room, but feeling as they do that they represent the views of a large number of Fellows, your determination to hold office until "a President is found who will command the unanimous suffrages of the Fellows" will cause this unhappy strife to be prolonged with increasing bitterness until you vacate the President's Chair two years' hence.

> Believe me to remain,
> Very sincerely yours,
> WILLIAM CROOKES.

This letter prevailed upon Ramsay to retire gracefully from the position in which he had been placed by his supporters. For a time, only, for he was elected to the Presidency he so richly deserved a few years later.

The next distinction offered to Crookes was the Presidency of the British Association for the Advancement of Science for 1898. He replied :

> *March 30th,* 1897.

MY DEAR HARCOURT,

I have given the subject of the Presidency of the British Association next year my serious consideration, and it is with much inward misgivings that I now write to say it is impossible to refuse an honour so far beyond my aspirations and one which is offered to me in so flattering a manner. I shall therefore be happy* to be your President in 1898 at Bristol, and only hope I shall not disappoint the expectations of my too confiding friends.

Kindly convey the substance of this letter to the Council, and believe me,

> Very sincerely yours,
> WILLIAM CROOKES.

P.S.—* "Happy" is not the right word, but it can stand.

The year 1897 was the year of Queen Victoria's Diamond Jubilee, and Crookes applied for a special pass for the day of the procession :

June 12th, 1897.

THE SECRETARY,
    Metropolitan Police Office,
       New Scotland Yard, S.W.
    DEAR SIR,
       I write to apply for the privilege of a special pass for self and friend to walk down the streets on the line of Procession on the 22nd inst. after they are cleared of traffic. I have been favoured on similar occasions—such as the entry of the (now) Princess of Wales into London, the Thanksgiving procession of Her Majesty to St. Paul's after the recovery of the Prince of Wales, and the Jubilee celebrations ten years ago—on these occasions I have had tickets, printed and signed " Pass the Bearer Everywhere," and I hope on the forthcoming occasion you will continue the privilege.

                 I remain, truly yours,
                 WILLIAM CROOKES.

He also applied for facilities to see the Naval Review :

June 20th, 1897.

DEAR ADMIRAL WHARTON,
       I wrote a day or two ago accepting with many thanks your kind invitation for my wife and myself to see the Naval Review on board the *Gladiator.* Feeling that travelling from London and back the same day would be very fatiguing I have tried to get a bed at Portsmouth for the night, but find there is no chance, every place being full. Would it be possible for us to get a berth on board one of the ships ? for payment of course. I would not trouble you in the matter, but I do not know anyone else to whom I could apply.

    Shortly before the great day arrived, Crookes was agreeably surprised to receive the following letter from the Premier :

[*Private.*]

MY DEAR SIR,

It affords me great satisfaction to be authorised to inform you that the Queen has been pleased to direct that you should receive the Honour of Knighthood on the occasion of Her approaching Jubilee, in recognition of the eminent services you have rendered to the advance of scientific knowledge during Her Majesty's reign.

It is very gratifying to me to be the instrument of making you acquainted with Her Majesty's gracious intention.

Believe me,
Yours truly,
SALISBURY.

PROFESSOR CROOKES, F.R.S.

Crookes replied :

To the MOST HONOURABLE THE MARQUESS OF SALISBURY,
Foreign Office, S.W.

MY LORD,

I find it difficult to fitly express my sense of the high and unexpected honour you have just told me Her Majesty will be pleased to confer on me on the occasion of her approaching Jubilee. I cannot deny it is a distinction I shall highly esteem as the reward of a life's work in the pursuit of abstract scientific research, but I prize it still more, feeling as I do that it is also a recognition of the part played by science in the material progress of this great Empire.

Allow me also, my Lord, to say how much I appreciate the kind words in which you have conveyed to me the intimation of Her Majesty's gracious intention.

I have the honour to remain,
Your obedient servant,
(*Signed*) WILLIAM CROOKES.

The split infinitive in the first few words shows that " Lallah's " blue pencil had not passed across this letter. He wrote to her on the same day :

June 20th, 1897.

MY DEAR LALLAH,

I don't want you and George to hear the news from anyone but myself, so I write to tell you that I have had a letter from Lord Salisbury saying he is authorised to inform me that the Queen has been pleased to direct that I should receive the honour of Knighthood on the occasion of her approaching Jubilee " in recognition of the eminent services I have rendered to the advance of Scientific knowledge during Her Majesty's reign." This came a few days ago, and was quite unexpected, as I never thought honours would fall to the share of one who has cultivated almost entirely abstract science.

It is very gratifying to find such recognition of the part played by scientific research in the material progress of the Empire. I suppose the news will come out in Tuesday papers.

Nellie sends love to you and George, and with the same from myself.

Believe me, ever affectionately yours,

WILLIAM CROOKES.

And again later :

June 30th, 1897.

MY DEAR LALLAH,

I know you will excuse me for not having written before now. Ever since Jubilee day I have scarcely done anything but scribble away as fast as possible autograph replies to letters and telegrams of congratulations which came pouring in from all parts of the country. I determined to answer as soon as possible, and have neglected almost everything else for this. The bulk are now finished. I have only a dozen or so arrived this morning which will not take long.

But I am not going to put off you and George, my oldest and best friends, with a letter. I want to come and thank you in person. Shall you be in at lunch time to-morrow ? I am deeply touched by the overwhelming extent of congratulations I have received, and by the very flattering way in which my work is referred to in all quarters. I hope I shall be able to preserve a level head !

With much love to you both, in which Nellie joins.

Believe me, affectionately yours,

WILLIAM CROOKES.

A LOST ESSAY

At Crookes's request, four prisms—symbols of the spectroscope—as well as a radiometer were embodied in his arms. The device was a sort of heraldic pun :

Ubi Crux Ibi Lux.

Towards the end of the year Crookes seems to have written an " Essay on Immortality," but he refrained from publishing it owing to the opposition of Dr. Bird and his sister. He wrote to the former :

Dec. 8th, 1897.

MY DEAR GEORGE,

I said I would try and improve my " Essay on Immortality," and have therefore written it out again, and have altered the conclusion. In spite of your and Lallah's condemnation I plead for a new trial. Perhaps on reading it in its new dress you may be able to recommend some parts worthy of rescue from the flames. If, however, you still condemn, I will be a Spartan mother and sacrifice my child for the good of my audience.

I will write soon about that lunch. So sorry you cannot join our dinner, but it is perhaps wisest.

With love to Lallah and yourself,

Believe me, affectionately yours,

WILLIAM CROOKES.

The evolution of the great Presidential Address for the British Association can be traced by the letters written to Lallah. The subject chosen for the address was the problem of the world's wheat supply, but a man with Crookes's mind would naturally be expected to make some pronouncement on the psychic problems which had created such a stir in the seventies. There are eight letters from Crookes to Miss Bird bearing on the preparation of the address. They are given below :

July 3rd, 1898.

MY DEAR LALLAH,

I have added to, corrected, and polished the address up to the last pitch of my unaided powers, and now I want you and

357

George to help me a little with the finalities. I have received all your proofs and adopted the corrections, and Bernard, my mathematical son, has been going over the figures, which did not agree in different places. Then I have written a couple of slips about psychical matters, and one on other branches of science.

Where are you, and when do you return home? I pine for an interview. Tell me when lunch will be ready!

Ever affectionately,
WILLIAM CROOKES.

July 9th, 1898.

My dear Lallah,

I send the rest of the Wheat section. Do I bore you? My conscience pricks me, but your and George's goodness have spoilt me. All will be the better for a polish, but if you would not mind looking at the latter half of slip 8 I think much good would be done. The par. about bananas is a tip from Stanley.

My mathematical son Bernard is carefully going over all my figures and calculations, so as to keep me right in that respect. I will bring the psychical part with me on Monday. Oliver Lodge has looked over that.

Nellie wants to see you. May I bring her on Monday?

With love, believe me,

Ever yours affectionately,
WILLIAM CROOKES.

July 20th, 1898.

My dear Lallah,

I enclose one or two pars. which are new, and therefore in need of artistic polish, which only you and G. can give! Also I send the two pars. cut out of slip 10. I want you to look at them again, and let me come and plead their cause at a quiet lunch. I may induce you to bless them, instead of cursing them twice over! At all events I want if possible to save a portion of them from the general wreck.

Shall you be at home on Saturday?

With love, believe me,

Affectionately yours,
WILLIAM CROOKES.

LITERARY POLISHING

August 2nd, 1898.

My dear Lallah,

I have improved the address in my opinion very much, by cutting out all the statistical part and relegating it to an appendix. Would you mind looking over a little bit I wrote yesterday on a new discovery I have made, and one I want to introduce to the audience with a flourish of trumpets ? Enter, the younger brothers to Thallium, with about 35 years' difference of age ! I also send the troublesome Psychical part, in which I have adopted all your and George's suggestions. I have marked a part which does not read at all smoothly. I had struck out the " foul mine " simile, but Professor Lodge begs me to give it, as he thinks it is very apt. I have tried to bring it in, but it does not fit very well.

When is George returning ? Can I catch you both at home before I leave this day week ?

With love, in which Nellie joins,

Believe me, affectionately yours,

William Crookes.

August 5th, 1898.

My dear Lallah,

I have just wired to you asking you not to trouble about the proof I sent a few days ago, as I have altered it very much, and will send the revise to-night. I want just a little time to speak to you about one or two things before you leave. I am going to Sark on Wednesday next, and expect to remain till about the 25th. Then we rush back to prepare for Bristol.

If possible I want to break the neck of the address before I leave for Sark, so as to only leave final corrections in minor points open. The Secretaries want me to send them the copy of Address soon.

When at Sark I may want to consult you on some points. Where will you be ? And can you be bothered ? I am not happy about the psychical part. The matter cut out makes some of my allusions and remarks unintelligible. I want to see if I cannot bring in again some allusions to the tabooed passages so as to make all consistent. Is George staying at Windmill Hill while you are away ?

With love, in which Nellie joins,

Believe me, ever affectionately yours,

William Crookes.

THE END OF THE CENTURY

DIXCART HOTEL,
SARK,
CHANNEL ISLANDS,
Aug. 12th, 1898.

MY DEAR LALLAH,

Here is the *Magnum opus* practically complete. I still want suggestions on some of the more recondite scientific points from two friends, but their emendations will only be in figures and such-like. If you have time to run your eye over it as a connected whole I shall be oh ! so grateful ! The brain wave part I have decided to leave out. Oliver Lodge's criticism has mainly induced me. He is right. It is only a guess, and one that will raise hostility among many who otherwise would be friendly. This is as beautiful as ever. The sun is brilliant, but its rays are tempered by a cool breeze, and there is shade among the rocks. We have had a long walk this morning, and have come in tired. Nellie is sleeping the sleep of the just on the sofa, surrounded with a circle of light literature and newspapers she has attempted to read. I feel as if I should like to do the same. With much love,

Believe me, ever affectionately yours,
WILLIAM CROOKES.

Aug. 7th, 1898.

MY DEAR LALLAH,

Back from Sark for a week ! Last night I made a little addition to one of the slips, in consequence of a new and important physiological discovery a friend told me. I gave instructions to the printers to send you a proof. If possible, let me have it by return of post. The chief alteration is in the second paragraph. I shall be here all next week. What are your movements ? Nellie sends best love. She is very well. With love, believe me,

Ever yours affectionately,
WILLIAM CROOKES.

Sept. 1st, 1898.

MY DEAR LALLAH,

At last the fateful month has commenced, the Address has gone to the official printer, and I am indulging in nightmare dreams of hisses and failure. If I succeed in escaping, it will mainly

360

be on account of the invaluable assistance rendered me by you and George. I will send you some papers if there is anything in them worth reading. We are very busy as you may suppose, preparing for our journey to Bristol, which will be next Tuesday morning.

We are much upset by the awful death of Hopkinson and his three children. We knew them well. Also our next-door neighbour, Mr. Huggins, has just died on his holiday, and I have this morning heard that another good man of science, Dr. Rose, holiday making, has just died. Truly holiday making is a dangerous pastime, and I wish we were safe home again. My address at Bristol will be : c/o The Worshipful the Mayor, Mansion House, Bristol. A word of sympathy amid the chorus of execration will be strength to bear it, and if the chorus is one of praise I shall value your encouragement more than that of all the newspapers in the world.

Nellie is well, but anxious. She sends love to you and George, and with the same from myself, believe me,

> Always affectionately yours,
> WILLIAM CROOKES.

What will be your address from 6th to 15th ?

Some of the salient passages in the address may be quoted here :

Are we to go hungry and to know the trial of scarcity ? That is the poignant question. Thirty years is but a day in the life of a nation. Those present who may attend the meeting of the British Association thirty years hence will judge how far my forecasts are justified.

If bread fails—not only us, but all the bread-eaters of the world —what are we to do ? We are born wheat-eaters. Other races, vastly superior to us in numbers, but differing widely in material and intellectual progress, are eaters of Indian corn, rice, millet, and other grains ; but none of these grains have the food value, the concentrated health-sustaining power of wheat, and it is on this account that the accumulated experience of civilised mankind has set wheat apart as the fit and proper food for the development of muscle and brains.

It is said that when other wheat-exporting countries realise that the States can no longer keep pace with the demand, these countries will extend their area of cultivation, and struggle to keep up the supply *pari passu* with the falling off in other quarters. But will this comfortable and cherished doctrine bear the test of examination ?

Cheap production of wheat depends on a variety of causes, varying greatly in different countries. Taking the cost of producing a given quantity of wheat in the United Kingdom at 100s., the cost for the same amount in the United States is 67s., in India 66s., and in Russia 54s. We require cheap labour, fertile soil, easy transportation to market, low taxation and rent, and no export or import duties. Labour will rise in price, and fertility diminish as the requisite manurial constituents in the virgin soil become exhausted. Facility of transportation to market will be aided by railways, but these are slow and costly to construct, and it will not pay to carry wheat by rail beyond a certain distance. These considerations show that the price of wheat tends to increase. On the other hand, the artificial impediments of taxation and customs duties tend to diminish as demand increases and prices rise.

* * * * *

For years past attempts have been made to effect the fixation of atmospheric nitrogen, and some of the processes have met with sufficient partial success to warrant experimentalists in pushing their trials still further ; but I think I am right in saying that no process has yet been brought to the notice of scientific or commercial men which can be considered successful either as regards cost or yield of product. It is possible, by several methods, to fix a certain amount of atmospheric nitrogen ; but to the best of my knowledge no process has hitherto converted more than a small amount, and this at a cost largely in excess of the present market value of fixed nitrogen.

* * * * *

I have dwelt on the value and importance of nitrogen, but I must not omit to bring to your notice those little known and curiously related elements which during the past twelve months have been discovered and partly described by Professor Ramsay and Dr. Travers. For many years my own work has been among what I may call the waste heaps of the mineral elements. Professor Ramsay is dealing with vagrant atoms of an astral nature. During the course of the present year he has announced the existence of no fewer than three new gases—krypton, neon, and metargon. Whether these gases, chiefly known by their spectra, are true unalterable elements, or whether they are compounded of other known or unknown bodies, has yet to be proved. Fellow-workers freely pay tribute to the painstaking zeal with which Professor Ramsay has conducted a difficult research, and to the philosophic subtlety brought to bear on his investigations. But, like most discoverers, he has not escaped the flail of severe criticism.

*　　*　　*　　*　　*

The phenomenon discovered by Zeeman, that a source of radiation is affected by a strong magnetic field in such a way that light of one refrangibility becomes divided usually into three components, two of which are displaced by diffraction analysis on either side of the mean position and are oppositely polarised to the third or residual constituent, has been examined by many observers in all countries. The phenomenon has been subjected to photography with conspicuously successful results by Professor T. Preston in Dublin and by Professor Michelson and Dr. Ames and others in America.

*　　*　　*　　*　　*

Quite recently M. and Mme. Curie have announced a discovery which, if confirmed, cannot fail to assist the investigation of this obscure branch of physics. They have brought to notice a new constituent of the uranium mineral pitchblende, which in a 400-fold degree possesses uranium's mysterious power of emitting a form of energy capable of impressing a photographic plate and of discharging electricity by rendering air a conductor. It also appears that the radiant activity of the new body, to which the discoverers have given the name of Polonium, needs neither the excitation of light nor the stimulus of electricity ; like uranium, it draws its energy from some constantly regenerating and hitherto unsuspected store, exhaustless in amount.

*　　*　　*　　*　　*

These, then, are some of the subjects, weighty and far-reaching, on which my own attention has been chiefly concentrated. Upon one other interest I have not yet touched—to me the weightiest and the farthest reaching of all.

No incident in my scientific career is more widely known than the part I took many years ago in certain psychic researches.

363

THE END OF THE CENTURY

Thirty years have passed since I published an account of experiments tending to show that outside our scientific knowledge there exists a Force exercised by intelligence differing from the ordinary intelligence common to mortals. This fact in my life is of course well understood by those who honoured me with the invitation to become your President. Perhaps among my audience some may feel curious as to whether I shall speak out or be silent. I elect to speak, although briefly. To enter at length on a still debatable subject would be unduly to insist on a topic which—as Wallace, Lodge, and Barrett have already shown—though not unfitted for discussion at these meetings, does not yet enlist the interest of the majority of my scientific brethren. To ignore the subject would be an act of cowardice—an act of cowardice I feel no temptation to commit.

To stop short in any research that bids fair to widen the gates of knowledge, to recoil from fear of difficulty or adverse criticism, is to bring reproach on Science. There is nothing for the investigator to do but to go straight on, " to explore up and down, inch by inch, with the taper his reason " ; to follow the light wherever it may lead, even should it at times resemble a will-o'-the-wisp. I have nothing to retract. I adhere to my already published statements. Indeed, I might add much thereto. I regret only a certain crudity in those early expositions which, no doubt justly, militated against their acceptance by the scientific world. My own knowledge at that time scarcely extended beyond the fact that certain phenomena new to science had assuredly occurred, and were attested by my own sober senses, and better still, by automatic record. I was like some two-dimensional being who might stand at the singular point of a Riemann's surface, and thus find himself in infinitesimal and inexplicable contact with a plane of existence not his own.

I think I see a little farther now. I have glimpses of something like coherence among the strange elusive phenomena ; of something like continuity between those unexplained forces and laws. This advance is largely due to the labours of another Association of which I have also this year the honour to be President—the Society for Psychical Research. And were I now introducing for the first time these inquiries to the world of science I should choose a starting-point different from that of old. It would be well to begin with *telepathy* ; with the fundamental law, as I believe it to be, that thoughts and images may be trans-

364

ferred from one mind to another without the agency of the recognised organs of sense—that knowledge may enter the human mind without being communicated in any hitherto known or recognised ways.

* * * * *

An eminent predecessor in this chair declared that "by an intellectual necessity he crossed the boundary of experimental evidence, and discerned in that matter, which we in our ignorance of its latent powers, and notwithstanding our professed reverence for its Creator, have hitherto covered with opprobrium, the potency and promise of all terrestrial life." I should prefer to reverse the apophthegm, and to say that in life I see the promise and potency of all forms of matter.

In old Egyptian days a well-known inscription was carved over the portal of the temple of Isis : "I am whatever hath been, is, or ever will be ; and my veil no man hath yet lifted." Not thus do modern seekers after truth confront Nature—the word that stands for the baffling mysteries of the universe. Steadily, unflinchingly, we strive to pierce the inmost heart of Nature, from what she is to reconstruct what she has been, and to prophesy what she yet shall be. Veil after veil we have lifted, and her face grows more beautiful, august, and wonderful, with every barrier that is withdrawn.

After delivering the address, Crookes wrote to Miss Bird :

c/o SIR R. SYMES,

Mayor,

THE MANSION HOUSE,

BRISTOL,

Sept. 8th, 1898.

MY DEAR LALLAH,

The eventful night has come and gone, and it was a *brilliant success,* and I am overwhelmed with compliments. Several old stagers saying it was the best address that they had ever heard. I spoke for an hour and 20 minutes, and was heard in the farthest parts of the hall.

The Wheat (curtailed) did not bore, and the psychic part did not shock the audience. The latter part, indeed, got most of the applause. It was all very gratifying. Now the real work begins, the after-dinner speeches and such like. Write to me here. I want to know the impression in the London papers. I am too busy to read anything but my own notes. Believe me and Nellie to remain ever very affectionately

Yours,
WILLIAM CROOKES.

P.S.—Our love also to George.

And a month later :

Oct. 10*th,* 1898.

MY DEAR LALLAH,

We have at last returned home, and not at all sorry to do so as the weather has turned cold. I have an enormous accumulation of arrears in front of me—excuse the Bull—and I am laboriously struggling with it. We should so like to see you. Shall you be at home on Friday ? That is about the first day we are free. After the B.A. meeting at Bristol we took a tour in Cornwall and enjoyed the quiet time all by ourselves immensely after the bustle of Bristol. The pace there was forced too much to be endurable for more than a week, but it was very brilliant all the same.

I hope you and George are well. With our united love to you both, believe me,

Ever affectionately yours,
WILLIAM CROOKES.

The following letters show further developments in connection with the address :

Dec. 16*th,* 1898.

MY DEAR LALLAH,

I am sorry not to see you before Christmas, but I must not neglect my opportunities so long when you return to Hampstead. I am thinking about a general reply to all the criticisms on my Address. I have collected over 100 of them, all unfavourable, and shall have no difficulty in showing their utter futility. I am convinced I am right, and can fortify myself in all ways by quotations from recognised authorities. But my difficulty is to know what to do with my reply when finished. I cannot ask *The Times* to insert it, for it would occupy too much space, and they might think it too dry. I shall put it in type for convenience of correction, and then send it to the *XIXth Century* on spec. I have just received a serious attack in an American journal, and I am given to understand it will be reprinted in *The Times*. If so, I shall rejoice, as it will give me the wished-for opening. I wish you and George a bright and happy time with your friends at this Xmas time. We shall be very quiet, but the grandchildren will liven us up a little bit.

With love to you and George, in which Nellie joins.

Believe me, very affectionately yours,

WILLIAM CROOKES.

Jan. 18*th*, 1899.

MY DEAR LALLAH,

I received your letter some time ago, but did not answer it as I knew you were moving about, and it was not worth while bothering you. Now I am overjoyed to find that you are at home again ; I hope in very good health. Yes ! I am on the war-path again, and have a complete answer to all the attacks on my address. I have put it all in type so as to correct and cut about as I like. But the difficulty is that no one will attack the address in a really good paper or periodical. I have had a slashing attack in the (American) *Monthly of Science*, but I want some English journal of repute to copy it and then I shall have a chance of delivering my reply, which will be convincing I think. But I want to talk to you about all these things and others in general. It seems ages since I saw you and George, and as I have always said "To know you is a liberal education !" I feel as if I was neglecting my opportunities if I do not occasionally see and talk with you. Nellie sends kind love to you both, and with much from myself to you and George,

Believe me, very sincerely yours,

WILLIAM CROOKES

March 1*st*, 1899.

MY DEAR LALLAH,

A thousand thanks for your corrections and emendations. As you have not recorded any fines of 5/- each, I hope there are not many split infinitives ! May I send a revise ? I hardly like to do so, as I respect your eyes too much to allow me to do anything that is likely to strain them.

The *North American Review* has an article in its current number written by the Chief Statistician to the Agricultural Dept. of the U.S. Government, in which he goes very fully into the wheat question, and practically endorses all I said. It is quite cheering to read. I will bring a copy the next time I come, and read it to you, if the infliction will not be too great. (I hope my offer will not render me less welcome at lunch than before.) Now I hope you and George are very well and are able to come to dinner with us on *Tuesday next, March 7th.* We shall have Mrs. Tweedy, Miss Kingsley (whose book I am now studying), and some other pleasant people. Nellie sends kind love, and with the same from myself to you and George.

Believe me,
Very affectionately yours,
WILLIAM CROOKES.

April 13th, 1899.

MY DEAR LALLAH,

I have arranged to be at home on the afternoon of Wednesday next, the 19th inst., to help Nellie pour out tea for visitors. We shall be delighted to see you then, and if you can induce Miss Harraden to accompany you our pleasure will be increased.

" Put not your trust in " Newspapers ! *The Times* has proved a deceiver, and I am mourning. And after such a nice letter from the Editor, too.

With our united love to George and yourself, believe me,

Affectionately yours,
WILLIAM CROOKES.

July 10th, 1899.

MY DEAR LALLAH,

We have been spending a delightful " week-end " at Lord Brownlow's at Ashridge, and on returning I find your card saying you have already gone and will not return for a fortnight. I hope you will enjoy yourselves much. On your return I will do my best to see you. The little speech will keep till then ; but alas ! The Preface will not, so I enclose it in the hopes that you or George may be able to find a few minutes to look at it.

You will think I am very obstinate, but I cannot help thinking " prophet " is the best word, so I have altered the expression at the commencement so as to remove the apparent contradiction. But if you still strike out " false prophet " and put in " misleading pessimist " I will bow and adopt it.

Nellie joins in best love to you both.

Believe me,

Ever affectionately yours,

WILLIAM CROOKES.

7, KENSINGTON PARK GARDENS,

LONDON, W.,

Oct. 4th, 1899.

MY DEAR LALLAH,

At last we are back to our home, and never have I been so tempted to sing " Home, Sweet Home " as on this occasion. Six weeks is too long for old folks like me to be living in Hotels and strange houses away from our friends. I could not get to see you on my hurried visit to London in the middle of the holiday When I left, the " Wheat Problem " was not published, and on my return I find several copies waiting for me. I gave Murrays a list of friends to whom I wanted copies sent, but they have not sent all I desired, and I am afraid the copy for you was one of those omitted. I therefore hasten to remedy the omission. Do not think that because the copy has been delayed that you have been forgotten. Man's ingratitude could not reach to so low a depth, so I hope you will ascribe all shortcomings to the publisher.

How are you both ? Well, I hope. We are pretty well— not exactly in rude health, but passably so. If you are now at home I will run over some day and recount our various doings since I saw you last, or we will " take them as read," as we say of a dry paper in the sections of the B.A. Nellie sends kindest love to you and George, in which I cordially join.

Believe me,

Affectionately yours,

WILLIAM CROOKES

Excuse mess of typewriter. It has got out of order.

THE END OF THE CENTURY

The account of Crookes's great experience at Bristol would be incomplete without quoting the following letter, written to Sir Oliver Lodge before the meeting :

August 10th, 1898.

My dear Lodge,

        I am truly grateful for your plain speaking. If every writer had as candid a friend there would be much less nonsense written.

The whole thing goes out at once, of course. Now you point out all the nonsense, I see the absurdities myself. You say you prefer the soberer and more historic style. So do I for general use, but I commenced my task by reading carefully all the Addresses for the last 20 years, and there is a general character running through all but two or three, which gives the tone to the address. The President allows himself more latitude than he would in a paper before the R.S., and always indulges in bold speculations put forward in a more or less picturesque style. But I am glad you have pulled me up. Where you say "hypothesis" others would say "sheer nonsense."

I hope what I sent last night will pass muster.

Just off to Sark.

Believe me, very truly yours,
William Crookes.

CHAPTER XVIII

DIAMONDS

(1897–1903)

THE NEW CENTURY came upon Crookes at the age of sixty-eight. His greatest work was done, and the remaining nineteen years of his life were mainly spent in reaping what he had sown. He was given, in profusion, University degrees which in his younger days would have been most valuable to him, but which, as he once said to a friend, were but " dust and ashes " in his old age. The Universities of Dublin, Oxford, the Cape of Good Hope, Cambridge, Sheffield, and Durham conferred upon him their honorary degrees of Doctor of Science, and Birmingham gave him an LL.D. on the occasion of the opening of the new University buildings at Bournbrook. It is probably true that most of the honorary graduates on such occasions feel much as Crookes did about it. The honour is supposed to be bestowed by the university conferring the degree. But the governing boards of the university do not confer such degrees upon unknown men, but upon men whose names will add lustre to the university. There is no evidence to show that Crookes ever sought such degrees. An honour he did covet and work for, that of Correspondent to the Institut de France, he obtained at last in 1906. But even that was some twenty years after the date on which such an election would have been of vital importance to him.

His public reputation had been growing apace, and any striking discovery made, or alleged to have been made, about the beginning of the century was referred to him for endorsement, or otherwise, by innumerable publicists and others throughout the world. Thus, when in 1897, a certain Dr. Emmens, an American chemist, claimed to have converted Mexican silver dollars into gold by a process of prolonged hammering, Crookes was overwhelmed with inquiries. His attitude on the question was very correct. While not denying the ultimate possibility of such conversion, he gave it as his opinion that transmutation, if ever accomplished, was most unlikely to begin by the conversion of silver into gold. To the editor of the *National Review* he wrote that he had had some correspondence with Dr. Emmens in the matter, but had as yet seen nothing to convince him that the thing had been actually done. To Professor Barker he wrote :

I should have thought that sulphur into selenium, or chlorine into bromine, or nickel into cobalt would be the first to be transmuted. But I do not think transmutation will be effected on those lines. More probably we shall decompose the so-called elements by the application of some undiscovered means, and then recompose them as we like, working by the light of the knowledge gained in their dissection. But these are dreams, only to be indulged in in strict privacy to one's friends at Christmas time.

It was inevitable that a man in Crookes's position should often be approached by people anxious to find employment for their sons. To one such applicant he put the question :

What are his tendencies and bent of his mind ? I do not mean how does he occupy himself from 10 to 4, but what does he do after hours. No one can get on in England by simply working as a machine six hours a day. All the best work of our leading men has been done during the other 18 hours, too often, alas ! devoted to play and sleep.

Crookes's persistent and largely unsuccessful efforts to find remunerative enployment for his own sons may have made him somewhat pessimistic.

It was somewhat unfortunate that William Crooks, the Poplar labour leader, should have had a name so closely resembling Crookes's. The latter felt it necessary on one occasion to write to *The Times* to point out the difference in spelling, among other differences. This annoyed the Labour leader, but Crookes wrote him a very courteous and friendly letter which removed the sting from the incident. It is quite possible that, even now, the name William Crookes or Crooks represents a champion of the horny hand rather than a wizard of science to the majority of English people who are familiar with the sound of the name.

Crookes's chief assistant at his private laboratory was Mr. James H. Gardiner (who was with him till his death in 1919). Mr. Charles Gimingham had left in 1882. For secretarial work in connection with the *Chemical News* he engaged a lady assistant of the name of Miss Heather. Before making this selection he had written to Sir William Ramsay :

Thanks. I think a woman would be if anything better than a man, always providing the qualifications are equal. A woman is more conscientious than a man in many things, and is not always trying to get another appointment to "better" herself. But on the other hand she goes and gets married, which is quite as fatal !

Ramsay appears to have suggested engaging Dr. Donnan, but Crookes said he was "too big a man for my modest needs." (He now fills the Chair at University College, London, vacated by Ramsay's death.)

In the later days of his life Crookes turned his attention to diamonds, probably under the inspiration of his visit to Kimberley in 1896, when he took numerous photographs.

DIAMONDS

His lecture on diamonds at the Royal Institution in 1897 was very carefully prepared, and attracted a brilliant audience, but it was marred by the failure of the chief experiment, that of making diamonds artificially. Crookes tried, in fact, to repeat Moissan's experiment of manufacturing diamonds in an electric furnace consisting of a hollow block of magnesite containing a graphite crucible filled with pure iron and sugar charcoal, through which a current of 1,000 amperes was sent in order to produce an enormously high temperature.

The plan was to withdraw the crucible when white hot and to plunge it into water, when the congealing would subject the charcoal to an enormous pressure and thus produce small diamonds. One assistant was to take the lid off the furnace, while another withdrew the electrodes, Crookes himself took out the crucible with a special pair of tongs. At the critical moment, however, the crucible stuck, and Crookes broke it in his efforts to get it out of the furnace.[1]

One of the morning papers headed its account of the lecture with the facetious line, " How Diamonds are Not Made." But in reality the lecture was very remarkable and enjoyable. Discussing the possible modes of formation of natural diamonds, Crookes said :

It is conclusively proved that the diamond has not been formed *in situ* in the blue ground. The diamond genesis must have taken place at great depths under enormous pressure. The explosion of large diamonds on coming to the surface shows extreme tension. More diamonds are found in fragments and splinters than in perfect crystals ; and it is noteworthy that although many of these splinters and fragments are derived from the breaking up of a large crystal, yet in no instance have pieces been found which could be fitted

---

[1] Crookes was anxious to repeat the experiment at the Royal Institution to retrieve his prestige, but Dewar told him it was too dangerous, as the men had to keep the belt on the dynamo by hand to produce the great current required.

together Does not this fact point to the conclusion that the blue ground is not their true matrix ? Nature does not make fragments of crystals. As the edges of the crystals are still sharp and unabraded, the *locus* of formation cannot have been very distant from the present sites. There were probably many sites of crystallisation differing in place and time, or we should not see such distinctive characters in the gems from different mines, nor indeed in the diamonds from different parts of the same mine.

How the great diamond pipes originally came into existence is not difficult to understand, in the light of the foregoing facts. They certainly were not burst through in the ordinary manner of volcanic eruption ; the surrounding and enclosing walls show no signs of igneous action, and are not shattered nor broken even when touching the " blue ground." These pipes after they were pierced were filled from below, and the diamonds formed at some previous epoch too remote to imagine were erupted by a mud volcano, together with all kinds of debris eroded from the adjacent rocks. The direction of flow is seen in the upturned edges of some of the strata of shale in the walls, although I was unable at great depths to see any up-turning in most parts of the walls of the De Beers mine.

As regards the actual process of formation, Crookes sketched it out as follows :

It must be borne in mind I start with the reasonable supposition that at a sufficient depth there were masses of molten iron at a great pressure and high temperature, holding carbon in solution, ready to crystallise out on cooling. In illustration I may cite the masses of erupted iron in Greenland. Far back in time the cooling from above caused cracks in superjacent strata through which water found its way. On reaching the iron the water would be converted into gas, and this gas would rapidly disintegrate and erode the channels through which it passed, grooving a passage more and more vertical in the endeavour to find the quickest vent to the surface. But steam in the presence of molten or even red-hot iron rapidly attacks it, oxidises the metal and liberates large volumes of hydrogen gas, together with less quantities of hydrocarbons of all kinds—liquid, gaseous, and solid. Erosion commenced by steam would be continued by the other gases, and it would be no difficult task for pipes, large as any found in South Africa, to be scored out in this manner. Sir

DIAMONDS

Andrew Noble has shown that when the screw stopper of his steel cylinders in which gunpowder explodes under pressure is not absolutely perfect, gas finds its way out with a rush so overpowering as to score a wide channel in the metal ; some of these stoppers and vents are on the table. To illustrate my argument Sir Andrew Noble has been kind enough to try a special experiment Through a cylinder of granite is drilled a hole o 2 inch diameter, the size of a small vent. This is made the stopper of an explosion chamber, in which a quantity of cordite is fired, the gases escaping through the granite vent. The pressure is about 1,500 atmospheres, and the whole time of escape is less than half a second. Notice the erosion produced by the escaping gases and by the heat of friction, which have scored out a channel over half an inch diameter and melted the granite along their course. If steel and granite are thus vulnerable at comparatively moderate gaseous pressure, is it not easy to imagine the destructive upburst of hydrogen and water gas grooving for itself a channel in the diabase and quartzite, tearing fragments from resisting rocks, covering the country with debris, and finally at the subsidence of the great rush, filling the self-made pipe with a water-borne magma in which rocks, minerals, iron oxide, shale, petroleum, and diamonds are churned together in a veritable witch's cauldron ? As the heat abated the water vapour would gradually give place to hot water, which, forced through the magma, would change some of the mineral fragments into the now existing forms.

Crookes also mentioned an alternative hypothesis, viz. that all diamonds had been originally contained in meteorites, which had fallen from the sky, boring huge " pipes " in the ground and scattering diamonds broadcast. He thought that there was much to support this hypothesis, especially the frequent occurrence of diamonds in the fragments of the Cañon Diablo meteorite which fall in Arizona. Crookes dissolved a piece of this meteorite weighing 5 lbs. in acids in order to extract diamonds, " an act of vandalism in the course of science for which I hope mineralogists will forgive me," and exhibited photographs of small diamonds thus obtained. His final conclusion was that both these origins of diamonds were true.

In 1900, Crookes became an Hon. Secretary of the Royal Institution of Great Britain. He held this office for thirteen years, devoting much time to the onerous and responsible task of arranging the programme of lectures every year. It brought him into close association with Sir James Dewar, the renowned chemist who was also in collaboration with Crookes over the inspection of the Metropolitan water supply. It was one of the tragedies of Crookes's life that this association, at one time amounting to a close friendship, with Sir James Dewar was finally shattered by a legal dispute between them over the discovery of certain colloid substances by Crookes's eldest son.

The water supply work began in 1880. It was before the time of the Metropolitan Water Board, and London was supplied by a number of companies such as the New River, East London, Chelsea, West Middlesex, Lambeth, Grand Junction, and Southwark and Vauxhall Water Companies. From 1892, Dewar and Crookes were directors of the Water Inspection Laboratory in Colville Road, with a staff of assistants. Their relations with the various companies were often anything but friendly, and at one time their claims against the companies amounted to well over £6,000. On June 13, 1904, Crookes wrote to Dewar :

What are we to do with these Lambeth people ? It is positively disgraceful the way in which they are begging and praying for something off our account. If it was a poor company and they had lost an expensive case I could understand their anxiety to get a reduction, but they are a rich Company, and owing to our exertions they are about half a million in pocket !

On the other hand we must remember that we have been willing to make a reduction of our charges in the case of other Companies to the extent of £500 or more. But we have pocketed the cash, and it is not pleasant to disgorge.

The income which Crookes derived from this source was about £400 a year on the average. It was nearly the same as the profit from the *Chemical News*. His total income at this time must have been between £3,000 and £4,000 a year.

Since 1898, Crookes had been a member of the Explosives Committee of the Ordnance Board, of which Lord Rayleigh, O.M., F.R.S., was President, and Sir A. Noble and Lord Haldane were members. The Committee had a laboratory for research at Woolwich. The head of this laboratory was a German, Dr. Silberrad. About this man Crooks wrote in 1904 to Lord Rayleigh : " Silberrad is very anxious to make our laboratory and establishment second to none in Europe, and he always keeps the example of the great Berlin establishment before his eyes." Crookes had a high opinion of Silberrad's " enormous store of information," and was anxious that his salary of £1,000 a year should be increased so as to retain him against certain temptations from America.

Crookes had no close relations with Germany, and did not speak German. His knowledge of German was insufficient for any close understanding. In 1903 he was invited to deliver an address in English before the Congress of Applied Chemistry at Berlin. He chose the subject, " Modern Views on Matter : The Realisation of a Dream." He claimed that the dream just realised was essentially a British dream, comprising the resolution of chemical elements into simpler forms of matter or even refining them altogether away into ethereal vibrations or electrical energy. He referred to his own hypothesis of the origin of the elements from a " formless mist " which he called " protyle " [1] in 1886.

This, as Crookes himself afterwards acknowledged, was a wrong spelling. It should have been "prothyle"—πρώτη ὕλη. The ignoring of the *spiritus*

After giving a short account of the electronic conceptions of matter then just dawning upon mankind, he said that the end of the nineteenth century " saw the birth of new views of atoms, electricity, and ether. Our views to-day of the constitution of matter may appear satisfactory to us, but how will it be at the close of the twentieth century ? Are we not incessantly learning the lesson that our researches have only a provisional value ? A hundred years hence shall we acquiesce in the resolution of the material universe into a swarm of rushing electrons ? "

The German expedition was rather a hurried one. He went with Lady Crookes via Harwich and the Hook of Holland, and stayed in Berlin at the Continental Hotel, where he paid forty-two marks a day for a suite of rooms. His address was delivered on June 5, 1903, and was reprinted in the Report of the Smithsonian Institute for 1903.

In 1904, Crookes was awarded the Copley Medal of the Royal Society for his researches on the action of radiant matter upon various substances.

In 1905, Crookes was invited to deliver a lecture on Diamonds at Kimberley on the occasion of the meeting of the British Association in South Africa. He had just discovered small diamonds in the ashes of cordite exploded in a closed vessel by Sir Andrew Noble, at a pressure probably amounting to 50 tons per square inch and an absolute temperature estimated at 5,400° C.[1]

The lecture at Kimberley was prepared with great elaboration. From photographs and measurements of the lecture theatre communicated to him by post, Crookes prepared a model of the lecture theatre to scale and arranged

---

asper is a common fault among English Greek scholars. Faraday was free from this fault.

[1] See *Proceedings* of the Royal Society, 76A, p. 458, 1905

the disposal of the lantern and other apparatus. He specified the weight to be sustained by the platform, the current to be used, and even the points of entrance and exit on the platform. He brought all his influence to bear to secure the loan of large diamonds for the lecture, and was somewhat impatient of the questions of Wernher, Beit & Co. as to the precise use to which the stones were to be subjected. He sent his assistant, Mr. J. H. Gardiner, to Kimberley a week before his own departure to prepare things on the spot.

On arrival in Cape Town, he received the degree of Doctor of Science from the University of the Cape of Good Hope, while similar degrees were conferred upon Professor G. H. Darwin, Sir Richard Jebb, Sir David Gill, Dr. Kapteyn, Dr. Bohr, and other men of science.

The lecture itself was a brilliant success. It was an elaboration of the Royal Institution lecture of 1897, with some additional information and experiments. He exhibited the blackening of diamond under the impact of cathode rays, and finally converted a diamond into graphite by the heat of the electric arc before his delighted audience. The lecture was so popular that it had to be twice repeated.

Crookes returned to England in October. On November 13, 1905, he wrote to Miss Alice Bird :

MY DEAR LALLAH,

Forgive me for so long delaying an answer to your kind letter of welcome home. Since returning, I have been overwhelmed with arrears of correspondence which had to be attended to immediately, and everything else had to give way.

We have been back about a fortnight, and are only just beginning to realise the wonderful journey we have had. I have a lot of photographs illustrating the chief points of interest on our travels, and you must come here some day and see them all. Are you at home for some time ? I should like to see you and tell you what a fine success the lecture was, and how it had to be repeated, and a third repetition was wanted, but time would not admit of it. Then the Matoppos and Cecil Rhodes's grave, and the Victoria Falls, will be dreams of beauty and sublimity all my life. They cannot be described adequately. Nothing but seeing them will suffice. I am glad to say Nellie and I both stood the intense heat of the Red Sea, the East Coast, and Cairo well, and have had no relapse from the cold we were suddenly hurried into as we came out of the Mediterranean.

We send our love, and believe me,

Very affectionately yours,

WILLIAM CROOKES.

CHAPTER XIX

RADIUM AND ITS RAYS

ON APRIL 10, 1906, Sir William and Lady Crookes celebrated their golden wedding. The celebration took place at the Empress Rooms, Kensington, and was arranged by the staff of the Royal Palace Hotel There was no band, but there was solo singing according to the following programme :

" During the evening Miss Marta Cunningham will sing :—

1 (*a*) ' Maytime ' ....*Cowdell.*
   (*b*) ' Love is meant to make us glad ' ....*German.*
2. (*a*) ' Sealed Orders ' ....*Willeby.*
   (*b*) ' Filles de Cadiz ' ....*Delibes.*
3. (*a*) ' World that once was a Garden ' ....*Lohr.*
   (*b*) ' Chanson Bohême ' ....*Bizet.*

M. Stirling Mackinlay, Esq., will sing :

1. (*a*) ' If I built a World for you ' ....*Liza Lehmann.*
   (*b*) ' The two Grenadiers ' ....*Schumann.*
2. (*a*) ' Mattinata ' ....*Tosti.*
   (*b*) ' Jest her way ' ....*Aitkin.*"

It was a day of triumph for the aged and extremely united couple. Over eighty telegrams of congratulation were received from all parts of the world. Some four hundred guests attended the celebration, and as Crookes had ordered twelve dozen of champagne—he was something of a connoisseur and chose " Perrier Jouet, 1895 "—there was much good fellowship and hilarity. Supper was served in two relays. Among those present was Crookes's old friend and later critic, John Spiller. To Professor Maskelyne he wrote afterwards :

> I am sorry to hear that illness prevented your coming to our reception on Tuesday. So many of my old friends were prevented by illness—chiefly influenza—from attending, that it was quite sad to read the letters and telegrams of regret.
>
> I think with one exception you would have been the oldest friend, in point of time, present, and when I look back to the Oxford days, and think of the kindness you, one of the honoured Dons, showed to me a raw undergraduate, I still more regret that illness was the cause of your non-appearance.
>
> When you are in Town again I should like to know, as I want to show you a diamond that will interest you. It started a year ago as a cut brilliant of the first water, but by means of Radium treatment it has become of a rich bluish-green colour very similar to a green tourmaline, but retaining all the fire of the diamond.

All who knew the Crookeses were much impressed by the extraordinary affection existing between Crookes and his wife. And, indeed, Lady Crookes had deserved well of her husband. She had been a devoted wife, and had borne him ten children. Of these seven survived childhood, and one—the second girl—died at the age of thirteen. Henry, the eldest son, was his mother's favourite, but Crookes considered none of his children as being up to his own standard with the exception of Bernard, and his friends were disposed to agree with him. In addition to her activities as a mother, Lady Crookes was a wise and intelligent companion and friend in all Crookes's affairs, and in the earlier days she frequently acted as his secretary and amanuensis. She was a great talker, and on many occasions, it is said, she ruled the conversation to the total extinction of her rather retiring husband, and kept it on a level far below the high altitudes which her

383

guests expected from the presence of a world-famous savant.

Crookes's tender devotion to his wife and his long-suffering patience with her mother, who had joined the household on their marriage, wove a halo of saintliness round Crookes's head which never faded. If he exacted the toll of parenthood to the uttermost, it must be remembered that he came of a long-lived and very prolific stock, and that he did his best to live up to his responsibilities.

Just about the time of Crookes's golden wedding the War Office dissolved the Explosives Committee and created a new body called the Ordnance Research Board. Crookes was invited to join this Board as a salaried officer, and accepted with alacrity.

At this time the great revolution in electrical and chemical theory inaugurated by the discovery of Röntgen rays and radium was accomplishing itself. When Henri Becquerel, the discoverer of the radium rays named after him, died of heart failure on August 25, 1908, it was Crookes who wrote the obituary for the Royal Society. He described how Becquerel followed up his father's researches on phosphorescence, and continued :

> When Lenard and Röntgen published their memorable papers on new radiations he attacked the subject with renewed vigour, and zealously followed up his father's work of 1872 on the light emitted by phophorescent uranium compounds.
> The writer visited Henri Becquerel's laboratory one memorable morning when experiments were in progress which culminated in the discovery of the " Becquerel rays," and of " Spontaneous radio-activity." Uranium salts of all kinds were seen in glass cells, inverted on photographic plates enclosed in black paper, and also the resulting images automatically impressed on the sensitive plates. Becquerel was working on the phosphorescence of uranium compounds after insolation ; starting with the discovery that sun-excited uranium nitrate gave out rays capable of penetrating opaque paper and then

acting photographically, he had devised another experiment in which, between the plate and the uranium salt, he interposed a sheet of black paper and a small cross of thin copper. On bringing the apparatus into daylight the sun had gone in, so it was put back into the dark cupboard and there left for another opportunity of insolation. But the sun persistently kept behind clouds for several days, and, tired of waiting (or with the unconscious prevision of genius), Becquerel developed the plate. To his astonishment, instead of a blank, as expected, the plate had darkened under the uranium as strongly as if the uranium had been previously exposed to sunlight, the image of the copper cross shining out white against the black background. This was the foundation of the long series of experiments which led to the remarkable discoveries which have made " Becquerel rays " a standard expression in Science. This property of uranium compounds was found not to fade with lapse of time ; it does not require starting by exposure to the sun's rays, but is an inherent property of the atoms of uranium itself. Faraday's saying comes to mind—" One good failure often teaches more than the most successful experiment."

The discovery in 1898 of radium gave a powerful impetus to all radiographic work, and profoundly modified all previous ideas on that subject. Henri Becquerel was one of the first to correlate the new facts brought forth by radium with his old discoveries in connection with uranium. He was the first to obtain a small specimen of radium bromide from M. and Mme. Curie, and one of the earliest experiments he tried with it was to expose the bundle of rays emitted by the radium salt to the influence of an electromagnet. By this means he separated them into three categories : (1) Rays which are positively electrified particles of the dimensions of a chemical atom ; these rays are scarcely at all deflected by the magnet ; (2) rays (now recognised as *electrons*) which are deflected considerably by a magnet in the opposite direction to the first-named rays ; and (3) rays which are not at all deflected by a magnet and consist probably of ethereal vibrations, the same as Röntgen or X-rays.

Crookes was too conservative a chemist to accept at once Curie's demonstration of the spontaneous heat permanently developed by radium. He wrote to *The Times* and also to the Royal Society, quoting some experiments to show that the energy emitted by radium was somehow extracted from the molecules of the air, in which case it would cease to be evolved in a perfect vacuum. He withdrew the Royal Society communication when the spontaneous breaking up of the radium atom was finally established, and it was never printed. In writing to a fellow F.R.S. he put the matter as follows : " A few deci-grammes of radium have undermined the atomic theory of chemistry, revolutionised the foundations of physics, revived the ideas of the alchemists, and given some chemists a bad attack of ' swelled head ' ! " He also wrote to his financial adviser (June 20, 1903) about a mine con-taining a little pitchblende :

All this excitement about radium has caused a big boom in pitchblende, and the mine is worth ten times as much as it was in the time of North.   One German firm, Buchler & Co., of Brunswick, are scooping in riches to a fabulous extent by selling radium at about £12,000 an ounce, and they are calling for more pitchblende, and will pay anything for it.   Everybody in the scientific world is buying radium, and learned Societies are being asked in all directions to supply funds for the purchase of radium for experiment and investiga-tion.   I know of several hundreds of pounds likely to be supplied in that way.   Were I younger I would pay all off and buy the mine myself and work up the radium and become a rival to Rothschild in wealth.   As it is, however, I suppose I must let a German firm have it all !   If we could find some chemical manufacturer worth his salt I would give him all the information necessary for the extraction of radium from pitchblende   I am the only man in England who knows it.   Under these circumstances can you not squeeze our German friends a little more ?   But I put it in your hands knowing you will do the best you can for me.

He also wrote to the same " German friends " : " Can you tell me what is the best way to exhibit the lumi-nosity either of a radium compound or the action of radium on a screen, so that it can remain permanent ? " Buchler & Co. recommended zinc blende, and Crookes painted twelve milligrammes of radium bromide over the blende screen and obtained enough light to read print with it. He then mounted a pin-point of radium bromide in front of a blende screen and a lens in front of the radium, and thus obtained the Spinthariscope, a little instrument which exhibited the effect of bombardment by the " Alpha rays " of radium very beautifully, each impact of an Alpha particle giving rise to a tiny flash, and the whole effect resembling a mass of tiny scintillating stars.

In 1908, Crookes was elected Foreign Secretary of the Royal Society, an office which he only relinquished in 1913 on his election as President. A cloud was cast over that year by Lady Crookes's serious illness with pneumonia, which left her permanently enfeebled.

In 1909 the *Chemical News* completed its hundredth volume, and its fiftieth year of existence, " and I think," Crookes wrote to Miss Bird, " it will be legitimate to indulge in a little cock-crowing on the occasion." About the end of that year the " daylight comet " appeared, and Crookes paid frequent visits to Windmill House, Hampstead (the Birds' residence), to catch a glimpse of it. " If no comet is visible," he wrote on one occasion, " we shall at all events have the pleasure of seeing you, and that is worth a wilderness of comets."

In 1910 the King bestowed upon Crookes the Order of Merit. Crookes appreciated this very high distinction the more as he had been disappointed in 1897 at being made a simple Knight Bachelor " like any little local mayor." He had now attained his heart's desire, and there only remained the Presidency of the Royal Society yet to be achieved.

On November 11, 1910, Crookes made a remarkable speech at a banquet given to Past Presidents by the Chemical Society. He drew attention to the scientific importance of anomalies, which were frequently responsible for what Huxley called " The Great Tragedy of Science "—the slaying of a beautiful hypothesis by an ugly fact.

I fear (he continued) the radiometer has been guilty of more than one tragedy of this kind. But the whirligig of time brings its revenges, and my friend Sir James Dewar, by his researches on high vacua and low temperature, now threatens to kill the radiometer ! I will take a brief glance backwards. My old master and dear friend Hofmann first started me on the road of research. About the year 1850 he proposed I should commence an investigation on the " Selenocyanides," using for this purpose some selenium I had extracted from seleniferous residues he had brought from Tilkerode. The result of this research was published in 1851 in the *Quarterly Journal of the Chemical Society*.

When first I entered the Royal College of Chemistry, Liebig was in the height of his fame ; his *Chemistry in its Application to Agriculture* would be a useful textbook to-day. In this book was first declared the importance—nay, the absolute necessity—of mineral matter and nitrogen to the well-being of the vegetal organism. Long and fierce were the controversies aroused by Liebig's views of the absorption of nitrogen from the air. Partisanship was keen ; as a pupil of Hofmann I naturally studied the subject in connection with our food supply—and my interest continues to this day.

Our food depends on agriculture, and as the available wheat lands of the world inevitably contract, the economical cultivation of the soil becomes more and more a question of chemistry. At the Bristol meeting of the British Association in 1898 I re-opened the nitrogen question, not with a view to produce what has been called a " cosmic scare," but to show how by the application of known physical and chemical reactions, the shortage of fixed nitrogen, and consequently of wheat, can be overcome.

One of the first analyses I performed at the College of Chemistry was that of the mineral cerite. The rare earths—my first, but not my only love—were then represented by ceria, lanthana, didymia, yttria, erbia, and terbia, and a specimen of yttria extracted from this cerite first gave me a key, thirty-three years later, to the citron band which then was haunting my laboratory.

Soon after Schrötter's discovery of amorphous phosphorus the great Professor came to the College of Chemistry to demonstrate to Hofmann the conversion of yellow phosphorus into the red variety. At that time I had acquired some skill in blowpipe manipulation, and was asked to help fit up the apparatus under Schrötter's directions. The new art of photography early attracted my attention. I remember the excitement caused by the announcement of Daguerre's and Fox Talbot's discoveries. I was working at photography in 1848, and not long after I had the privilege of being shown the Talbotype process by the master himself. One of my most highly prized relics is a copy of Talbot's *Pencil of Nature*. In these snapshot days it may be of interest to remember that portraiture was an affair of minutes, not seconds. I think there is one at least of my audience whom I victimised by a five-minute sitting in sunlight with his face chalked !

Niepce de St. Victor had discovered that uranium salts possessed the property of storing up light and giving it out in the dark, and in 1858 I took what was perhaps the first radium photograph in this country, by writing with solution of uranium nitrate on a card, insolating it, and then putting it face to face in the dark with a sheet of photographic paper ; the image of the writing was reproduced on the paper. Also among my earliest recollections in photography is a photograph of the solar spectrum in its true natural colours by Edmond Becquerel. Those who had the advantage of examining Becquerel's coloured photographs in their early perfection will agree with me that they hardly fell short of the recent productions of Lippman.

I took some of the earliest stereoscopic photographs for Wheatstone, and in 1852 I commenced work on the photography of solar and metallic spectra—work which has increasingly fascinated me to the present day.

A few days afterwards, at the Institution of Electrical Engineers, he indulged in some further lively recollections.

My memory goes far back. I have seen the old dandy-horse change into a velocipede worked with arms and legs ; this soon developed into a bicycle ; power was added, and the bicycle became a motor-car, and now the motor-car has taken to itself wings and has become an aeroplane.

I have heard " God Save the King " before Victoria's reign, and have seen *The Times* newspaper in mourning for the Crown on three separate occasions. But if old age has its drawbacks it also has compensations. It has enabled me to re-make the acquaintance of an old friend, Halley's Comet, which I saw on its appearance in 1835.

In my early days flint and steel were still used for producing fire, and as a child I used to make sulphur-tipped splints of wood, and occasionally graze my knuckles with the flint and steel.

When Schönbein in 1846 announced his discovery of gun-cotton I set to work at once and made some. So successful were my efforts that I nearly blew up myself and my young school companions.

He wound up a most entertaining speech by a *tour de force* :

It has been said that Englishmen cannot meet together for the inauguration of any new departure without celebrating the occasion by a dinner. This I suppose is one of the virtues we have inherited from our ancestors, but if their good qualities have been transmitted to us, I fear we must plead guilty to some share in their bad attributes. We still retain at our feasts the skeleton so dear to ancient Egyptian banquets, only we put him into a tail-coat, prop him on his hind legs, and make him utter platitudes. My friends, I feel myself that skeleton.

On February 16, 1911, a formal presentation was made to the Royal Society of a portrait of Crookes by E. A. Walton, acquired for the Society by subscription.

That year Crookes was troubled by frequent attacks of influenza, a disease which carried off his old friend, Georges Fournier, in Paris. On receiving the Medal of the Society of Chemical Industry (October 23, 1911), he said :

During my long life I have striven loyally with all my might for the truth. No one can be more fully conscious than I am how little I have won compared with what remains to be solved and achieved. Those of us who have attempted to penetrate from the known to the unknown, from the vague to the definite, are surprised not so much by the extent of our knowledge as by the depth of our ignorance. As I have proceeded in my investigations I have discerned new and unexplored regions opening out to the right and to the left. To some of these mysterious tracks I may return and endeavour to capture them and to bring them within the domain of science. Who shall dare to say what new treasures of truth and even of practical utility may there await the patient inquirer ? I hope and believe I shall still be able to do good work in my research laboratory, and this notwithstanding the ominous assurances of friends that I grow younger every year !

A few days afterwards, while out driving in his carriage with Lady Crookes, he unfortunately knocked down a boy, who threatened proceedings for injury. Crookes wrote to Mr. Morriss :

> Thanks for your letter and suggested settlement. I would not on any account have Lady Crookes put in the witness box. In her present state of health the consequences might be serious. If you can get the thing settled for the small sums you name I shall be very glad. You must not, however, believe all they say on the other side. Is it likely my wife and I would tell our coachman to drive away without ascertaining if the boy was injured ? But I do not want to bother about rebutting evidence, although were it necessary I could show I was not such a brute as the boy thinks.
>
> Please settle the matter as soon as you can.

On February 24, 1912, he told Miss Heather that his son, Henry, who was then working at the Water Supply Laboratory, had discovered a method of getting colloid metals in strong solution and in stable form. " The silver colloid," he added, " is doing wonders, and medical men are taking it up generally."

Crookes's satisfaction at the success of his son was, however, short-lived. Crookes had been Hon. Secretary of the Royal Institution since 1900, and had spent much time arranging the numerous afternoon and evening lectures of each session. Sir James Dewar was head of the Research Laboratory. In spite of their close collaboration over the Water Companies, and Crookes's championship of Dewar in the Ramsay controversy in the Chemical Society, there were numerous cases of friction, which culminated in a rupture on the matter of Henry Crookes's discovery. Dewar brought an action against Crookes and his son for infringement of certain rights. The matter, after causing some commotion in the Institution, ended in Crookes's resignation in 1912, after which he never visited the Royal Institution again. The legal dispute ended in a compromise, which, however, finally involved the bankruptcy of Henry Crookes, a calamity from which his father was unable (or even unwilling ?) to save him.

Crookes's election as President of the Royal Society in 1913 came as a balm upon the wound inflicted by this unfortunate occurrence. That Crookes met with some opposition is shown by the following letter to Professor Silvanus Thompson .

July 7, 1913.

My dear Silvanus,

    I understand that some printed documents were brought before the Council of the R.S. at their last meeting but one in order to show that I was a patentee of a " Quack Medicine " and was therefore unsuited for the high Office under discussion. As you were there, can you tell me anything about these documents ? They must be false documents got up for the occasion, for it is impossible that any genuine prints can be in existence to warrant such a statement.

My son Henry is a man between 50 and 60 years of age, and he never consults me on business matters. I have known for some years that he was at work on the scientific preparation of metals in the colloid form, and I have taken some interest in the bactericidal action of silver and mercury in this form. Hearing from a medical friend that he had cured some microbic diseases with these colloids, I drew Dr. Plimmer's attention to them, and gave my son an introduction to him.

I knew *absolutely nothing* of Henry's patenting the discovery until some time after the patent was taken out, and I may say the same about the Company which was subsequently formed. The first intimation I had that such a Co. was being got up was an application from Henry for me to advance a little money for expenses, telling me I could have some shares in his Company to the amount I would advance him. Before this I knew absolutely nothing about the Company or that any such thing was in contemplation.

I expressed my astonishment to Henry for keeping me in the dark about such an important step, whereupon he said he had kept me purposely in ignorance, as he thought, from previous remarks I had made to him, that my opinion as to the importance of his discovery was not very high. I asked him particulars about the Company, and he told me it would be quite a private one and the total capital would not be more than about £1,500. I said I would not be a shareholder in the Company if it was going to be advertised and boomed before the public He told me there were very few shareholders and they were chiefly private friends of his own. His father-in-law had taken £200 worth of shares, and other friends had put in the same amount, so I gave him a cheque for £200, not to be behind the others. I am not a director and know nothing of the business of the Co. I only have to wait for dividends which I don't think will come.

When Henry showed me the Prospectus of the Co. I objected strongly to the name, "Crookes's Collosols." I said it would be thought that I was the patentee and getter-up of the Company, and it would seriously damage my scientific position. Henry did not see the force of my objection and told me his co-patentee and partner insisted on the wording being retained. I thereupon consulted my lawyer as to whether my name and scientific position could thus be exploited to advertise a company. I was advised that I could legally stop them using the term "CROOKES's COLLOSOLS," and make them use "H. CROOKES's COLLOSOLS" instead. I felt that an action by me against my son could not be entertained, it would injure him, so I decided to take the risk. My sinister prediction has now come true !

Here you have the plain unvarnished tale. I know that I have a bitter and unscrupulous enemy who will take every unfair means of damaging my prospect of succeeding to the high honour of the Presidency, and I would not on any account let anyone think I was exerting myself actively to secure supporters. But when such lies are being told I feel I can rely on my friends to see that they do not prevail.

Crookes was, however, triumphantly elected, and occupied the august Presidency for three years. The same year saw his election as President of the Society of Chemical Industry, and as an Associate Member of the Washington Academy of Sciences. Henceforth the scientific world had no more honours to bestow.

APPENDIX TO CHAPTER XIX

List of Titles of Papers by Sir William Crookes on Radium and its Radiations.

"Sur la Source de l'Energie dans les Corps Radio-actifs." *Comptes Rendus*, vol. 128, p. 176–8 (1899).

"Source de l'Énergie que présentent les Corps Radio-actifs." *Revue Générale des Sciences*, vol. 10, p. 120 (1899).

"Source of Energy in Radio-active Bodies." *Chemical News*, vol. 79, p. 105 (1899).

"Source of Energy in Radio-active Bodies." *Nature*, vol. 59, p. 311 (1899).

"Radio-activity of Uranium." (A paper read before the Royal Society, May 10, 1900.) *Roy. Soc. Proc.*, vol. 66, pp.409–23, and one plate (1900).
Chemical News, vol. 81, pp. 253–5 and 265–7 (1900).
Chem. Soc. Journ., 1900, p. 586.
Journ. Soc. Chem. Ind., vol. 18, p. 862 (1900).
Am. Journ. of Science, vol. 10, Ser. I, pp. 318–19 (1900).

"La Radio-activité de l'Uranium." · *Revue Générale des Sciences*, vol. 11, p. 949 (1900).

"Radio-activity and the Electron Theory." (A paper read before the Royal Society, Feb. 6, 1902.) *Roy. Soc. Proc.*, vol. 69, pp. 413–22 (1902).
Chemical News, vol. 85, pp. 109–12 (1902).
Nature, vol. 65, pp. 400–2 (1902).
Chem. Soc. Journ., 1902, p. 374.
Electrician, vol. 48, pp. 777–9 (1902).

"The Stratification of Hydrogen." (A paper read before the Royal Society, Feb. 6, 1902. *Roy. Soc. Proc.*, vol. 69, pp. 399–413 (1902).

Chemical News, vol. 85, pp. 85–7 and 97–100 (1902).

Nature, vol. 65, pp. 375–8 (1902).

Chem. Soc. Journ., 1902, p. 374.

Electrician, vol. 48, pp. 702–3 and 739–42 (1902).

Pharmaceutical Journal, vol. 14, Ser. IV, p. 140 (1902).

Science Abstracts, vol. 5, pp. 570–1 (1902).

"Radium." *Science*, New Ser., vol. 17, pp. 675–6 (1903).

"The Mystery of Radium." *The Times*, March 26 and April 2, 7, and 14, 1903.

Chemical News, vol. 87, pp. 158 and 184 (1903).

"List of Objects exhibited by Sir William Crookes, F.R.S., illustrating certain properties of the emanations of Radium." Soirée of the Royal Society, May 15, 1903, 3 pp.

Chemical News, vol. 87, p. 245 (1903).

"The Emanations of Radium." (A paper read before the Royal Society on March 19, 1903.) *Roy. Soc. Proc.*, vol. 71, pp. 405–8 (1903).

Chemical News, vol. 87, pp. 157–8 (1903).

Nature, vol. 67, pp. 522–4 (1903).

Chem. Soc. Journ., 1903, p. 461.

Electrician, vol. 50, pp. 986–7 (1903).

"Certain Properties of the Emanations of Radium." *Chemical News*, vol. 87, p. 241 (1903).

"Modern Views on Matter : The Realisation of a Dream." (An Address delivered before the Congress of Applied Chemistry at Berlin, June 5, 1903.) London. 16 pp.

Chemical News, vol. 87, pp. 277–81 (1903).

Pharmaceutical Journal, vol. 16, Ser. IV, p. 815 (1903).

Journal of the Franklin Institute, vol. 156, p. 316 (1903).

Science, New Ser., vol. 17, pp. 993–1003 (1903).

Scientific American Supplement, vol. 56, pp. 23014–16 (1903).

"Moderne Anschauungen über Materie. Die Verwirklichung eines Traumes," von William Crookes. *Zeitschrift für Naturwissenschaften*, Bd. 76, pp. 292–314 (1904).

RADIUM AND ITS RAYS

"The Spinthariscope." The Royal Society's Soirée, May 15, 1903.
Chemical News, vol. 87, p. 241, and pp. 280–1 (1903).
Nature, vol. 68, p. 303 (1903).
American Journal of Science, vol. 16, 4th Ser., p. 99 (1903).
Société Française de Physique. *Resumé des Communications faites dans la Séance du Juillet,* 3, 1903, No. 200, pp. 1–2.
"The Ultra-violet Spectrum of Radium." (A paper read before the Royal Society, August 1, 1903.). *Roy. Soc. Proc.,* vol. 72, pp. 295–304, plates 16–18 (1903).
Chemical News, vol. 88, pp. 202–5, plates 1–3 (1903).
Journ. Chem. Soc., p. 3 (1904).
"The Ultra-violet Spectrum of Radium." Correction to paper. *Roy. Soc. Proc.,* vol. 72, p. 413 (1903).
"The Action of Radium Emanations on Diamond." (A paper read before the Royal Society, June 16, 1904.) *Roy. Soc. Proc.,* vol. 74, pp. 47–9 (1904).
Chemical News, vol. 90, pp. 1–2 (1904).
Nature, vol. 70, pp. 209–10 (1904).

Sir William Crookes and Sir James Dewar.

"Note on the Effect of Extreme Cold on the Emanations of Radium." (A paper read before the Royal Society, May 28, 1903.) *Roy. Soc. Proc.,* vol. 72, pp. 69–71 (1903).
Chemical News, vol. 88, pp. 25–26 (1903).
Nature, vol. 68, pp. 213 and 611 (1903).

CHAPTER XX

THE LAST STAGE

(1913-19)

THE LAST SIX YEARS of Crookes's life were spent in the warm radiance of public appreciation and regard. While his health permitted, he attended many dinners and other functions, and delighted his hearers and disciples with many witty or inspiring words. Representing the Royal Society at a meeting of the Fishmongers' Company, he said in reply to a toast :

A year ago the Royal Society celebrated its 250th anniversary. I have heard many interesting conjectures about the beginnings of scientific knowledge, and many comparisons between man's ignorance of natural phenomena then and his increased knowledge of them now. It is impossible to refrain from wondering how the Society's semi-millennium will be celebrated. What will the Fellows have to say about the gross ignorance of the twentieth century. But unless some of our present researchers add to their approaching discovery of the Philosopher's Stone that of the Elixir Vitæ, none of us will have a chance of joining in these festivities !

After giving my life to Science for sixty-five years I have accumulated a mass of memories of eminent men, epoch-making discoveries, and notable experiments. Indeed, I should feel that I was growing an old man if my imagination could no longer fling itself forward and get a glimpse of the future—of dazzling moments when one feels "like some watcher of the skies, when a new planet swims into his ken." The power of realising events before they occur is the most valuable endowment of the scientific man. It is the most exquisite faculty in our mental equipment, and it is through this vision splendid that we claim kinship with the Divine. I cannot

397

express this better than in the words of a modern poet who also is a scientific professor :

> Visions of beauty and splendour,
> Forms of a long lost race,
> Sounds and faces and voices,
> From the fourth dimension of space;
> And on through the Universe boundless
> Our thoughts go lightning shod,
> Some call it Imagination,
> And others call it God.

At the Annual Dinner of the Old Students' Association of the Royal College of Science (incorporating the Royal College of Chemistry) on January 25, 1913, he indulged in some further reminiscences :

To exceed the Psalmist's limit of fourscore years is to be left solitary as regards early friendships. My old friend John Spiller, so far as I can ascertain, is the last of the band of researchers in the early fifties who worked with me at the Royal College of Chemistry.

The state of Science when I was a student was not quite so backward as is generally supposed. Wireless telegraphy, bacteriology, spectrum analysis, and radium were unknown, but other discoveries of world-wide importance were at least in embryo. Wheatstone was at work on the electric telegraph ; and Faraday, earnestly on the track of magneto-crystallic force, was grappling with the magneto-electric machine—parent of the modern dynamo and of electric lighting. Faraday was first to draw attention of physicists to the highly insulating properties of gutta-percha. He said it was equal to that of the transparent cuticle left behind when an ethereal solution of gun-cotton was allowed to evaporate on a glass plate. The same great chemist was experimenting in low temperatures ; I remember seeing him freeze mercury in a red-hot crucible, throw the solid lump on to an anvil, and hammer it out before it melted. Grove, also in the field, had decomposed water by heat alone into its constituent gases, and Joule was determining the mechanical equivalent of heat ; Ebelman was making synthetic rubies by crystallising alumina under great heat from a glassy solvent, and by suitable modification of flux and coloured oxides, either ruby, sapphire, or Oriental emerald could be made, each with its own true crystalline and optical characteristics.

One memorable incident stands out prominently. I was present at the Royal Society's rooms in Somerset House in 1850, when Faraday demonstrated the magnetic character of oxygen gas. Few Royal Society announcements excited so much enthusiasm as did that paper. In the same year I was at one of Mr. Barlow's " At Homes " after a Friday evening lecture at the Royal Institution. My host introduced me to a young man who had made some interesting discoveries in the high ultra-violet end of the spectrum. He had recently been elected to Newton's Chair of Mathematics in the University of Cambridge, and as we both had been working on similar spectrum phenomena, he visually and I photographically, the lifelong friendship between Sir George Stokes and myself dates from that evening.

Pasteur had already discovered and separated dextro and lævo-tartaric acid, and he had started that brilliant series of researches on microbic life—researches which have revolutionised the science of medicine. Indirectly Pasteur solved the famous mediæval problem : " How many angels can stand on the point of a needle ? " Altering the word " angels " to " devils," I have found that, of one of the deadliest diseases that ever scourged mankind, no less than 500 of the maleficent microbes—veritable devils—could, without over-crowding, find place on the point of the finest needle.

In June 1914, Sir William and Lady Crookes for the first time received the Fellows and guests at the Royal Society's Soirée. It was a memorable occasion, and I remember it clearly.

The President stood remarkably erect for his eighty-two years, and wore the Order of Merit on its blue ribbon. He was, as a matter of fact, 5 ft. 8½ inches in height, having lost about half an inch in the preceding ten years. Lady Crookes had no airs of the *grande dame*, but her greeting seemed to convey a genuine welcome. They seemed both to be enjoying the novel dignity of the occasion. Crookes wrote to Miss Bird afterwards :

THE LAST STAGE

June 20th, 1914.

My dear Lallah,

I am so happy to be able to report that Nellie was able to go to the Soirée on Tuesday. She stood by my side for more than half an hour shaking hands with the guests and their ladies. She was not much tired afterwards, and to-day she has been to the White City and walked about for some hours. As soon as you get back to Windmill we will come over to see you. Your emendations and condensations to the Silicon paper were in good time, and the paper is now announced for reading at the R.S. next Thursday. H.R.H. Prince Arthur will be there to be admitted a F.R.S. by me, so he will be present at the reading of the paper. I hope he will not be too much bored by it ! I hope your stay with your friends at Painswick will set you up in good health again.

With kindest love from Nellie and myself, believe me,

Very affectionately yours,

WILLIAM CROOKES.

The outbreak of the European War took Crookes quite unawares, and even those who pretend that England prepared for war " of malice prepense " would hardly think that the President of England's most important scientific Society would have been left in ignorance of the plot. He had made arrangements to spend the month of September, as usual, in Sark, and had engaged rooms for the 24th and 25th of August at Weymouth to wait for the boat. The early days of August he had planned to spend with Mr. Arthur Franklin and his family at Chesham.

It was there that he heard of the outbreak of hostilities against France. He wrote to Mr. Gardiner

I am sorry to have to cut short our stay here, for the change is doing Lady C. much good.

Political news and the war are awful, and our host, being a Foreign Banker, is very anxious as to the prospect of England being dragged in.

400

On August 7th he wrote to cancel his engagement to visit Sark :

August 7th, 1914.

MY DEAR MRS. STOCK,

It is with very great regret on the part of Lady Crookes and myself that I write to say we cannot pay you a visit this summer. This wretched war into which we have been forced by no fault of our own makes it too serious a risk for us to venture taking a holiday in Sark. Even were we able to get there, which I am advised by good authorities is doubtful, it might be practically impossible for us to get home again.

I am writing to Weymouth to see if there is likely to be any difficulty about visitors there, and if I am advised that all is safe I think we shall go there for a week or two, and then, if all danger is over, we might run across to Sark. But I fear it will be twelve months before it will be safe to sail about the Channel.

I am very sorry it is necessary for us to come to this decision, as Lady Crookes, not being well, has been looking forward to the visit to Sark to set her up in good health, as our visits there always have done.

Crookes's address at the Anniversary Meeting of the Royal Society on November 30, 1914, referred in moving terms to the death of his friend, Sir Joseph Wilson Swan, the inventor of the bromide printing process and co-inventor of the Edison and Swan incandescent lamp. Referring to the war, he struck a note of moderation :

At a time like the overwhelming present, when our existence is at stake, it behoves those of us who cannot actively share in the great upheaval to take stock of our position among the nations of the civilised world. It is our duty, without morbid regret for the past, or craven fear of the future, to think of our own aims, and to reflect upon the aims of our adversaries as far as we are able to understand them.

And further :

The Royal Society is always cosmopolitan in spirit, and we should be unworthy of our heritage could we not, even at this desperate moment, take a dispassionate view of events, and preserve a well balanced and rational attitude towards our contemporaries of German nationality.

The Royal Society appointed four War Committees to advise the Government on scientific questions, and Crookes exercised a general supervision over these Committees without taking a very active part. The following remarkable letter to Miss Bird shows Crookes in a very attractive light :

May 29th, 1915.

My very dear Lallah,

I am going to advocate Dewar's claim to the Copley Medal of the R.S. Notwithstanding the bitter enmity between us, I do not wish my personal feelings to stand in the way of the reward which I think his good scientific work merits. Will you be so kind as to look over the enclosed and criticise it severely, for I do not wish it to be thought that I was trying to appease his wrath or suppress my own.

I am happy to say my dear Nellie is decidedly better.

With united love,

Ever affectionately yours,

William Crookes.

In 1915 Crookes devised a signalling lamp by which signals could be sent without disclosing the fact that there was any signalling going on. No details are available, but it was probably based on the use of polarised light. The Admiralty, however, after a test, rejected the device as impracticable.

On November 30, 1915, Crookes delivered his last address as President of the Royal Society, and gracefully ceded the Chair to his successor, Sir J. J. Thomson.

Crookes was appointed a member of the Admiralty Board of Invention and Research. This Board, like its military companion, the Munitions Inventions Department, was a sort of defensive bulwark against the importunities of inventors who daily bombarded the Admiralty and War Office with futile inventions for which they demanded fabulous remunerations. It worked well for its purpose, and even discovered and brought out a few positive results. Its deliberations were a curious mixture of inadequate theory and ill-informed practice. The university men on the Board had all the principles at their fingers' ends, but sometimes failed in war-time applications. They often proposed things which were impossible to carry out at sea or in the field, and sometimes turned down proposals which were quite practical for some unimportant theoretical qualm. On the whole, however, they enabled the Government to " muddle through," and it was they who finally overcame the appalling menace of the submarine. It cannot be said that Crookes was of any particular use in such positions. He was failing in health, and his heart showed frequent irregularity of action. His visits to the Ghost Club—a society of inquirers founded by W. Stainton Moses, M.A.—became fewer, and even the Royal Society Dining Club, at which he had for many years been the senior Fellow and therefore Chairman, began to miss him. He declined all invitations to public dinners, describing them humorously as " the most lethal of functions," and generally reduced his pace of living.

On April 10, 1916, Sir William and Lady Crookes celebrated their diamond wedding by a modest family gathering, at which his sons, Bernard and Lewis, and his grandson, Willie Crookes, were present. In reply to a congratulatory message from Alice Bird, Crookes wrote :

Your very nice friendly letter gave dear Nellie much pleasure, and we got talking about the time when we first saw you, at Yoxford.

THE LAST STAGE

On May 10th, Crookes wrote to an artist friend, Miss Sartoni :

I am very anxious about the condition of my Wife. She is very ill. I wish I had as good a portrait of her as my own portrait which you were so good as to paint from photographs.

Lady Crookes died on May 17th, and Crookes was prostrated with grief. When Miss Bird and her sister called, he broke out in a passionate fit of weeping. After sixty years of the closest companionship and loving friendship, such a loss was irreparable. He went down to Broadstairs for a month, staying very quietly at the Carlton Hotel. He rallied bravely for a time, and tried to take up the tangled thread of his life. He got Miss Sartoni to paint a life-sized portrait of Lady Crookes and hung it in his library. After a while he resumed his Sunday evening gathering of men friends at his house, to whom Lady Crookes used to dispense a supper of anchovy sandwiches and wine while she lived. He wore Lady Crookes's expanding bracelet and sat by her portrait very silently. He wrote to his friend :

Nov. 4th, 1916.

MY DEAREST LALLAH,

You are very good to inquire how my cold is getting on. I feel decidedly better, but the cold blustering weather will, I fear, keep it hanging about a little longer.

I must come over and have a talk to you about Sir Conan Doyle's article.

In the meantime let me beg you to read Sir Oliver Lodge's book *Raymond.* You will find it of absorbing interest, and most convincing.

Wishing you are well, and continuing to have good news of your Sister,

Believe me, affectionately yours,

WILLIAM CROOKES.

Under the stress of his bereavement Crookes resumed his visits to mediums, and also held séances in his house. A medium and her daughter sat with Crookes in the dark room off the laboratory. A musical box tinkled out its tune, and the old man sat patiently waiting for the " signs and miracles " which would tell him of the presence of his beloved partner. The words " Is that you, Willie ? " sounding out of the darkness would galvanise him into an attitude of strained attention, but he retained enough of his critical faculty to guard himself against a hasty acceptance of a gift offered to him as genuine, and left the séance room in a state of depression and bewilderment. At a photographic séance with another medium he obtained a photograph of himself with a female face as what spiritualists would call " an extra," but as it showed no resemblance to Lady Crookes he attached no importance to it.

He also visited the " Hope Circle " at Crewe, taking his own plates, fitting them into slides in the dark room in Mr Hope's presence and signing his name on the plates. He thus obtained an unmistakable likeness of Lady Crookes. The negative, according to Mr. Gardiner, showed clear signs of double exposure, but Crookes clung to the conviction that this was a real " spirit photograph " of his dead wife, and treasured it accordingly. He wrote to Miss Bird :

Jan. 24th, 1917.

MY DEAREST LALLAH,

Here are two letters from a lady whose name I cannot exactly read. She asks for one of the Spirit photographs. I do not like to send any copy to other than friends who knew my dear wife. I look upon the picture as a Sacred Trust, and do not like the idea that one is in stranger's hands, to be shown about to anyone as a curiosity.

If you know the lady sufficiently well to think she would keep it sacred, and would like me to send her a copy I will do so.

I am so sorry the weather is too bad, and I am so ailing, to permit me to take a drive over to see you. I feel deeply sympathetic for you in your loss, for I have only to turn my thought inward, to realise your grief.

The photograph must be a slight consolation, as it proves the continuity of the Self after passing through the change called Death.

" If one survives all survive."

I think I am getting a little better, but I feel the want of getting out and taking more exercise than I get in the house. I hope you are getting well again.

With much love, believe me,

Ever affectionately yours,

WILLIAM CROOKES.

In a subsequent letter, Crookes warned Miss Bird against publishing anything about his " sacred relic." All that winter Crookes was poorly. In April 1917 he severed his connection with an institution calling itself the British College of Psychic Science. In that month he had a fall which injured his arm and leg. By June he had, however, quite recovered, and was well enough to go to Bournemouth, accompanied by his nurse. He recovered so far as to be able to work in perfecting his " eye-preserving glass " for spectacles, which he had primarily worked out for the protection of glass-workers. He also worked at the rare metal, scandium, a research on which he presented to the Royal Society just before his death.

In August 1917 he was able to write to " Lallah " as follows :

I would much rather have no allusion by Mr. Clodd, in print, about the Spirit photograph of dear Nellie. I look upon it as too sacred to be made public. Also, when any public mention of any discovery of mine is made, I am deluged with private letters about it, and I fear that if Mr. Clodd prints anything about the picture I should receive letters innumerable asking for information on the subject and all about it. These I could not undertake to reply to. The first sunny morning, if I have no engagement otherwise, I will do myself the great pleasure of driving over to see you, when I hope I shall find you in good health, as I am at present, I am glad to say.

It was the winter of the great air raids on London, and Crookes's activities were hampered by the difficulty of getting about after dark.

In 1918, Crookes's circles were drawing closer. He, however, exhibited preparations of scandium compounds at the British Scientific Products Exhibition.

His last letter to the Royal Society was addressed to Dr. Schuster on December 10, 1918, enclosing the paper on scandium, with numerous photographs of its spectrum pasted on cardboard.

His last recorded letter consisted of instructions to his house agent to put things right at 20, Mornington Road (the house at which he had lived many years) for the tenant who occupied it.

Crookes died at 6 a.m. on April 4, 1919, in the presence of his second nurse. None of his children or other relations were present, though his daughter had seen him the night before, and his faithful assistant, Mr. J. H. Gardiner, was about him every day. The funeral service took place on April 10th, not at the parish church of St. Peter's, where he used to worship for many years, but at St. John's, Ladbroke Grove.

The principal mourners were Sir William's sons, Mr. Bernard, Mr. Walter, and Mr. Lewis Crookes, Mr. and Mrs. Jack Cowland (son-in-law and daughter), Mrs. Henry Crookes (daughter-in-law), and Mr. Charles Crookes (nephew). Others present in the congregation were :

Mr. Hepworth Collins, Sir Joseph Thomson, O.M. (Master of Trinity College, Cambridge, representing the Royal Society), the Hon. R. Clere Parsons (representing the Royal Institution of Great Britain), Sir Boverton Redwood (representing the British Science Guild), Professor John Perry (representing the British Association), Sir Herbert Jackson (President of and representing the Institute of Chemistry), Mr. W. M. Mordey (representing the Institute of Electrical Engineers), Sir William Tilden, Sir Lawrence Jones (representing the Society of Psychical Research), Professor Frank Clowes, D.Sc. (Past-President of and representing the Society of Chemical Industry), Dr. Alex. Scott (Vice-President of and representing the Chemical Society), Mr. W. F. Reid (Chairman of the Institute of Inventors), Sir David Prain, Sir J. Larmor, M.P., Sir William Davison (Mayor of Kensington), Dr. Abraham Wallace, Dr. Francis Fox, Lady and Miss Macdonell, Dr. Donald Hood, Dr. Emerson Reynolds, Dr. and Mrs. Henry Armstrong, Sir John Snell, Professor A. Schuster, and a number of representatives of the Notting Hill Electric Lighting Company. Sir William Barrett was unavoidably prevented from being present.

The burial took place later at Brompton Cemetery.

Thus passed one of England's greatest and most representative men of science, a man of the people, English of the English, a searching and painstaking student and investigator of Nature's secrets, an inspired prophet of future progress, a fearless expounder of new truths, a loyal friend, a chivalrous antagonist, a devoted husband, son, and father, a sturdy patriot, who yet had room in his great heart for fellow-creatures of all nations. Well might those gathered round his grave recall the wonderful lines penned about his greatest precursor :

> He lay, a little child, at Nature's breast
> And drew with reverent lips the pure white truth
> With which she fed his soul from earliest youth
> Until she lulled the grey-haired babe to rest.

Great mind and simple heart ! whose life expressed
The grand humility of wisdom's ways ;
Who sought the truth for love, and not for praise,
And, in the light of knowledge, deemed faith best.

For his was not the cold philosophy
Which, finding Law throughout the Universe,
Believes the world drives on beneath the curse
Of soulless Force and blind Necessity ;
But reading still above the unfolded Law
Love's revelation touched his soul with awe.

INDEX

INDEX

Printed in Great Britain by
UNWIN BROTHERS, LIMITED, THE GRESHAM PRESS, LONDON AND WOKING